T0305981

Behavior Dynamics in Media-Sharing Social Networks

In large-scale media-sharing social networks, where millions of users create, share, link, and reuse media content, there are clear challenges in protecting content security and intellectual property, and in designing scalable and reliable networks capable of handling high levels of traffic.

This comprehensive resource demonstrates how game theory can be used to model user dynamics and optimize design of media-sharing networks. It reviews the fundamental methodologies used to model and analyze human behavior, using examples from real-world multimedia social networks. With a thorough investigation of the impact of human factors on multimedia system design, this accessible book shows how an understanding of human behavior can be used to improve system performance.

Bringing together mathematical tools and engineering concepts with ideas from sociology and human behavior analysis, this one-stop guide will enable researchers to explore this emerging field further and ultimately design media-sharing systems with more efficient, secure, and personalized services.

H. Vicky Zhao is an Assistant Professor in the Department of Electrical and Computer Engineering at the University of Alberta. The recipient of the IEEE Signal Processing Society Young Author Best Paper Award 2008, she is an Associate Editor for the *IEEE Signal Processing Letters* and the *Journal of Visual Communication and Image Representation*.

W. Sabrina Lin is a Research Associate in the Department of Electrical and Computer Engineering at the University of Maryland. She received the University of Maryland Future Faculty Fellowship in 2007.

K. J. Ray Liu is a Distinguished Scholar-Teacher of the University of Maryland, where he is Christine Kim Eminent Professor in Information Technology. He received the IEEE Signal Processing Society Technical Achievement Award in 2009, and was Editor-in-Chief of the *IEEE Signal Processing Magazine* and the founding Editor-in-Chief of the *EURASIP Journal on Advances in Signal Processing*.

Behavior Dynamics in Media-Sharing Social Networks

H. VICKY ZHAO
University of Alberta, Canada

W. SABRINA LIN
University of Maryland, College Park

K. J. RAY LIU
University of Maryland, College Park

CAMBRIDGE
UNIVERSITY PRESS

CAMBRIDGE
UNIVERSITY PRESS

University Printing House, Cambridge CB2 8BS, United Kingdom

One Liberty Plaza, 20th Floor, New York, NY 10006, USA

477 Williamstown Road, Port Melbourne, VIC 3207, Australia

314-321, 3rd Floor, Plot 3, Splendor Forum, Jasola District Centre, New Delhi - 110025, India

103 Penang Road, #05-06/07, Visioncrest Commercial, Singapore 238467

Cambridge University Press is part of the University of Cambridge.

It furthers the University's mission by disseminating knowledge in the pursuit of education, learning and research at the highest international levels of excellence.

www.cambridge.org
Information on this title: www.cambridge.org/9780521197274

© Cambridge University Press 2011

First published 2011

A catalogue record for this publication is available from the British Library

Library of Congress Cataloging in Publication data
Zhao, H. Vicky, 1976–
Behavior dynamics in media-sharing social networks / H. Vicky Zhao, W. Sabrina Lin, K. J. Ray Liu.
   p.   cm.
Includes bibliographical references and index.
ISBN 978-0-521-19727-4 (hardback)
1. Social networks. 2. Consumer behavior. 3. Human behavior. 4. Game theory. I. Lin, W. Sabrina, 1981– II. Liu, K. J. Ray, 1961– III. Title.
HM742.Z46 2011
302.30285'675 – dc22   2011006139

ISBN 978-0-521-19727-4 Hardback

To Our Families

Contents

Preface

In the past decade, we have witnessed the emergence of large-scale media-sharing social network communities such as Napster, Facebook, and YouTube, in which millions of users form a dynamically changing infrastructure to share multimedia content. This proliferation of multimedia data has created a technological revolution in the entertainment and media industries, bringing new experiences to users and introducing the new concept of web-based social networking communities. The massive production and use of multimedia also pose new challenges to the scalable and reliable sharing of multimedia over large and heterogeneous networks; demand effective management of enormous amounts of unstructured media objects that users create, share, link, and reuse; and raise critical issues of protecting the intellectual property of multimedia.

In large-scale media-sharing social networks, millions of users actively interact with one another; such user dynamics not only influence each individual user but also affect the system performance. An example is peer-to-peer (P2P) file sharing systems, in which users cooperate with one another to provide an inexpensive, scalable, and robust platform for distributed data sharing. Because of the voluntary and unregulated participation nature of these systems, user cooperation cannot be guaranteed in P2P networks, and recent studies showed that many users are free riders, sharing no files at all. To provide a predictable and satisfactory level of service, it is important to analyze the impact of human factors on media-sharing social networks, and to provide important guidelines for better design of multimedia systems. The area of human and social dynamics has recently been identified by the US National Science Foundation (NSF) as one of its five priority areas, which also demonstrates the importance of this emerging interdisciplinary research area.

This book, *Behavior Dynamics in Media-Sharing Social Networks*, aims to illustrate why human factors are important, to show that signal processing can be used effectively to model user dynamics, and to demonstrate how such understanding of human behavior can help improve system performance. We cover recent advances in media-sharing social networks, and study two different types of media-sharing social networks, multimedia fingerprinting and P2P live streaming social networks. We review the fundamental methodologies for modeling and analyzing human behavior, and investigate the impact of human dynamics on multimedia system design. Our goal is to encourage researchers from different areas to further explore the emerging research field of behavior modeling and forensics, to improve our understanding of user dynamics in media-sharing social networks, and ultimately to design systems with more efficient, secure, and personalized services.

We partition the book into five parts. In Part I, we illustrate the fundamental issues of media-sharing social networks, including quantitative social network analysis and media semantics, in Chapter 1; provide overviews on multimedia fingerprinting and P2P video streaming in Chapters 2 and 3, respectively; and offer an introduction to game theory that will be used throughout the later chapters in Chapter 4.

In Part II, the focus is on user dynamics in media-sharing social networks. The notion of equal-risk fairness in multimedia fingerprinting colluder social networks is addressed in Chapter 5, followed by the study of how to leverage side information to reach a better equilibrium via game-theoretical analysis in Chapter 6. The concept of risk–distortion tradeoff is considered in Chapter 7 to understand how optimal strategies may vary and depend on decisions of both attackers and detectors.

Because of the constant user interactions in social networks, cooperation becomes a major issue. Therefore, Part III is dedicated to the consideration of cooperation stimulation with the notion of fairness. Game-theoretic models with different bargaining strategies and fairness criteria are developed in Chapter 8 to study optimal strategies of feasible attacks in multimedia fingerprinting colluder social networks. In Chapter 9, an optimal cooperation strategy of cooperative stimulation in P2P video streaming is considered, followed by the study of the optimal price setting for mobile P2P video streaming in Chapter 10.

In Part IV, we turn our attention to the identification of misbehaving users. In multimedia fingerprinting social networks, even when colluders agree on a strategy, they may not execute accordingly. Instead, for example, they may cheat to take more advantage in further minimizing their own risk. In Chapter 11, such a traitor-within-traitor phenomenon is investigated. Similarly, the presence of the malicious attack will discourage nonmalicious users to join the social network. In Chapter 12, the design of methodologies to identify hostile users and the cheat-proof cooperation strategies against malicious attacks are considered.

Finally, in Part V, the impact of social network structure on the performance of social networking is discussed. In Chapter 13, the impact of centralized social networks with trusted ringleaders and distributed peer-structured social networks on multimedia fingerprinting collusion is considered, and in Chapter 14, forming a social structure with a group lead agent is investigated for P2P streaming social networks.

This book is intended to be a reference book or textbook for graduate-level courses such as social computing and networking, image and video communications and networking, and network/information security. We hope that the comprehensive coverage and a holistic treatment of media-sharing social networking will make this book a useful resource for readers who want to understand this emerging technology, as well as for those who conduct research and development in this field.

We would like to thank Mr. Yan Chen for his research contributions that are included in this book.

Part I

Introduction

1 Introduction to media-sharing social networks

With recent advances in communications, networking, and computer technologies, we have witnessed the emergence of large-scale user-centered web 2.0 applications that facilitate interactive information sharing and user collaboration via Internet – for example, blogs; wikis; media-sharing websites such as Napster, Flickr, and YouTube; social networking services such as Facebook, LinkedIn, and Twitter; and many others. Different from traditional web applications that allow only *passive* information viewing, these web 2.0 sites offer a platform for users to *actively* participate in and contribute to the content/service provided. The resulting trend toward social learning and networking creates a technological revolution for industries, and brings new experience to users.

The emergence of these websites has significant social impact and has profoundly changed our daily life. Increasingly, people use the Internet as a social medium to interact with one another and expand their social circles, to share information and experiences, and to organize communities and activities. For example, YouTube is a popular video-sharing website on which users upload, share, and view a wide variety of user-generated video content. It targets ordinary people who have Internet access but who may not have a technical background on computers and networking, and enables them to upload short video clips that are viewable to the worldwide audience within a few minutes. Its simplicity of use and the large variety of content offered on the website attract more than one billion views per day, according to a blog by Chad Hurley (cofounder of YouTube) on October 9, 2009, and make video sharing an important part of the new Internet culture.

According to the Alexa Global traffic ranking, among the 20 hottest websites, many of them are social networking and media-sharing websites – for example, Facebook, MySpace, Twitter, and YouTube. The increasing popularity of these interactive and user-centered websites also means new business opportunities. In July 2009, eMarketer projected that even with the current world economy hurdle, the amount of money US marketers spend on online social network advertising will reach $1.3 billion in 2010, a 13.2 percent increase compared with 2009 [1]. In a later report in December 2009, eMarketer predicted that in 2010, worldwide online advertising spending on Facebook would reach $605 million, corresponding to a 39 percent increase compared with 2009 [2]. In particular, it was predicted that the non-U.S. advertising spending on Facebook would increase by 65 percent in 2010 [2].

With recent advances in wireless communication technologies, mobile social networks have become increasingly popular. Internet-based social networks such as MySpace and Facebook have turned mobile; they enable mobile phone users to access their websites, upload mobile photos and videos to their profiles, and so forth. In addition, new mobile social networks that are designed specifically for mobile applications – for example, Foursquare, Loopt, and Gowalla, which enable users to explore and discover their local vicinity – have also emerged. A report released by Informa on March 25, 2010 forecast that US mobile social networking ad revenue would rise by 50 percent to $421 million in 2010, and would continue its robust growth into 2013 with a breakthrough of the $1 billion revenue mark [3]. Other potential applications for integrated sensor and social networks include traffic monitoring, human movement and behavior analysis, collaborative rehabilitation for seniors, and many others [4,5].

However, this emerging trend of social learning and networking also poses new challenges and raises critical issues that need to be addressed for further prolif-eration and development of such social networks. Listed below are just a few of them.

• With the resulting avalanche of information that users create, share, and distribute over networks, it is crucial to effectively manage these data and to support accurate and fast searching of information. This is particularly challenging for media objects (audio, image, and video), as the same audio or video clip (or its portions) may be processed in many different ways and appear in a variety of different contexts and formats [6]. As an example, a recent study of traffic flow in FastTrack, one of the largest peer-to-peer (P2P) file-sharing networks, showed that there were 26,715 different versions and 637,381 different copies of the song "Naughty Girl" on FastTrack. Among them, 62 percent of the versions and 73 percent of the copies were "polluted" – that is, either they were nondecodable or their time durations were significantly longer or shorter than the official CD release [7].

• Social networks facilitate easy information sharing among users, and the same easy access of such networks enables *anyone* to view the shared content [8]. Information sharing at such an unprecedented scale in social networks may post serious security and privacy concerns. For example, in medical and scientific research, collecting human behavior and health information requires very strict scrutiny, whereas social networks make it possible to collect such information much more easily without contacting the subjects [9]. Using and republishing such public information in research without informed consent may be considered as an invasion of privacy [9].

• Social networks may be misused and manipulated by people for defamation, profit, and many other purposes [10]. For example, researchers from Harvard University recently discovered that scammers created sophisticated programs to mimic legitimate YouTube traffic and to provide automated feedback for videos and other content they wished to promote [11]. A recent study by researchers at the University of California at Berkeley found that some eBay users artificially boost their reputations by buying and selling feedbacks, so they can seek higher prices on items that they sell [12].

From these examples, it can be seen that users play an active and important role in social networks, and the way in which they behave and use the huge amount of information available on social networks has a significant impact on system performance. These new challenges call for novel solutions to model user interactions and study the impact of human behavior on social networks, to analyze how users learn from each other as well as from past experiences, and to understand people's cognitive and social abilities. These solutions will facilitate the design of future societies and networks with enhanced performance, security, privacy, availability, and manageability. This is an interdisciplinary research area, covering signal processing, social signal processing, information science, sociology, psychology, and economics, in which signal and information processing plays a critical role. The advanced signal and information processing technologies will enable us to better characterize, understand, and ultimately influence human behaviors as desired.

This book focuses on an important class of social networks, media-sharing networks, in which users form a dynamically changing infrastructure to upload, exchange, distribute, and share images, videos, audio, games, and other media. Famous examples include YouTube, Napster, and Flickr. Also, many P2P file sharing systems – for example, BitTorrent and KaZaa – have been used to share digital media. Catching the current trend of delivering TV programs over the Internet, we have also seen many successful deployments of P2P live streaming, sometimes called P2PTV, in which video streams (typically TV programs) are delivered in real time on a P2P network. Examples of such P2PTV applications include PPLive, PPStream, SopCast, QQLive from China, Abroadcasting from the United States, and LiveStation from the United Kingdom. They attract millions of viewers, and the aggregated bandwidth consumption may reach hundreds of gigabits per second [13]. In this book, we study user behavior in media-sharing social networks and analyze the impact of human factors on multimedia signal design. We use two different types of media-sharing social networks, multimedia fingerprinting and P2P live streaming, as examples.

Before we move on to the modeling and analysis of user behavior in media-sharing social networks, we first quickly review recent advances in other research areas in media-sharing social networks, including social network analysis and media semantics in social networks.

1.1 Quantitative analysis of social networks

Social networks are defined as "social structures that can be represented as *networks* – as sets of *nodes* (for social system members) and sets of *ties* depicting their interconnections" [14]. The two elements, actors (or nodes) and relations, jointly form a social network. Actors can be individual persons, small groups, formal organizations, or even countries, who are connected to each other via certain relationships, such as friendship, trade, or colleagues. In addition to describing how a set of actors are connected to each other, *social network analysis* describes the underlying patterns of social structure and investigates their impact on individual behavior, as well as analyzing them on the system level [15].

1.1.1 Social network representation, notations, and relationship measures

1.1.1.1 Representation

There are two different methods to represent and analyze a social network, *sociogram* and *sociomatrix* [15]. In a sociogram, graphs and graph theory are used to visually represent and analyze social networks. Here, actors are denoted as points (also called *nodes* or *vertices*) and a relation (tie) is represented using a line (also called an *arc* or *edge*). A sociomatrix uses tabular matrices to depict social networks and to facilitate complex mathematical analysis. Here, an $N \times N$ matrix x is used to represent a social network with N actors, and the element x_{ij} at row i and column j represents the relationship between the ith and jth actors, in which actor i is the initiator and j is the recipient. These two representation methods are equivalent and contain the same information.

Different networks represent different relations, and there are many different types of social networks. If the relation is *nondirected*, or mutual – for example, classmates and colleagues – all lines in the graph representation have no arrowheads, and in the matrix representation, we have a symmetric matrix with $x_{ij} = x_{ji}$. Other types of relations are *directed*, and we have directed graphs with arrowheads in the lines. For example, A trusts B, whereas B may not trust A; therefore, there is one link from A to B but not vice versa.

In addition to directionality, the lines (arcs) in a social network may be binary or measured with different value scales. For example, in simple relations such as classmates or colleagues, there is either a line (presence) or no line (absence) between two nodes, which corresponds to binary networks. For other types of relations, each line not only indicates the *existence* of the relation, but also values the *intensity* of the relation. For example, one actor can rank other actors in the network as "friends," "acquaintances," or "strangers," indicating different levels of relations [15].

To summarize, there are four basic types of social networks: binary nondirected, binary directed, valued nondirected, and valued directed.

1.1.1.2 Notations

Graph theory is often used in social network analysis; thus, we first introduce some basic concepts in graph theory.

Given a social network represented using a graph, a *subgraph* is a subset of nodes and lines, in which all lines in the subgraph must be between pairs in the subgraph. A *walk* is an alternating sequence of incident nodes and lines, which connects the starting and the ending nodes; the *length* of a walk is the number of lines contained in the walk. A *path* is a walk with distinct nodes and lines – that is, every node and every line are visited only once in the walk.

A graph is *connected* if there is a path between every pair of nodes in the graph, and is called *disconnected* otherwise. A node that is not connected to any other nodes is called an *isolate*. A *graph component* is a maximal subgraph that forms a connected graph. In a connected graph, a node is a *cutpoint* if its removal would disconnect the graph, and a line is a *bridge* if its removal would disconnect the graph into two or more components. The notion of cutpoint and bridge is important in network analysis, as networks with cutpoints and bridges are more vulnerable to disruptions than those with many redundant paths to sustain information and resource flows [15].

1.1.1.3   Relationship measures

There are many important relationship measurements in graph theory – for example, nodal degree, geodesic distances, and density – which we will briefly introduce here.

Nodal degree: For a node i in a binary nondirected graph, its *nodal degree* is the total number of lines that are incident with it. With directed graphs, *nodal indegree* and *nodal outdegree* should be distinguished. Nodal indegree is the number of lines received by node i, and nodal outdegree is the number of lines sent by node i. For valued graphs, we can use the mean values of the lines connected to node i to represent its nodal degree. Nodal degree reflects the node's level of involvement in network activities [15], and the mean nodal degree averaged over all nodes shows the aggregate level of activity in the network.

Geodesic distance: The *geodesic distance* between a pair of nodes is the length of the shortest path that connects them. If there is no path between two nodes, then their geodesic distance is infinite or undefined. For directed graphs, the geodesic distance from node i to node j may be different from the geodesic distance from j to i. For example, node i may be able to send a message to j, but not vice versa. Geodesic distance measures the closeness of two nodes and plays an important role in distance-based analysis, such as in clustering analysis [15].

Density of a graph: *Density* measures the extent to which nodes in a graph are connected among themselves. For a binary nondirected graph with N nodes, its density D is the number of lines in the graph (L) divided by the maximum possible lines ($\binom{N}{2}$ when there is a direct link between any pair of nodes in the graph) – that is, $D = L/\binom{N}{2}$. For directed graphs, the denominator is changed to $2\binom{N}{2}$, because for each pair of nodes in a directed graph, there are two possible lines with different directions. For valued graphs, the numerator is replaced by the summation of all lines' values.

1.1.2   Centrality and prestige

An important usage of graph theory is to identify the "most important" actors and/or groups in social networks [15,16]; the concepts of *centrality* and *prestige* quantify an actor (or group)'s prominence (involvement in the network activities) in a network. An individual actor's prominence reflects its visibility to other actors in the networks, and at the group level, it evaluates the divergence of all group members' prominence [15]. The difference between centrality and prestige is whether the direction of lines counts.

In centrality, a prominent actor has many direct links with other actors regardless of the direction, whereas in prestige, a prominent actor receives many incoming lines but does not initiate many outgoing ties.

The most widely used centrality measures are *degree, closeness,* and *betweenness*. In this section, we use binary nondirected graphs to illustrate these concepts, and the definitions for directed and valued graphs are available in references [15,16].

Degree centrality: At the individual level, node i's *degree centrality* is defined as its nodal degree normalized by the total number of actors in the graph, and is a real number between 0 and 1. Actors with high degree centralities have more connections with others and higher visibility in the networks.

At the group level, group degree centrality measures the extend to which actors differ in terms of their individual degree centralities, and resembles the standard deviation of the individual degree centralities among group members [15]. When all group members have the same degree centrality, the group degree centrality is zero. In the other extreme case of a star graph, in which one node is connected to all other nodes but there is no connection between any other two nodes, the group degree centrality achieves the maximum possible value.

Closeness centrality: For node i, its *closeness* reflects how quickly it can interact with other actors, such as by communicating directly or via few intermediaries [15]. For node i, its actor closeness centrality index is the inverse of the mean geodesic distances between i and all other nodes in the graph, and it takes the smallest value when node i is directly linked to all others in the network. At the group level, the group closeness centralization index measures the extent to which actors differ in their individual closeness centralities.

Betweenness centrality: The communication between two nonadjacent nodes depend on other actors, especially those who are on the paths between these two. These "other actors" may potentially have some control over the interaction between these two nonadjacent actors, and the *betweenness centrality* concept quantifies how other actors control or mediate the relations between connected nodes [15]. Actor betweenness centrality measures the extent to which an actor lies on the shortest path between pairs of other actors, and the group-level betweenness measures the extend to which this value varies across group members. Detailed definitions and explanations of betweenness centrality can be found in references [15,16].

Prestige is used when it is much more important to specify the initiators and the recipients of relations than just giving mere participation [15]. It measures the extent to which an actor "receives" relations sent by others, and emphasizes inequality in control over information and/or resources. For node i in a directed graph, its actor degree prestige is its normalized indegree, and it takes a larger value when node i is more prestigious. A detailed discussion can be found in reference [16].

1.1.3   Cohesive subgroups

Another important task in social network analysis is to identify *cohesive subgroups* of actors who are connected via many direct, reciprocated choice relations, and who share information, achieve homogeneity of thoughts and behavior, and act collectively [15,16].

In graph theory, *clique* is an important concept for analyzing group structures and for understanding how cohesion benefits group members as well as restricts the range of social contacts [17,18]. A clique is "a maximal complete subgraph of three or more nodes, all of which are directly connected to one another, with no other node in the network having direct ties to every member of the clique" [15]. Thus, every pair of nodes in a clique is connected by a direct link, and their geodesic distance is 1. Furthermore, it rigorously separates members inside a cohesive subgroup from outsiders. Because of this very strict requirement, large cliques are seldom found in real networks [16].

To address this rigid definition, the *n-clique* concept is introduced, in which the geodesic distance between any pair of nodes cannot exceed n and no node can be more than n links away from any others [15]. A larger value of n makes the clique more inclusive (with more nodes) but less cohesive (among its members). Another possible solution is the *k-core* concept, based on nodal degrees; in a k-core subgroup, each node is adjacent to at least k other nodes in the subgroup [16]. There have also been definitions of cohesive subgroups derived from the relative closeness of ties within the subgroup as well as the relative distance of ties from subgroup members to outsiders [16]. Readers who are interested are referred to reference [16] for more discussions and detailed explanations.

1.1.4   Structural equivalence

Many works on social network analysis focus on the network role and position analysis – that is, study of actors' structural similarities and patterns of relations. Two actors are perfectly *structurally equivalent* if they have exactly identical patterns of links sent to and received from all other actors [15]. That is, node i and j are equivalent if and only if the following conditions hold: (1) if node i receives a link from node k, then there is also a link from node k to node j; and (2) if node i sends a link to node k, node j also sends a link to node k. Structurally equivalence often causes fierce competition, as one actor can be easily replaced by another without affecting the network structure.

In real networks, the above definition is often too rigorous to be useful. A more practical scenario is that some nodes may be *approximately* structurally equivalent – that is, their relations to other nodes are similar but not identical. Many works have been done to measure the relation similarity between two nodes – for example, the Euclidean distance-based measurement, the correlation-based definition, automorphic and isomorphic equivalence, and regular equivalence [16].

Given these relation similarity measurements, the next step is to partition actors into subsets (also called *positions*), in which actors in one subset are closer to being equivalent than those in different subsets. There are many different ways to partition actors, including convergence of iterated correlations (CONCOR), hierarchical cluster-ing, and multidimensional scaling. The last step is to describe the ties between and within positions – that is, how positions are related to each other. The commonly used methods – density tables, image matrices, reduced graphs, blockmodels, and relational algebras – have also been used for algebraic analysis of role systems. Details can be found in reference [16].

1.1.5   Other methods for network analysis

In the preceding sections, we focused on the study of "one-mode" networks linking actors to actors. *Affiliation networks*, also called *membership networks*, represent the involvement of a set of actors in a set of social events. An affiliation network is a "two-mode" network containing two types of nodes, actors, and events, and a set of relations between each nodal type. Research on affiliation network analysis aims to uncover the relational structures among actors through their joint involvement in events, and to reveal the relational structure of events attracting common participants [15].

A binary affiliation network can be represented using an *affiliation matrix x*, where $x_{ij} = 1$ if actor i participates in event j and $x_{ij} = 0$ otherwise. It can also be represented using a *bipartite graph*, where nodes are partitioned into two subsets, one including all the actors and the other with all the events, and one line in the graph links one actor to one event [16]. Galois lattices and correspondence analysis are often used to analyze affiliation networks; interested readers are referred to references [14,15] for detailed discussions.

In addition to the preceding analysis of deterministic (also called descriptive) social networks, probability and statistics have also been introduced in social network analysis [16]. Research topics in this area include statistical analysis of reciprocity and mutuality, structure inference, modeling and prediction for processes on network graphs, analysis of network flow data, and many others [16,19,20]. Readers who are interested are referred to references [16,19,20] for recent advances in this area.

Traditional social network analysis treats the network as a *static* graph, which is generated either from data aggregated over a long period of time or from data collected at a specific time instance. Such analysis ignores the temporal evolution of social networks and communities. To address this issue that has been overlooked, recently, there has been a growing trend to analyze how communities evolve over time in *dynamic* networks [21–24]. Lin *et al.* [25] proposed a unified framework to analyze communities and their evolution, in which the community structure provides evidence about how they evolve, and at the same time, the evolutionary history suggests which community structure is more appropriate.

1.2   Understanding media semantics in media-sharing networks

With the increasing popularity of media-sharing social networks, an important issue is to effectively manage these "billion-scale" social media and support accurate and efficient search of media objects. It requires accurate interpretation and understanding of media semantics. A promising approach is to annotate digital media with a set of keywords, such as "bridge" and "airplane" (sometimes called *labels*, *concepts*, or *tags*, in different contexts), to facilitate searching and browsing [26]. In this section, we first quickly review recent advances in social media annotation. We then focus on recent study on the emergent and evolutionary aspects of semantics and on leveraging social processes to understand media semantics.

1.2.1   Social media annotation

Depending on the flexibility of the keywords used to annotate digital media, media annotation methods can be classified into two categories, *labeling* and *tagging* [26]. In labeling, given a fixed concept set often called *ontology*, annotators decides whether a media object is relevant or irrelevant to a concept, whereas in tagging, which is ontology-free, users can freely choose a few keywords to annotate a media object.

The labeling process can be either manual or automatic. Manual labeling by users is tedious, labor-intensive, and time-consuming; Hua and Qi [27] proposed that the future trend for large-scale annotation is to leverage Internet users to contribute efforts. Ontology-driven automatic media annotation (also called *concept detection* or *high-level feature extraction*) and the closely related research area of content-based image/video retrieval have attracted much research activity in the past decades. These methods extract low-level content-based features that can be easily computed from digital media (for example, color histograms) and map them to high-level concepts that are meaningful and intuitive to humans. The mapping of low-level numerical features to high-level concepts (labels) is often done using learning algorithms – for example, neural networks, support vector machines, manifold learning, and feedback learning [26,28]. Recently, many automatic annotation systems have also used external information – for example, location information– to further improve the accuracy. Interested readers are referred to references [29,30] for review of recent works in this area.

Tagging enables users to freely choose the keywords (tags), and arguably provides better user experience [26]. Many social media websites, including Flickr and YouTube, have adopted this approach and encouraged users to provide tags to facilitate data management. In addition, the recent ESP Game motivates users to compete in annotating photos with freely chosen keywords in a gaming environment [31]. However, the free-form nature of tagging also poses new challenges: tags are often inaccurate, wrong, or ambiguous, and often may not reflect the content of the media [26]. To address the issue of "noisy" user-contributed tags, other context cues, such as time, geography-tags, and visual features, are fused with user-contributed tags to improve the search result and to recommend relevant tags [32]. Another approach to social tagging is to rank tags – for example, according to their relevance to the media content [33] and/or their clarity in content description [34] – to improve the visual search performance.

1.2.2   Semantic diversity

There are a few implicit assumptions in this concept detection framework – that is, the concept semantics is stable and the context is consistent. However, social media objects shared online originate from an unlimited number of sources [29], and there are an extraordinary large number of concepts that may not be shared universally [35]. The same keyword (concept) may have totally different meanings in different contexts. For example, on Flickr, the tag "yamagata" may refer to the Japanese town, the visual artist Hiro Yamagata, or the singer Rachel Yamagata [25]. Therefore, it is unrealistic to build one single classifier to learn all concepts for all media available on the networks, and it is of crucial importance to address this "semantic diversity" or "domain diversity" issue in media semantics in social networks [29,35].

To address this issue, some adaptive and cross-domain learning methods have been proposed, which efficiently and effectively adapt the concept detectors to new domains [36–38]. Zunjarwad *et al.* [39] proposed a framework that combines three forms of knowledge – global (feature-based distance), personal (tag co-occurrence probability), and social trust (finding people with correlated experience). The basic idea there is to use social trust and personal knowledge, and to recommend annotations only from people who share similar experience and opinions.

1.2.3   Emergent semantics

Social networks are highly dynamic and time-evolving, and so are media semantics. In real-world social networks, the visual representations of abstract concepts may change over time, and new concepts may emerge. In addition, some transient concepts that are relevant only to a specific event may exist for only a short period time [35]. In media-sharing social networks, it is important to consider and analyze the dynamic, emergent, and evolutionary aspects of media semantics owing to (explicit or implicit) collaborative activity, which is often ignored in media computing society [25]. Such investigation helps study how human beings interact with, consume, and share media data, and opens new views to the understanding of the relationship between digital media and human activities [25].

Emergent semantics has been studied in distributed cognition and sociology, and is defined by Cudré-Mauroux [40] as "a set of principles and techniques analyzing the evolution of decentralized semantic structures in large scale distributed information systems." It not only addresses how semantics are represented, but also analyzes how self-organizing and distributed agents discover the proper representation of symbols (concepts) via active interaction among themselves [40]. Lin *et al.* [25] provided a review of recent works in emergent and evolutionary semantics in media-sharing social networks, interested readers are referred to reference [25] and the references therein for detailed discussion.

One challenging problem in emergent media semantics is *community discovery* – that is, how to extract human communities that collaborate on certain topics or activities [25]. For example, Flickr allows people to connect their images to communities via image "groups," in which images shared by a group of users are organized under a coherent theme [41]. But the challenge is to find the *right* community that will ensure *reachability* to other users for useful comments. Lin *et al.* [42] used the concept of "mutual awareness" to discover and model the dynamics of thematic communities. That is, users are aware of each other's presence through observable interaction (e.g., comments, hyperlinks, trackbacks), and the expansion of mutual awareness leads to community formation. The work of Lin and colleagues [43] extracted grammatical properties (the triplets of *people*, *actions*, and *media artifacts*) of interactions within communities, which will help generalize descriptors of communities. Lin *et al.* [44] analyzed the temporal evolving patterns of visual content and context in Flickr image groups, in an effort to understand the changing interest of users and to infer the genres of images shared in the group.

Another challenge is to characterize the *information flow* (or communication flow) in social networks and the roles that individuals play within the networks [25]. Such analyses are important in information source ranking and quality assessment, and identification of suitable time periods for marketing [25]. Choudhury and co-workers [45] proposed a temporal prediction framework to determine communication flow between members in a network, including the *intent to communicate* (the probability that one user wants to talk to another person) and the *communication delay* (the time taken to send a message). Their work showed that social context greatly affects information flow in social networks, in which *social context* refers to the patterns of participation (information roles) and the degree of overlap of friends between people (strength of ties). Choudhury *et al.* [46] developed a multiscale (individual, group, and community) characterization of communication dynamics; their analysis of the technology blogs (Engadget) showed that communication dynamics can be a strong predictor of future events in the stock market. Choudhury and colleagues [47] studied the temporal phenomenon *social synchrony*, in which a large number of users mimic a certain action over a period of time with sustained participation from early users, and a computational framework was proposed to predict synchrony of actions in online social networks.

2 Overview of multimedia fingerprinting

During the past decade, increasingly advanced technologies have made it easier to compress, distribute, and store multimedia content. Multimedia standards, such as JPEG, MPEG, and H.26x [48–51], have been adopted internationally for various multimedia applications. Simultaneously, advances in wireless and networking technologies, along with a significant decrease in the cost for storage media, has led to the proliferation of multimedia data. This convergence of networking, computing, and multimedia technologies has collapsed the distance separating the ability to create content and the ability to consume content.

The alteration, repackaging, and redistribution of multimedia content pose a serious threat to both governmental security and commercial markets. The ability to securely and reliably exchange multimedia information is of strategic importance in fighting an unprecedented war against terrorism. A recent high-profile leak involved a classified video of Osama bin Laden's camp captured by an unmanned aerial surveillance vehicle, when one copy of the tapes shared between the Pentagon and CIA officials was leaked to the news media [52]. Without effective traitor-tracing tools, different agencies would still be reluctant to share critical information, which jeopardizes the mission of fighting terrorism and defending national and global security. To prevent information from leaking out of an authorized circle, it is essential that the governments have the forensic capability to track and identify entities involved in unauthorized redistribution of multimedia information.

In addition to the demands from homeland security, preventing the leak of multimedia information is also crucial to the economy. The US copyright industries, which includes prerecorded CD/DVDs and tapes as well as videos, motion pictures, and periodicals, accounts for about 5.2 percent of the US gross domestic product (GDP), or \$531.1 billion, and are responsible for close to 6 percent of all US employment [53]. The copyright industries, however, are experiencing a substantial decline in income and job positions, which is largely attributed to piracy. For example, US music sales by unit were reported to having dropped 31 percent from mid-2000 to 2003. Hollywood is actively seeking technologies whereby each preview copy of a new movie is individually and invisibly labeled prior to sending to Academy Award-voting members to prevent the leak to the market. A preliminary technology based on robust watermarking was adopted in the 2004 Academy Award season and successfully captured a few pirates [54]. As with other information security and forensics research, this "cat-and-mouse" game between

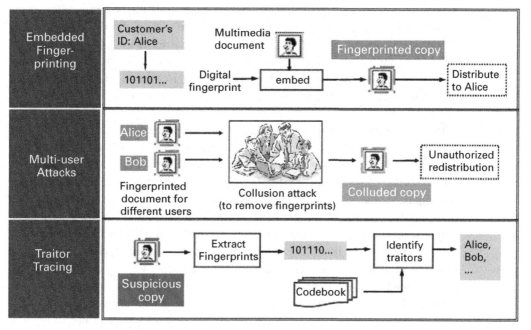

Fig. 2.1 Embedded fingerprinting for traitor tracing

technology developers and adversaries is becoming more intense, as smarter attacks pose serious challenges to the existing technologies for media security and rights management.

To protect the value of multimedia, content providers must have a suite of forensic tools that will allow them to track and identify traitors involved in the fraudulent use of media. Traitor-tracing fingerprinting is an emerging technology to enforce digital rights polices, whereby unique labels, known as *digital fingerprints*, are inserted into content prior to distribution to assist in investigating how unauthorized content was created, and what entities were involved in forming the fraudulent media.

This chapter reviews the basics of traitor-tracing multimedia fingerprinting. After a brief overview of traitor-tracing multimedia fingerprinting, we steer our attention to scalable video fingerprinting, in which users receive fingerprinted copies of different resolutions because of network and device heterogeneity. Detailed formulation of the fingerprint embedding and multiuser collusion attacks establishes a foundation to unveil our technical discussion on behavior modeling and analysis in the subsequent chapters.

2.1   Traitor-tracing multimedia fingerprinting

As shown in Figure 2.1, digital fingerprinting labels each distributed copy with the corresponding user's identification information, known as a *fingerprint*, which can be used to trace culprits who use their copies illegally. Fingerprints are embedded into the host signal using traditional data-hiding techniques [55–57], and human visual/audio models [58,59] are used to control the energy and achieve the imperceptibility of the embedded fingerprints. When the digital rights enforcer discovers the existence of an illegally redistributed copy of multimedia, he or she extracts the fingerprint from the suspicious copy. Correlation-based detection statistics [60–62] are often used to measure the similarity between the extracted fingerprint and each of the original fingerprints, and users whose detection statistics are above a predetermined threshold are identified as suspicious attackers.

However, protecting digital fingerprints is no longer a traditional security issue with a single adversary. The global nature of the Internet has enabled a group of attackers (colluders) to work together and collectively mount attacks to remove the fingerprints. These attacks, known as *multiuser collusion*, pose serious threats to intellectual property rights. Analysis of the strategies, capabilities, and limitations of attackers is an indispensable and crucial part of research in multimedia security.

Linear collusion is one of the most feasible collusion attacks that may be employed against multimedia fingerprinting [63–65]. Given K different fingerprinted signals $\{\mathbf{X}^{(i)}\}$ of the same content, attackers generate a colluded copy $\mathbf{Y} = \sum_k a_k \mathbf{X}^{(k)}$, where the weights satisfy $\sum_k a_k = 1$ to maintain the average intensity of the original multimedia signal (thus the perceptual quality of the attacked copy). With *orthogonal* fingerprinting, such an averaging attenuates the energy of the kth contributing fingerprint by a factor of a_k^2 and thus reduces colluder k's probability of being detected. Ergun *et al.* [63] modeled collusion attacks as averaging differently fingerprinted copies with equal weights (that is, $a_k = 1/K$) followed by the addition of noise. Their work showed that $O\left(\sqrt{N/\log N}\right)$ colluders are sufficient to defeat the underlying fingerprinting system, where N is the fingerprint length.

In addition to linear averaging, another important class of collusion attacks is based on operations as taking the minimum, maximum, and median of corresponding components of the fingerprinted signals [66]. For example, given K fingerprinted signals $\{\mathbf{X}^{(i)}\}$, to generate the jth component of the colluded copy $\mathbf{Y}(j)$, colluders use the minimum value of $\mathbf{X}^{(1)}(j)$, $\mathbf{X}^{(2)}(j)$, ..., $\mathbf{X}^{(K)}(j)$ and let $\mathbf{Y}(j) = \min\left(\{\mathbf{X}^{(k)}(j)\}\right)$. Because each fingerprinted copy is expected to have high perceptual quality, colluders have high confidence that $\mathbf{Y}(j)$ is within the JND range. Similarly, colluders can also let $\mathbf{Y}(j) = \max\left(\{\mathbf{X}^{(k)}(j)\}\right)$ and take the maximum value of $\{\mathbf{X}^{(i)}(j)\}$. They can also use the median value and select $\mathbf{Y}(j) = \text{median}\left(\{\mathbf{X}^{(k)}(j)\}\right)$. Detailed analysis of linear and nonlinear collusion attacks on orthogonal fingerprints was provided in reference [67]. The gradient attack was proposed by Kirovski and Mihcak [68]; this uses the combination of several basic nonlinear collusion attacks given in reference [67]. The work of Wang *et al.* [61] evaluated the collusion resistance of multimedia fingerprints as a function of system parameters, including fingerprint length, total number of users, and system requirements.

Collusion attacks pose serious threats to multimedia intellectual property rights. To provide reliable and trustworthy traitor-tracing performance, it is important to design anticollusion fingerprints. In the literature, techniques from a wide range of disciplines were used to improve the fingerprinting system's collusion resistance. A two-layer fingerprint design scheme was proposed by Zane [69], in which the inner code from spread spectrum embedding [58,59] is combined with an outer error-correcting code

(ECC) [70]. A permuted subsegment embedding technique and a group-based joint coding and embedding technique were proposed by He and Wu [71] to improve the collusion resistance of ECC-based multimedia fingerprinting while maintaining the detection efficiency. Dittmann *et al.* [72] used finite projective geometry to generate codes whose overlap with each other can identify colluding users. The anticollusion code based on combinatorial theories was proposed by Trappe *et al.* [62]. Wang *et al.* [73] used prior knowledge of the possible collusion patterns to improve the collusion resistance of the fingerprinting systems. The anticollusion dithering technique was proposed by Varna and colleagues [74] to resist multiuser collusion attacks for compressed multimedia. Readers who are interested in anticollusion fingerprint design are referred to reference [75] for a detailed discussion of current research in this area.

2.2   Scalable video coding system

Most prior works on multimedia fingerprinting focused on the scenario in which all colluders receive fingerprinted copies of the same resolution. In reality, as we experience the convergence of networks, communications, and multimedia, scalability in multimedia coding becomes a critical issue to support universal media access and to provide rich media access from anywhere using any devices [76]. Thus, it is of immediate importance to study multimedia fingerprinting with scalable coding and investigate collusion-resistant traitor tracing techniques when users receive fingerprinted copies of different resolutions [77].

To achieve scalability, we use layered video coding and decompose the video content into nonoverlapping streams (layers) with different priorities [76]. The base layer contains the most important information of the video sequence and is received by all users in the system. The enhancement layers gradually refine the resolution of the reconstructed copy at the decoder's side and are received only by those who have sufficient bandwidth.

Figure 2.2 shows the block diagrams of a three-layer scalable codec. The encoder downsamples the raw video and performs lossy compression to generate the base layer bit stream. The encoder then calculates the difference between the original video sequence and the upsampled base layer, and applies lossy compression to this residue to generate the enhancement layer bit streams. At the receiver's side, to reconstruct a high-resolution video, the decoder must first receive and decode both the base layer and the enhancement layer bit streams. The upsampled base layer is then combined with the enhancement layer refinements to form the high-resolution decoded video.

As an example, we use temporally scalable video coding, which provides multiple versions of the same video with different frame rates. Our analysis can also be applied to other types of scalability, as the scalable codec in Figure 2.2 is generic and can be used to achieve different types of scalability. The simplest way to perform temporal decimation and temporal interpolation is by frame skipping and frame copying, respectively. For example, temporal decimation with a ratio of 2:1 can be achieved by discarding one frame from every two frames, and temporal interpolation with a ratio of 1:2 can be

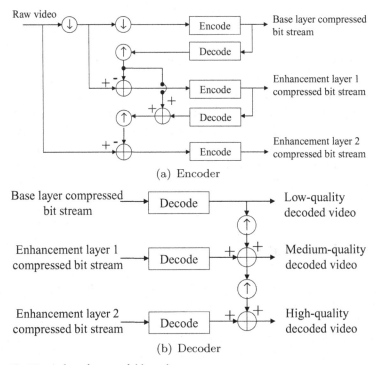

Fig. 2.2   A three-layer scalable codec

realized by making a copy of each frame and transmitting the two frames to the next stage.

We consider a temporally scalable video coding system with three-layer scalability, and use frame skipping and frame copying to implement temporal decimation and interpolation, respectively. In such a video coding system, different frames in the video sequence are encoded in different layers.

Define F_b, F_{e1}, and F_{e2} as the sets containing the indices of the frames that are encoded in the base layer, enhancement layer 1, and enhancement layer 2, respectively. Define $F^{(i)}$ as the set containing the indices of the frames that user i receives. Define $\mathbf{U}^b \triangleq \{i : F^{(i)} = F_b\}$ as the subgroup of users who subscribe to the lowest resolution and receive the base layer bit stream only; $\mathbf{U}^{b,e1} \triangleq \{i : F^{(i)} = F_b \cup F_{e1}\}$ is the subgroup of users who subscribe to the medium resolution and receive both the base layer and the enhancement layer 1; and $\mathbf{U}^{all} \triangleq \{i : F^{(i)} = F_b \cup F_{e1} \cup F_{e2}\}$ is the subgroup of users who subscribe to the highest resolution and receive all three layers. $\mathbf{U}^b$, $\mathbf{U}^{b,e1}$, and $\mathbf{U}^{all}$ are mutually exclusive, and $M = |\mathbf{U}^b| + |\mathbf{U}^{b,e1}| + |\mathbf{U}^{all}|$ is the total number of users.

2.3   Scalable video fingerprinting

We consider a digital fingerprinting system that consists of three parts: fingerprint embedding, collusion attacks, and fingerprint detection. We use temporal scalability as an example and analyze the fairness issue during collusion. In this scenario, fingerprints embedded at different layers will not interfere with one another. Our model can also be applied to other types of scalability, such as spatial and signal-to-noise ratio (SNR) scalability. However, with spatial or SNR scalability, the content owner must take special care during fingerprint design and embedding to prevent fingerprints at different layers from interfering with one another. This issue of fingerprint design and embedding is beyond the scope of this book.

2.3.1 Fingerprint embedding

Spread spectrum embedding is a popular data hiding technique owing to its robustness against many attacks. Here, we briefly go through the embedding process and clarify the notations used in the following chapters.

For the jth frame in the video sequence represented by a vector $\mathbf{S}_j$ of length N_j, and for each user i who subscribes to frame j, the content owner generates a unique fingerprint $\mathbf{W}_j^{(i)}$ of length N_j. The fingerprinted frame j that will be distributed to user i is $X_j^{(i)}(k) = S_j(k) + JND_j(k) \cdot W_j^{(i)}(k)$, where $X_j^{(i)}(k)$, $S_j(k)$, and $W_j^{(i)}(k)$ are the kth components of the fingerprinted frame $\mathbf{X}_j^{(i)}$, the host signal $\mathbf{S}_j$, and the fingerprint vector $\mathbf{W}_j^{(i)}$, respectively. JND_j is the just-noticeable difference from human visual models, and it is used to control the energy and achieve the imperceptibility of the embedded fingerprints. Finally, the content owner transmits to user i all the fingerprinted frames $\left\{\mathbf{X}_j^{(i)}\right\}$ to which he or she subscribes.

We apply orthogonal fingerprint modulation and assume that the total number of users is much smaller than the length of the embedded fingerprints. For each frame j in the video sequence, with orthogonal modulation, fingerprints for different users are orthogonal to each other and have the same energy; thus, for users i_1 and i_2,

$$\langle \mathbf{W}_j^{(i_1)}, \mathbf{W}_j^{(i_2)} \rangle = ||\mathbf{W}_j||^2 \delta_{i_1,i_2}, \tag{2.1}$$

where δ_{i_1,i_2} is the Dirac–Delta function. $||\mathbf{W}_j||^2 = N_j \cdot \sigma_w^2$, where σ_w^2 is the variance of the watermark $\mathbf{W}_j^{(i)}$.

2.3.2 Multiuser collusion attacks

Attackers apply multiuser collusion attacks to remove traces of the embedded fingerprints. As discussed by Wang et al. [61], with orthogonal fingerprint modulation, nonlinear collusion attacks can be modeled as the averaging attack followed by additive noise. Under the constraint that the colluded copies from different collusion attacks have the same perceptual quality, different collusion attacks have approximately identical performance. Therefore, it suffices to consider the averaging-based collusion only.

An important issue in collusion is to ensure its fairness. We consider in this chapter the simplest equal-risk fair collusion, in which all colluders have the same probability of being caught. There are also other definitions of fairness, which will be explored in Chapter 8.

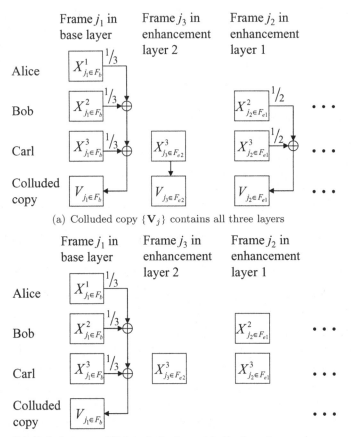

(a) Colluded copy $\{\mathbf{V}_j\}$ contains all three layers

(b) Colluded copy $\{\mathbf{V}_j\}$ includes frames in the base layer only

Fig. 2.3 Two trivial solutions of collusion by averaging all fingerprinted copies

When colluders receive copies of the same quality, averaging all copies with the same weight reduces the energy of each contributing fingerprint by an equal amount, and therefore gives each colluder the same probability of being detected. However, achieving equal-risk fairness is much more complicated when colluders receive copies of different resolutions owing to network and device heterogeneity, especially when the attackers wish to generate a copy of high resolution.

With the above temporally scalable fingerprinting system, we consider a simple example of collusion including three attackers: Alice, who receives the base layer only; Bob, who receives the base layer and enhancement layer 1; and Carl, who receives all three layers. Figure 2.3 shows two trivial solutions of collusion by averaging the three fingerprinted copies. In Figure 2.3(a), the colluded copy includes all three layers and is generated as follows:

- For each frame $j_1 \in F_b$ in the base layer, colluders average the three copies of fingerprinted frame j_1 that they have and generate $\mathbf{V}_{j_1 \in F_b} = \frac{1}{3}\left(\mathbf{X}_{j_1}^1 + \mathbf{X}_{j_1}^2 + \mathbf{X}_{j_1}^3\right)$.

- For each frame $j_2 \in F_{e1}$ in enhancement layer 1, colluders average the fingerprinted frame j_2 from Bob and Carl, respectively, and $\mathbf{V}_{j_2 \in F_{e1}} = \frac{1}{2}\left(\mathbf{X}_{j_2}^2 + \mathbf{X}_{j_2}^3\right)$.
- For each frame $j_3 \in F_{e2}$ in enhancement layer 2, frame j_3 in the colluded copy equals to that in the copy from Carl and $\mathbf{V}_{j_3 \in F_{e2}} = \mathbf{X}_{j_3}^3$.

In the colluded copy in Figure 2.3(a), the three fingerprints corresponding to the three attackers have the same energy in the base layer; whereas the enhancement layers contain only Bob and Carl's fingerprints, not the fingerprint identifying Alice. It is obvious that among the three, Carl has the largest probability of being caught and Alice takes the smallest risk. Consequently, the collusion in Figure 2.3(a) does not achieve equal-risk fairness.

In Figure 2.3(b), colluders generate an attacked copy consisting of the base layer only, and the colluded copy equals to $\mathbf{V}_{j_1 \in F_b} = \frac{1}{3}\left(\mathbf{X}_{j_1}^1 + \mathbf{X}_{j_1}^2 + \mathbf{X}_{j_1}^3\right)$ for each frame $j_1 \in F_b$ in the base layer. Under the collusion in Figure 2.3(b), the fingerprints corresponding to the three attackers have the same energy in the colluded copy; therefore, the three attackers have the same probability of being detected. Although the collusion in Figure 2.3(b) ensures equal risk for all colluders, the attacked copy has low resolution.

The question, therefore, is when there is difference in the resolution of fingerprinted copies owing to network and device heterogeneity, how colluders should conduct fair multiuser collusion that guarantees the collective equal risk among all attackers while still generating an attacked copy of high resolution. Assume that there are a total of K colluders, and SC is the set containing their indices. During collusion, colluders first divide themselves into three nonoverlapping subgroups: $SC^b \triangleq \{i \in SC : F^{(i)} = F_b\}$ contains the indices of colluders who receive the base layer only; $SC^{b,el} \triangleq \{i \in SC : F^{(i)} = F_b \cup F_{e1}\}$ contains the indices of colluders who receive the base layer and enhancement layer 1; and $SC^{all} \triangleq \{i \in SC : F^{(i)} = F_b \cup F_{e1} \cup F_{e2}\}$ contains the indices of colluders who receive all three layers. Define K^b, $K^{b,el}$, and K^{all} as the number of colluders in SC^b, $SC^{b,el}$, and SC^{all}, respectively.

Then, colluders apply the *intragroup* **collusion attacks:**

- For each frame $j \in F_b$ that they received, colluders in the subgroup SC^b generate $\mathbf{Z}_j^b = \sum_{i \in SC^b} \mathbf{X}_j^{(i)}/K^b$.
- For each frame $j \in F_b \cup F_{e1}$ that they received, colluders in the subgroup $SC^{b,el}$ generate $\mathbf{Z}_j^{b,el} = \sum_{i \in SC^{b,el}} \mathbf{X}_j^{(i)}/K^{b,el}$.
- For each frame $j \in F_b \cup F_{e1} \cup F_{e2}$ that they received, colluders in the subgroup SC^{all} generate $\mathbf{Z}_j^{all} = \sum_{i \in SC^{all}} \mathbf{X}_j^{(i)}/K^{all}$.

Define F^c as the set containing the indices of the frames that are in the colluded copy, and $F^c \in \{F_b, F_b \cup F_{e1}, F_b \cup F_{e1} \cup F_{e2}\}$. Then, colluders apply the *intergroup* **collusion attacks** to generate the colluded copy $\{\mathbf{V}_j\}_{j \in F^c}$:

- For each frame $j_1 \in F_b$ in the base layer, $\mathbf{V}_{j_1} = \beta_1 \mathbf{Z}_{j_1}^b + \beta_2 \mathbf{Z}_{j_1}^{b,el} + \beta_3 \mathbf{Z}_{j_1}^{all} + \mathbf{n}_{j_1}$. To maintain the average intensity of the original host signal and ensure the quality of the colluded copy, we let $\beta_1 + \beta_2 + \beta_3 = 1$. Our analysis can also be applied to other scenarios where $\beta_1 + \beta_2 + \beta_3 \neq 1$. To guarantee that the energy of each of the original

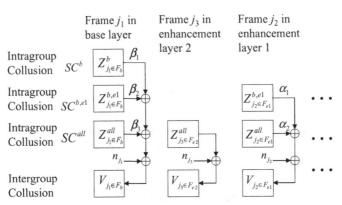

Fig. 2.4 Intragroup and the intergroup collusion attacks

fingerprints is reduced, we select $0 \leq \beta_1,\ \beta_2,\ \beta_3 \leq 1$. $\mathbf{n}_{j_1}$ is the additive noise that colluders add to $\mathbf{V}_{j_1}$ to further hinder detection.

- If $F_{e1} \subset F^c$ and the colluded copy contains frames in the enhancement layers, then for each frame $j_2 \in F_{e1}$ in the enhancement layer 1, $\mathbf{V}_{j_2} = \alpha_1 \mathbf{Z}_{j_2}^{b,e1} + \alpha_2 \mathbf{Z}_{j_2}^{all} + \mathbf{n}_{j_2}$, where $0 \leq \alpha_1,\ \alpha_2 \leq \alpha_1 + \alpha_2 = 1$, and $\mathbf{n}_{j_2}$ is additive noise. Our analysis can also be extended to the more general case of $\alpha_1 + \alpha_2 \neq 1$.
- If $F_{e2} \subset F^c$ and the colluded copy contains frames in all three layers, then for each frame $j_3 \in F_{e2}$ in enhancement layer 2, $\mathbf{V}_{j_3} = \mathbf{Z}_{j_3}^{all} + \mathbf{n}_{j_3}$, where $\mathbf{n}_{j_3}$ is additive noise.

Colluders adjust the energy of the additive noises to ensure that frames of similar content at different layers have approximately the same perceptual quality. We consider challenging scenarios with a large number of colluders (e.g., more than 100 attackers). In addition, we consider scenarios in which the energy of the additive noise $\mathbf{n}_j$ is comparable with that of the originally embedded fingerprints and the final colluded copy has good quality. For frame j_1 in the base layer, frame j_2 in enhancement layer 1, and frame j_3 in enhancement layer 2 that have similar content, we can show that this requirement can be simplified to $||\mathbf{n}_{j_1}||^2 \approx ||\mathbf{n}_{j_2}||^2 \approx ||\mathbf{n}_{j_3}||^2$ in the scenarios in which we are interested.

During collusion, colluders seek the *collusion parameters*, F^c, $\{\beta_k\}_{k=1,2,3}$, and $\{\alpha_l\}_{l=1,2}$, to ensure that all colluders have the same probability to be captured. The detailed analysis is in Section 5.1.

2.3.3   Fingerprint detection and colluder identification

When the content owner discovers the unauthorized redistribution of $\{\mathbf{V}_j\}_{j \in F^c}$, the owner applies a fingerprint detection process to identify the colluders.

There are two main detection scenarios, blind and nonblind detection. In the blind detection scenario, the host signal is not available to the detector and serves as additional noise during detection, whereas in the nonblind scenario, the host signal is available to the detector and is first removed from the test copy before detection. Different from other data-hiding applications in which blind detection is preferred or required, in many fingerprinting applications, the fingerprint verification and colluder identification process is usually handled by the content owner or an authorized forensic party who can have access to the original host signal. Therefore, a nonblind detection scenario is feasible and often preferred in multimedia fingerprinting applications.

For each frame $\mathbf{V}_j$ in the colluded copy, the detector first extracts the fingerprint $\mathbf{Y}_j = (\mathbf{V}_j - \mathbf{S}_j)/JND_j$. Then, the detector calculates the similarity between the extracted fingerprint $\{\mathbf{Y}_j\}_{j \in F^c}$ and each of the M original fingerprints $\{\mathbf{W}_j^{(i)}\}_{j \in F^{(i)}}$, compares with a threshold, and outputs a set $\widehat{SC}$ containing the estimated indices of the colluders. We use the correlation-based detection statistics to measure the similarity between the extracted fingerprint and the original fingerprint. The fingerprint detector can use fingerprints extracted from all layers collectively to identify colluders. With the collective fingerprint detector, for user i, the detector first calculates $\check{F}^{(i)} \triangleq F^{(i)} \cap F^c$, where $F^{(i)}$ contains the indices of the frames received by user i and F^c contains the indices of the frames in the colluded copy. Then the fingerprint detector calculates

$$TN^{(i)} = \left(\sum_{j \in \check{F}^{(i)}} \langle \mathbf{Y}_j, \mathbf{W}_j^{(i)} \rangle \right) \bigg/ \sqrt{\sum_{j \in \check{F}^{(i)}} ||\mathbf{W}_j^{(i)}||^2}, \tag{2.2}$$

where $||\mathbf{W}_j^{(i)}||$ is the Euclidean norm of $\mathbf{W}_j^{(i)}$. The fingerprint detector can also use fingerprints extracted from each individual layer to identify colluders; the details will be discussed in Chapter 6. Given the M detection statistics $\{TN^{(i)}\}_{i=1,\cdots,M}$ and a predetermined threshold h, the estimated colluder set is $\widehat{SC} = \{i : TN^{(i)} > h\}$.

2.3.4 Performance criteria

To measure the temporal resolution of the colluded copy, we use the total number of frames in the colluded copy $L^c = |F^c|$ (or, equivalently, the frame rate of the colluded copy). $L^c = |F_b|$, $L^c = |F_b| + |F_{e1}|$, and $L^c = |F_b| + |F_{e1}| + |F_{e2}|$ correspond to the three scenarios in which the colluded copy has the lowest, medium, and highest temporal resolution, respectively.

To measure the collusion resistance of a multimedia fingerprinting systems, the most commonly used criteria are

- P_d: the probability of capturing at least one colluder;
- P_{fp}: the probability of accusing at least one innocent user;
- $E[F_d]$: the expected fraction of colluders who are successfully captured; and
- $E[F_{fp}]$: the expected fraction of innocent users who are falsely accused.

Based on these criteria, both the fingerprint detector and colluders can evaluate the effectiveness of their strategies to make decisions.

3 Overview of mesh-pull peer-to-peer video streaming

With recent advances in networking, multimedia signal processing, and communication technologies, we have witnessed the emergence of large-scale video streaming social networks, in which millions of users form a distributed and dynamically changing infrastructure to share video streams. Statistics showed that more than 75 percent of the total US Internet audience have viewed online video, and the average online video viewer watched four hours of video per month [78]. With the fast deployment of high-speed residential network access, video is expected to be the dominating traffic on the Internet in the near future.

The traditional method for video streaming over the Internet is the client-server service model. A client sets up a connection with a video source server and video content is directly streamed to the client from either the video source server or a nearby content delivery server. The most popular client-server video stream service nowadays is YouTube, which drew 5 billion US video views in July 2008. However, client-server–based video streaming methods incur expensive bandwidth provision cost on the server. For example, the streaming rate for a TV-quality video is about 400 kilobits per second (kbps), which makes the client-server video streaming solution very expensive when more users join the system.

P2P video streaming encourages users to upload their downloaded data to other users in the network; each user acts as a server and a client at the same time. The system relies on voluntary contributions of resources from individual users to achieve high scalability and robustness and to provide satisfactory performance. Cooperation also enables users to access extra resources from their peers, and thus benefits each individual user as well. These video streaming users form one of the biggest multimedia social networks on the Internet, and P2P video streaming technology has enjoyed many successful deployments to date.

Based on the network structure, current P2P streaming systems can be classified into two categories: tree-push [79] and mesh-pull. Tree-based systems have well-organized overlay structures; mother peers proactively send or push the video streams to their children peers. The major drawback of tree-based streaming systems is their vulnerability to membership dynamics. When a peer leaves the system, the video delivery to all the peer's children in the tree will be temporarily disrupted, which causes significant quality degradation. Over decades, many tree-push systems have been tested and evaluated in academia, but they have seldom taken off commercially.

In mesh-pull systems, peers do not have to be structured as a static topology. Instead, a peer dynamically connects to some other peers, called *neighboring peers*, to pull videos from one another, and they form a mesh-shaped overlay. The peering relationships can be established or terminated dynamically based on the content and bandwidth availability of each peer, and they periodically exchange information about the availability of video chunks in their buffers. A recent simulation study [80] suggests that mesh-pull P2P video streaming systems have better performance than tree-push systems. The mesh-pull structure is also widely used in other P2P systems, including file sharing. BitTorrent, one of the most popular P2P file sharing systems, also employs mesh-pull structures.

Although the dynamic neighboring mechanism provides robustness to peer churns when users join and leave the system frequently, it makes the video distribution efficiency unpredictable. Also, neighbors can be any peers in the system, and different video chunks may go through different routes to reach the destination peers. Such a problem may cause video playback quality degradation, such as long startup delays or frequent playback freezes. Although mesh-pull P2P systems have these disadvantages, they still enjoy a large number of successful deployments to date, with tens of thousands of simultaneous users, because of their simple design principle and inherent robustness against a highly dynamic P2P environment. Mesh-pull systems include PPLive, PPStream, and many others. Each system has its own neighbor selection and peer cooperation policies, and the operation of the mesh-pull system relies on the bandwidth contribution from each peer.

In current P2P streaming systems, no reciprocity mechanisms, such as those used in BitTorrent, are deployed to encourage resource sharing between peers and to address the stringent need of receiving video chunks before the playback time. This encourages researchers to investigate how to stimulate user cooperation in P2P live streaming systems.

In this chapter, we explore the general design, challenges, and recent developments of mesh-pull P2P video streaming. Because user cooperation dominates the system performance, we then review current research on behavior modeling and strategy analysis of peer-to-peer systems.

3.1 Mesh-pull structure for P2P video streaming

In this section, we describe several key design components in mesh-based systems and the fundamental requirements for a successful mesh-pull P2P video streaming system.

3.1.1 Mesh formation

There are three major components of a P2P mesh-pull video streaming network: the track server, the video server, and streaming peers.

- The **track server** is to keep track of the active peers in each video session and the list of video streams. It provides information of streaming channels, buffer maps, and links to each peer, such that new peers who just join the network can download video data from multiple peers who are watching the same media content.

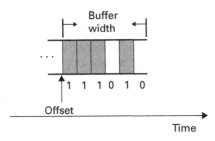

Fig. 3.1 A peer's buffer map

- *Video server:* A video stream is divided into media chunks and is made available from the video server for broadcast. All information about the video stream is available at the video server.
- The *streaming peer* contains a streaming engine and a media player in the same machine.

  The *streaming engine* exchanges video buffer map information with other peers, delivers chunks to other peers upon request, sends chunk requests to other peers, and downloads video chunks from other peer nodes or the video streaming server. The buffer map information includes the buffer offset (the time index of the first chunk in the buffer), the width of the buffer map controlled by the peer, and a binary string indicating which chunks are available in the buffer. An example of the buffer map is shown in Figure 3.1. A unique video ID is also contained in each buffer map message to differentiate between different video streams.

  Depending on the cooperation rule of each video streaming system, peers cooperatively deliver video chunks among themselves to some extent via the streaming engine. After peer 1 receives peer 2's buffer map information, peer 1 can request one or more chunks that peer 2 has advertised in its buffer map. Peers can download chunks from tens of other peers simultaneously. Video chunks are usually transmitted by TCP protocol, but in recent developments, chunks can also be delivered by UDP. Different mesh-pull systems may differ significantly on their neighbor-peer selection and chunk-requesting algorithms.

  The received chunks are decoded and reassembled into the raw video format, which are then forwarded to the *media player* for playing. When the client application software is started, the linked media player is launched, the address of the video stream is provided, and the media player sends an HTTP request to the streaming engine. Upon receiving the request from the media player, the peer's streaming engine assembles its chunks into a media file and delivers it to the media player. For video streaming, new chunks continuously arrive at the peer's streaming engine, and the streaming engine continuously adds data to the file. For chunks that do not come before the playback time, the media player uses the most recent available frames to replace them in the video stream. The media player first buffers the received data and starts to play the video once it has buffered a sufficient amount of continuous video stream. If too many chunks do not arrive on time, the media player will pause the video and wait until enough chunks are buffered.

When a new peer joins a mesh-pull video streaming system, it first connects to the track server. It then downloads a list of channels distributed by the streaming network from the track server. After the user selects the channel, this peer node registers itself in the track server and retrieves an initial list of peers that are currently watching the same video stream.

After receiving an initial list of active peers from the track server, the new peer will try to make partner relationships to a subset of the peers in the list, through TCP/UDP connections. The peering connection is established based on the mutual agreement between the requesting and the requested peers. If a connection request is accepted by a peer on the initial list, the new peer will add this peer to its neighbor list and obtain additional peer lists, which it aggregates with its existing peer list. After obtaining enough neighbors, the local peer starts to exchange video content with its neighbors. The definition of enough peers and when and how a peer refreshes the neighbor list differ from system to system. Several factors are usually considered when designing the connection establishment mechanism. For example, a peer has to consider its current number of connections, uploading and downloading bandwidth, and CPU and memory usage when responding to another peer's request for establishing relationships. The packet delay and loss characteristics on the network path between the two peers should also be taken into account to efficiently exchange video chunks between the peers.

Using this mechanism, each peer maintains and enlarges the list of other peers watching the same video. A peer on the list is identified by its UDP and TCP signaling port numbers and its IP address. The peer discovery and peer registration are usually running over UDP; when UDP fails, TCP can be used for the same purpose. The signaling overhead at the track server is considerably reduced by using the distributed peer discovery mechanism. Therefore, each track server is able to manage a large number of streaming users, in the order of millions.

To deal with peer churn, in which users frequently join and leave the system, a peer constantly updates its peer list during the session by sending small "ping" messages to the peers on the list, and finds new peers by exchanging its peer list with its neighbors through the TCP connections. A peer can also go to the track server to ask for a fresh list of active peers. If a peer wants to leave the session for good, it will notify the track server and its neighbors on the list to have its information removed there. However, when a peer disappears from the network unexpectedly – for instance, because of a computer crash – the leaving peer will still be on others' neighbor lists. To deal with such a problem, peers should regularly exchange pinging messages to make sure others are still in the network. A peer will be removed from other peers' lists if no pinging messages are received or responded within a predefined timeout period.

After establishing connections with enough number of neighboring peers, peers buffer and share video chunks with one another. Each peer first exchanges its buffer map with others, selects the video chunks that it needs, and requests the desired chunks from the video server or other peers. Clearly, when a peer requests chunks, it should give a higher priority to the missing chunks that are to be played first. Most likely, it also gives a higher priority to rare chunks – that is, chunks that do not appear in many of its partners' buffer maps [81]. Peers can also download chunks from the original video server.

3.1.2   Video streaming services

P2P video streaming applications can be divided into two categories: live streaming and video-on-demand service. In this subsection, we discuss the difference between these two video streaming services.

Live streaming service enables users to watch the live broadcast of video stream simultaneously, with a small playback time difference. Users watching the live video stream are synchronous and are viewing the same part of the video. As a result, video chunks downloaded by one peer are most likely to be useful to many other peers in the network. Therefore, the buffer length of the peers can be short and users can easily cooperate with one another.

Video-on-demand service allows a user to watch any part of the video. Unlike live streaming service, video-on-demand offers more flexibility and convenience to users. It also realizes the goal of video streaming systems: watch whatever you want whenever you want. Video-on-demand has been identified as the key feature to attract consumers to video streaming services such as PPLive. For example, many users would like to watch a live broadcast of the Super Bowl using a live streaming service. When some users want to revisit the game, they would like to use the video-on-demand service to freely choose the part of the game they would like to watch.

In video-on-demand service, although a large number of users are watching the same video, they are asynchronous and different users may watch different parts of the same video. In P2P video-on-demand, if video chunks are downloaded in the order of their playback time, a new peer can make little contribution because it is very likely that it does not have the chunks that other existing peers are looking for. Meanwhile, many existing peers can help the new peer, as they have watched the beginning part of the video and those chunks may be still in their buffers if their buffers are long enough. In such a case, tree-push peer-to-peer systems are not feasible for video-on-demand applications. This is because they are originally designed to implement multicasting at the application layer, and traditionally, users in a tree-push overlay are synchronized and receive the content in the order the server sends it out.

The challenges to offer video-on-demand services using mesh-pull P2P networks are at different levels. At the peer level, the video chunks must be received before their playback time, and ideally, they should be requested and downloaded in the same order as their playback time. However, the mesh-pull structure for video live streaming introduced in the previous section cannot solve this problem, because the track server keeps track of "all peers" watching the video but not their buffer map information. Therefore, when a new peer retrieves the peer list from the track server, it will be difficult to find the desired neighbors who have the chunks that it needs. To solve this problem, an extra record of each peer's viewpoint has to be kept in the tracking server to increase the efficiency of mesh formation. At the system level, the content sharing must be enabled among asynchronous peers, including keeping track of every user's playback in addition to the peer list. Hence, supporting video-on-demand using a mesh-pull P2P technique is not straightforward.

In a mesh-pull P2P video streaming system, to fully utilize users' upload bandwidth and to achieve the highest possible system throughput, it is better that video chunks at different users' buffers be different so there are always some chunks to exchange between users. Such an observation is called the *diversity requirement* in mesh-pull P2P systems. The content diversity improves the overall system throughput. However, if the system has high content diversity, video chunks are retrieved in a fairly random order, which is contradictory to the fact that the video chunks must be played in the sequential order of their playback time. As a result, the effective rate at which users can smoothly play a video stream may not be high. Because of the asynchronous nature of video-on-demand service, users are interested in different parts of content at a given moment, and the availability of different video chunks is also influenced by users' behavior. Thus, the challenge of designing a mesh-pull P2P video-on-demand scheme rests on the balance between the overall system efficiency and the conformation to the sequential playback requirement for asynchronous users.

A common solution is to set priorities for video chunks and increase the buffer length. First, in video-on-demand systems, peers should store all video chunks that they have received after they start watching the video. Such a large video buffer exists to increase the possibility of mutual interest among peers. Second, all missing video chunks are divided into two sets: a high-priority set and a low-priority set. The high-priority set contains missing video chunks that are close to their playback time, and the low-priority set includes missing chunks whose playback time is far away or has already passed. A selection process is used to decide which video chunk the peer should request in the next round. Each user sets a probability p such that a chunk in the high-priority set is requested with probability p, whereas a chunk in the low-priority set is downloaded with probability $1 - p$. By setting the value of p greater than $1/2$, chunks in the high-priority set will be requested earlier than those in the low-priority set. Intuitively, if a peer chooses a larger p, the chance of receiving a video chunk before its playback time is larger, which results in a smoother and higher-quality video. On the other hand, a smaller value of p increases the diversity of video chunks and would lead to better overall system efficiency, which means that peers help each other by a larger degree at a cost of their own video quality degradation.

Optimal allocation of the resources across different parts of the video and efficient management of the mesh overlay are important design issues. Annapureddy and colleagues [82] divided a video into segments, with each segment containing a set of video chunks. Segments close to the playback point are given high priorities to be requested first, which is very similar to the previous methodology. The difference is that there are more than two sets, which offers more freedom for users. Because different users have different concerns, increasing the number of sets increases the content diversity and the system throughput. Annapureddy *et al.* also used network coding to improve the resource utilization efficiency [82]. In the work of Dana *et al.* [83], the video server helps when the requested video chunks are not available and when the playback time is imminent. Unlike live streaming systems, both the video server and the track server keep record of each peer's current playback points. Upon receiving requests from peers, instead of delivering the requested chunks in a round-robin manner, the video server compares the playback time of the requested chunk with the requesting peer's current playback point. If they are very close, the video server will deliver that chunk first.

3.1.3   Challenges in peer-to-peer video streaming

We now explore the challenges and capacities of the mesh-pull P2P video streaming systems.

3.1.3.1   Peer heterogeneity

In P2P systems, users are heterogeneous, with different processing capacity and upload and download bandwidth, and different peers may choose to contribute different amount of upload bandwidth. In addition, peers can join and leave the system at any time, causing frequent membership updates. Peer heterogeneity and peer churn pose major challenges in designing high-quality P2P video streaming services; it is desirable to enable all peers to watch smooth high-quality videos without freezing or frame skipping, except for a short delay at the beginning.

One solution to address the peer heterogeneity issue is to use this heterogeneity – that is, to ask peers with larger upload bandwidth to help others by a greater amount. The intuition behind this design is that because the quality of service of a P2P video streaming system is determined largely by the total upload bandwidth contributed by all peers in the system, peers with higher upload bandwidth should be able to help more peers to fully use their upload bandwidth. The server divides the peers into two groups, ordinary nodes and super nodes; super nodes have much higher upload capacity than ordinary ones. If the upload capacity of some ordinary peers cannot sustain the video playback rate of their neighbors, super peers will be asked to contribute additional upload bandwidth to help these ordinary peers. Such a technique is called the *super node scheme* and is used widely in many recent systems. For example, in current P2P video streaming systems, most of the peers are broadband residential peers with DSL or cable connections, and they can be viewed as ordinary peers whose typical upload capacity is around or below 500 kbps. Some other peers are institutional peers with high-bandwidth Ethernet access, such as peers in campus networks. These institutional peers usually have upload capacity larger than a few mega-bps and can serve as super nodes.

However, the super node scheme has its limitations. The ratio between ordinary nodes and super nodes dominates the streaming system's capability to sustain satisfactory quality of service [84]. When the ratio of super nodes to ordinary nodes exceeds a critical threshold, the super node system can perform well. Otherwise, the system would perform very poorly.

3.1.3.2   Quality metrics

Because users in a mesh-pull video streaming system are watching videos, when evaluating the service quality they consider not only the traditional quality of service measurements, such as packet loss rate and delay, but also the reconstructed video equality. In the literature, the following criteria are often used to evaluate the quality of the streaming service.

- *Startup delay* is the interval from the time one peer selects a video to the time its media players starts playing the video. This time period is used to buffer a sufficiently large number of continuous chunks to address the dynamically changing network conditions and the peer churn. Although the startup delay is unavoidable for continuous playback, compared with the traditional TV services over cables that have negligible delays, it may cause a long waiting time and unhappy experience when a user switches the channel. For example, PPLive forces peers to buffer tens of seconds of nearly continuous video chunks before the video is played, which is much larger than that in traditional TV services.
- *Playback lag* is a serious issue in live-streaming service, when different users have different playback time and some peers are several minutes behind other peers. As a result, if users are watching the live Super Bowl, some peers might see a touchdown a few minutes later than others, which is an unpleasant experience. Additionally, because the video buffer in live-streaming services is usually short, peers with larger playback lags will not be able to contribute useful chunks to peers with smaller lags. As a result, huge playback lags will decrease the content diversity and the upload capacity of the system.
- *Video distortions* include freezing, frame skipping, rebooting, and low bit rate. Video streaming has real-time requirements for both live broadcast and video-on-demand services. The real-time requirement posts a stringent playback deadline for each video chunk. Although different users may have different playback deadlines for the same chunk as a result of the playback lag in live streaming or asynchronous playback in video-on-demand, the chunk still needs to be received before its playback deadline for each peer.

Depending on the system design, for a particular peer, there are two possible strategies when a chunk does not arrive before its playback deadline, which can also be adopted when a small group of continuous chunks do not arrive on time. One method is to have the media player freeze the playback with the most recently displayed video frame and wait for the late chunk to arrive. From the end user's perspective, freezing is the same as the video being paused by others. The other strategy is to skip frames encoded in the late chunk and to play the next available chunk. By doing so, the streaming engine must advance the deadlines for the subsequent chunks accordingly, which can be done by changing the buffer offset shown in Figure 3.1. If the system adopts frame skipping for late chunks, users will experience sudden movement of the video. In terms of video quality, both freezing and frame skipping are undesired, but different services would have different preferences of these two strategies. For example, the frame-skipping strategy is more desirable in live-streaming systems because the freezing strategy will increase the playtime lag to wait for late chunks.

When there are too many continuous chunks that do not arrive before their playback time, instead of freezing for a long time or skipping many frames, the streaming engine will terminate the connection with the player and reinitialize the entire streaming process. Such a phenomenon is called *rebooting*. The rebooting period directly corresponds to the startup delay, as it is equivalent to starting a new session.

Most video streams shared over P2P streaming systems are encoded into low bit rates and thus have low quality. This is a result of the large bandwidth consumption of video streaming applications, and the fact that video chunks must be received before strict playback deadlines. The super node method can help increase the video bit rate and improve the video quality by putting more loads onto the super peers. However, it is questionable whether super peers and their Internet service providers will have the capability and the incentives to provide the additional upload bandwidth.

The system designer of P2P video streaming service must take into account the preceding factors in addition to the network quality of service. For example, to address the issue of noticeable startup delay, many current P2P video streaming systems, such as PPStream, cache some advertisement videos in the end user's machine. When a user starts the application software and chooses a video, these commercials are played while the actual video streams are downloaded in the background. For example, in PPStream, these commercials are fifteen-seconds long and are played every time the user starts watching a new channel. Another possibility to reduce the startup delay and to maintain smooth playback is to include redundant downloading or network coding of video chunks. However, both schemes increase the amount of video traffic over the networks and the demand on each user's upload bandwidth.

3.1.3.3   Security threats

The distributed nature of mesh-pull P2P streaming makes the systems vulnerable to various attacks, in which malicious users aim to break the system or decrease the throughput as much as possible. The security threats can be divided into two categories, availability threat and integrity attack.

Availability threat: The availability of a system is violated if the system cannot respond to users' requests and deliver video streams to its users. An effective attack to the system availability is the denial-of-service attack [85]. For example, a malicious attacker may create a large number of nonexistent peers who pretend to be interested in the same channel and register all of them in the track server's peer list. Because each peer is identified only by its IP address and the UDP/TCP port, it is very easy to create many fake identities that will not respond to any chunk requests. As a result, a nonmalicious peer will have difficulties identifying other nonmalicious peers and downloading video chunks that it needs, because it wastes a great deal of time by sending chunk requests to these fake peers. In addition, malicious attackers may use the many fake identities that they generate and continuously send requests to the server to occupy its bandwidth and CPU processing time. It may disable the video server from answering requests from nonmalicious users and make the content unavailable to other users. This attack is particularly easy to employ if the video content server is a normal desktop/laptop computer with limited CPU power and bandwidth, which is often the case for user-generated videos.

Another example is that malicious users can waste other peers' local resources. The malicious peers can create a fake buffer map with the victim peer's ID and claim that the victim peer has many video chunks. By doing so, other peers would send chunk requests to the victim peer, because most of the chunks are "claimed" to be available at the victim peer's side. As a result, the victim peer's CPU power and network bandwidth would be largely consumed but not helping other nonmalicious peers.

Another concern regarding the system's availability occurs when streaming video over wireless networks. Wireless systems are particularly vulnerable to jamming attacks, eavesdropping, and active interference. Therefore, it is important to develop effective and mature security mechanisms for wireless networks. Many security mechanisms have been developed for WiFi and WiMax to address this issue, but Bluetooth is still in its developing phase with security flaws, which makes Bluetooth easy to attack.

Integrity threat: Malicious users can attack the P2P streaming system's integrity by modifying the video chunks themselves. They can mix the video stream with fake and useless chunks, which will significantly degrade the quality of the reconstructed media at the receiver's side. Such an attack is called a *pollution attack*. To address the stringent time constraint, peers will request a desired chunk as soon as they find it in the neighbor's buffer map. The chunk exchange is very frequent, and copies of the chunk can be spread out in the system very quickly. Such a fast-spreading feature is designed to enhance the content diversity and to improve the system throughput. However, the polluted chunks can also be propagated over the network quickly, not only by malicious users but also by nonmalicious users who are not aware of the existence of a polluted chunk in their buffers. After a polluted chunk is received and before it is decoded, the polluted chunk stays in the buffer and may be unintentionally forwarded by the nonmalicious user to other peers. A study by Liang and co-workers [7] demonstrated the potential damages that the pollution attack may cause to P2P streaming systems. In the experiment, when there were no attackers presented, one specific video channel had more than 3,300 viewers. When the polluted attack started, the video quality dropped drastically and became unacceptable. As a result, a large number of peers eventually left the system, and the number of viewers decreased to about 500 within thirty minutes.

To resist pollution attacks, it is important to detect polluted chunks and identify attackers (polluters) who intentionally send polluted chunks. Pollution detection schemes often use digital signatures and authentication tools, such as the chunk signing algorithms proposed by Liang *et al.* [7], to detect polluted chunks and remove them from the network as early as possible. To identify polluters, trust management [86] can be used to distinguish malicious attackers from nonmalicious users who unintentionally forward polluted chunks.

3.2 User dynamics in peer-to-peer video streaming

In the previous section, we discussed the challenges of mesh-pull P2P video streaming systems. One possible way to improve the quality of service is to change the system architecture, which may yield additional cost. For example, to reduce the startup delay without scarifying the smooth play of video, we can apply redundant downloading and network coding to video chunks. Such a scheme increases video traffic redundancy and consumes much more upload bandwidth, which are not desirable to resource-limited P2P video streaming systems.

Let us think from a different perspective of P2P systems. The gain of mesh-pull P2P design comes from user cooperation, and the robustness of the system comes from its independence from the underlying network structure. From this point of view, designing better peer cooperation mechanisms is very important. For example, better peer selection strategies and video chunk scheduling schemes can help reduce the playback lags. Therefore, the key to the success of mesh-pull P2P video streaming systems is to develop optimal cooperation strategies.

In this section, we introduce current work on peer selection and incentive mechanisms.

3.2.1   Peer selection

In P2P systems, peers seek collaboration from other peers in the network. Because of the limitation of peer upload capacity and high bandwidth demand for real-time video delivery, choosing the right peers with whom to collaborate is critical to the system performance. For example, after a peer has decided which video chunk to request, the peer must choose one or several peers from whom to request the chunk. The most straightforward solution is to ask a neighbor with the best connection. However, such a strategy may not always perform best. For example, different packets from the same sender to the same destination may be routed through different paths with different transmission delays, and some "popular" peers may receive multiple requests and may not be able to answer all of them on time. In addition, once the chunk to request is decided, a peer also needs to determine how many peers it contacts to request the chunk. Increasing this number may help increase the probability of receiving the chunk before its playback time, but it also increases the network load. This is particularly problematic when two or more copies of the chunk are routed through the same network bottleneck.

Nevertheless, because the mesh-pull P2P system lacks a central computing unit and usually contains a large number of users, computation of the optimal solutions for all users is not applicable. In addition, each peer has only partial information about the network and, therefore, the optimal strategy calculated by each peer will unlikely be the global optimal solution. Furthermore, given the fact that the connections between different pairs of peers are different and peers leave and join the system unexpectedly, using the same strategy for all users will not be optimal, and peer selection is still an open problem in P2P streaming systems.

One simple solution is random guesses and iterative updates of the peer selection strategy. Leveraging the fact that requesting the same video chunk from multiple peers does not always improve the throughput (e.g., video quality), iterative mechanisms can be used to identify a subset of senders that maximizes the overall throughput [87]. To request a certain video chunk, the requesting peer first randomly selects a peer from the list of peers who have the desired chunk, and puts it as the first peer in the active-sender set maintained by the requesting peer. The requesting peer then randomly and periodically adds another peer from the list of available neighboring peers to the active-sender set. Meanwhile, the requesting peer monitors the variations in the throughput of the entire active-sender set and that of each individual active sender. If the overall throughput of the active-sender set increases after adding the most recently selected sender peer, the new sender peer is kept in the set. Otherwise, it is replaced by another randomly selected peer.

The above iterative method of choosing the active-sender set can increase the diversity and reduce the probability that packets from two sender peers are congested at the same bottleneck link. However, if the access link from the Internet service provider (ISP) to the requesting peer is the bottleneck, changing the number of sender peers does not improve the overall throughput [87]. One way to address this issue is to introduce an approximate partial topology just sufficient for peer selection. The first step is to build the logical topology. A simple physical topology can be easily built using tools such as traceroute [88], in which the requesting peer asks all the candidate peers in the active list to perform traceroute toward the requesting peer. Then, the logical topology is constructed by merging consecutive links with no branching points into one segment. The approximated topology is now initialized and will be refined by the available bandwidth and the packet loss rate.

The available bandwidth of a path is the maximum rate that such a path can provide to a flow without reducing the rate of other traffic, and both available bandwidth and packet loss rate can be measured by sending probing packets. After obtaining the approximated graph to all peers on the neighbor list, the requesting peer finds the active-sender set that maximizes the expected aggregated rate at the requesting peer's side, under the constraint that the download bandwidth of the requesting peer is not exceeded. There are several peer-selection algorithms based on approximated topology. For example, because video chunks in the streaming network have strict playback deadlines, network delay should be considered in peer selection. Based on that, instead of considering the packet loss rate when constructing the estimated network graph, the requesting peer can probe the packet delay and select the active peers to minimize the product of the delay and the transmission rate. By doing so, the requesting peer optimizes bandwidth utilization with minimal end-to-end latencies based on its limited knowledge.

Up to now we have discussed several methods for a peer to construct its active-sender set, but we have not studied how much load the requesting peer should put on each sender peer. Most of the peer selection schemes need authentication of other peers and must keep track of other peers' reputations. An example is to introduce the pseudopayment mechanism, with which the peers must pay some credits to each other for chunk forwarding. In the pseudopayment mechanism, the requesting peer i "owes" the sender j the cost of transmission upon receiving the chunk, and when peer j sends a request in the future, peer i must pay back and send the requested video chunks. In this scheme, each peer i must first decide the cost c_i to transmit a chunk. When a requesting peer chooses which peers to ask for a chunk, it seeks the peers with the minimum total cost [89].

3.2.2   Incentives

A big difference between a traditional client–server streaming system and a P2P streaming system is that the capacity of a P2P system grows dynamically with the number of users. Ideally, if there are more users, the streaming system will have a larger capacity. However, such a statement has an implicit assumption – that users are always willing to help each other. This is not true, given the selfish nature of human beings. An evidence of user's selfishness is the presence of free riders who only want to download the video stream but do not share their video chunks with others.

As mentioned before, selecting peers only based on the network connection or streaming delay is not robust against free riders or attackers. A possible solution to address free riding is the tit-for-tat policy, under which peers copy each other's previous actions and punish noncooperative behavior by playing noncooperatively. That is, peer i answers peer j's chunk request only if j has previously sent chunks to i, and they send the same number of chunks to each other. Tit-for-tat is used in BitTorrent and has been shown to be effective in reducing the number of free riders. Still, for different P2P streaming applications with different requirements, it is important to design incentive mechanisms carefully to stimulate cooperation among peers. Here we introduce various aspects of incentive mechanisms in P2P streaming.

One popular category of incentive mechanisms is to introduce the concept of *differential service*. The idea is to give proper reward to cooperating peers so everyone has incentives to help others. There are several requirements that a successful differential service based incentive mechanism should meet.

(1) First, the system capacity should increase after a new peer joins the system. In other words, after the starting phase, the differential service should benefit all requesting peers and allow them to receive the video stream earlier when compared with an incentive mechanism that does not provide differential service. Low performance at the beginning is tolerable, but the differential mechanism should not deny newcomers.

(2) Next, ideally, it should be calculated and executed in a distributed fashion. Because of the fully distributed nature of a mesh-pull P2P system, the assumption of the existence of a central authority to enforce the cooperation policy may not be realistic and desirable. Although the track server keeps a list of all active peers and the channels they are currently viewing, when there are a large number of users with a huge amount of information flow among them, it will be challenging for the track server to monitor all users' interactions and enforce cooperation.

(3) Furthermore, if a peer cooperates at a larger amount and contributes more upload bandwidth, the probability that its requests are replied to should be larger, with shorter waiting times and smaller buffering delays.

If the differential service can satisfy these requirements, it gives peers incentives to contribute more upload bandwidth and to maximize the available resources in the network.

3.2.2.1 Peer selection based on differential service

Incentive mechanisms based on differential service have been used to design many cooperation strategies for mesh-pull P2P streaming systems. We start with the peer selection algorithms. The basic idea is to link a user's utility with its contribution to others so users have incentives to help others by a larger amount [90]. To get a quantitative measure of a user's contribution, the first step is to convert a user's contribution into a score, and map it into a percentile rank among all users. A user's rank determines whom it can ask for video chunks; a peer with a higher rank has more flexibility in peer selection and has a better chance to receive high quality videos. For example, a peer selection scheme may allow a peer to request video chunks only from peers with equal or lower ranks. Therefore, cooperative users who contribute more upload bandwidth can earn higher ranks and eventually receive high-quality videos. Meanwhile, free riders have very low ranks and thus limited choice in peer selection, resulting in low video quality.

By converting a user's contribution into a percentile rank, this peer selection scheme employs users' relative contributions instead of their actual contributions. Therefore, there will be more competition among peers and hence more incentives to contribute as many resources as possible. To address the issue that users join and leave the system frequently, users' ranks should be updated periodically.

When a peer first joins the system, it begins with a rank of zero and receives the best-effort service with the lowest quality offered by the system. The quality of this service may vary from system to system, and is highly unpredictable. If a user wishes to receive a better-quality streaming service, it must increase its rank by contributing to the system. A rational user will determine its optimal contribution level to maximize its utility, and consider both streaming quality and the cost of collaboration.

3.2.2.2 Chunk answering based on differential service

The preceding rank-based incentive mechanism addresses only the freedom of a requesting peer to choose the supply peer. Here we present an example of how a supply peer chooses which peers to serve upon receiving multiple chunk requests [91]. Each peer divides the requesting peers into two classes, depending on their contributed upload bandwidth. The higher class includes those who contribute larger upload bandwidth and whose requests will always be answered. The lower class contains peers who contribute smaller upload bandwidth, and whose requests will be answered with a probability smaller than 1. The differential chunk-answering policy favors higher-class requesting peers, leads to a faster amplification of the P2P system capacity, and ultimately benefits all requesting peers. This is reflected by the simplicity of the algorithm and the shorter average waiting time for all requesting peers. The class size is adaptively adjusted based on interactions with other peers to maximize the performance.

In the preceding algorithm, each peer makes its own decision on the partition of the requesting peers into two classes and on the class size; it is a fully distributed algorithm suitable for mesh-pull P2P system. Compared with the nondifferential chunk-answering mechanism, the differential mechanism can achieve a higher chunk-answering rate and a shorter average buffering delay. Furthermore, the chunk-answering mechanism distinguishes requesting peers who contribute more upload bandwidth from those who contribute less, and creates incentives for peers to contribute as much as possible.

3.2.2.3   Incentives for super nodes

The preceding rank-based peer selection mechanism has a serious problem when there are users with high bandwidth capacity. In that scheme, a user's rank is its relative, but not absolute, contribution, and users with high bandwidth capacity do not have incentives to contribute more than others as long as they can maintain their ranks. Meanwhile, because peers with lower ranks cannot send requests to these high-capacity peers, peers with higher ranks will receive fewer requests and end up contributing less to the network even when they have unused resources. Also, as mentioned earlier, one way to fully use the peer heterogeneity is to put more load on peers with higher capacity, which are referred as super nodes. But before the super-node scheme can be applied, incentives must be provided to the super nodes so that they are willing to contribute more upload bandwidth.

(1) *Centralized methodology:* An example of providing incentives for super nodes is the taxation model [92]. The basic idea is to encourage peers with rich resources to contribute more bandwidth and to subsidize peers who are poor in resources. By using the taxation model, the bandwidth resource is mapped to the wealth of peers, and the wealth will then be redistributed. To efficiently redistribute the wealth, the tax rate must be carefully designed so that even though peers with more resources must pay a higher tax, they are still willing to participate in the streaming service. Because peers are strategic, they individually own the bandwidth resources and are strategic in minimizing the cost of contributing resource and maximizing the benefit of the received video quality. The system must provide a reward mechanism to keep the resource-rich peers in the network.

In this taxation model, there must exist an asymmetry of roles, in which one user has the power to enforce the centralized rules. For example, one way to ask peers with higher capacity to contribute more is by taxing the users' download bandwidth. A tax schedule should be fixed and made public so peers can adjust their strategy to maximize their utility between the scheduled tax seasons. The central authority should not change the tax schedule, or if necessary, it should make the schedule change only at a very large time scale to minimize system instability owing to peers' reaction to the tax schedule change. The tax rate must to be fair in both horizontal and vertical ways. Horizontal fairness requires that individuals with similar wealth should bear similar tax liability. Similar to the tax rates in real world, vertical fairness enforces that people with more wealth have a higher tax rate, which in return enforces the super nodes to contribute more to the system. In such a scenario, there must be a central authority to collect the tax payment from individuals according to a predefined tax schedule. In P2P streaming, the video stream server is the natural empowered entity, as the video server owns the content and can determine the ways in which peers participate in the system via proprietary software. In other words, the video server has the freedom to design the rules, and the participating peers have the freedom to choose either to follow the rules and enjoy the video or to walk away.

(2) *Distributed methodology*: The taxation scheme described here relies on a central authority to control the tax schedule and enforce the payment. However, assuming the existence of a central authority may not be desirable in a distributed P2P system. Here we introduce a way to stimulate cooperation from resource-rich peers in a distributed manner.

The basic idea is that peers earn points by sending video chunks to others and pay points to receive chunks from others [93]. The time line is divided into fixed-length periods. In each period, peers compete with one another for good supply peers for the next period in a first-price auction procedure. In the first-price auction, requesting peers send bidding points to the supply peer, and the one with the highest bid gets the chunks at the price of his bid. When a peer fails in the competition for a certain supply peer, it must choose a new supply peer either from other supply peers or the winner of the bid, as the winner now can be considered as a suppler.

Peers who lose the bid can adopt two strategies to find the next supply peer: the shortest-path strategy and the balanced-tree strategy. In the shortest-path scheme, each requesting peer just looks for a supplier who has the shortest path. However, the shortest-path criterion makes many requesting peers end up competing for the same single well-located supply peer. As a result, the shortest-path scheme would very probably make some peers overloaded and other peers upload nothing. Under such a circumstance, the network resources will be severely wasted.

The balanced-tree strategy helps construct a short tree – that is, a small average overlay path length for all peers. A short delivery path has important implications for the system's performance, because the transience of peers is the dominant factor that affects the streaming stability in an overlay network, and a shortened delivery path can effectively reduce the probability of streaming disruptions. Therefore, the packet loss rate will be reduced, and the average utility of peers is increased.

In addition, the off-session points should be used to encourage peers who are no longer in the media sessions to continue contributing to the network. In return, the off-session peers earn points that can be used in later sessions to improve their media quality. To maximize its own utility, the off-session peer will seek to maximize its wealth instead of media quality, as it is not watching the video stream.

3.2.2.4   Incentives based on video quality

Because peers are heterogenous in terms of resources, scalable video coding is desirable for P2P video streaming service to allow resource-rich peers to have better video quality. With scalable coding, a video is coded into layers with nested dependency: the base layer is necessary to decode the video at the lowest resolution and a higher layer gradually refines the video quality. More received layers provide better video quality. Thus, one way to provide incentives is to reward cooperative peers with higher video quality [94].

Each peer first measures its download rates from its neighbors by accumulating the number of video chunks received from each neighbor over a certain time period. Then each peer reciprocates by providing a larger fraction of its upload rate to neighbors from which it is downloading at higher rates. These peers would also have a higher probability to be served if there are other competitors. In this scheme, each peer computes its neighbors' contributions in a distributed way. It avoids the centralized architecture and enhances the scalability of the system. In this scheme, a peer with higher contribution is more likely to obtain a larger share of its neighbors' upload rates, and thus may receive more layers and enjoy better video quality. On the other hand, a peer with a lower upload contribution is more likely to receive fewer layers, thus receiving a lower, but still acceptable, video quality. Such a scheme also gives no incentives to free-ride, because a free rider with no contribution is very unlikely to be served by its neighbors.

Furthermore, a peer periodically replaces a neighbor that contributed the least with a new peer to increase the system diversity and to maintain the service quality. To establish a stable relationship, it is desirable for a peer to build neighboring relationships with peers who have similar upload bandwidth. Instead of just looking at the upload bandwidth that can be falsely reported by neighbors, another approach is to locate a candidate neighbor peer based on its buffer map [95]. Before establishing a neighboring relationship, two peers exchange their buffer maps. Then they can evaluate whether the other peer has a higher chance to have the video chunks that it may need in the future. Also, because a peer that contributes more is supposed to receive more layers and vice versa, the upload bandwidth contribution of the peer and its cooperative behavior can be evaluated based on its buffer map. For example, a peer that is viewing more layers should be more cooperative, which also implies that it should have more upload/download bandwidth.

4   Game theory for social networks

Game theory is the mathematical study of cooperation and conflict. It provides a distinct and interdisciplinary approach to the study of human behavior and can be applied to any situation in which the choice of each player influences other players' utilities, and in which players take this mutual influence into consideration in their decision-making processes. Such strategic interaction is commonly used in the analysis of systems involving human beings, such as economy, sociology, politics, and anthropology. Game theory is a very powerful conceptual and procedural tool to investigate social interaction, such as the rules of the game, the informational structure of the interactions, and the payoffs associated with particular user decisions. Game theory can be applied to all behavioral disciplines in a unified analytical framework. In the later chapters of this book, game theory will be the main tool for modeling and analyzing human behavior in media-sharing social networks. In this chapter we introduce the basic and most important concepts of game theory that will be used extensively in this book.

The idea of game theory was first suggested by Emile Borel, who proposed a formal theory of games in 1921. Later in 1944, the mathematician, John von Neumann and the economist Oskar Morgenstern published *Theory of Games and Economic Behavior*, which provided most of the basic noncooperative game terminologies and problem setups that are still in use today, such as two-person zero-sum games. During the late 1940s, the developments of game theory focused on cooperative game theory, which analyzes optimal strategies for groups of individuals under the assumption that they can enforce collaboration among themselves to jointly improve their utilities. In 1950, the concept of equilibrium was introduced by John Nash, who demonstrated that finite games always have an equilibrium point, which is the best response of every player given all other players' choices. Ever since equilibrium was introduced, it has become the central focus of noncooperative game theory. Refinement of Nash equilibriums and extension of Nash equilibriums into Bayesian games, which consider incomplete information, were proposed later in the 1960s.

The 1950s were the most gorgeous years of game theory, when many important concepts were developed. Such concepts include the repeated games, game in extensive forms, and coalitional games. Within the following few decades, game theory was broadened theoretically and applied to problems of war, politics, economic theory, sociology, and psychology. Later on, game theory even established relationships with evolution and biology. Given the affluent applications in economy, a high-profile application of game theory was the design of auctions to allocate the resources more efficiently than traditional governmental practices at the end of the 1990s.

4.1   Noncooperative and cooperative games

Game theory is a formal model of interactive situations. A game typically involves more than one *player*, whereas a game with only one player is usually called a decision problem. Here we define N to be the number of players in the game. Players in a game make rational decisions by means of an abstract concept called *utility*. Utility of each player refers to the self-judged scale of the subjective welfare or the change in subjective welfare that this user derives from the outcome of the game, given the decision by all users. The possible decisions each player can make are called the *strategies* of the player. Different players can have different possible strategies and have different influences on others' or their own utility. Let player i's set of strategies be A_i; then user i's utility can be denoted as $\pi_i(\mathbf{A})$, where $\pi_i : A \to \mathbb{R}$ and $\mathbf{A} = \times_{i \in N} A_i$. A strategic game with the these parameters is usually denoted as $\langle N, A_i, \pi_i \rangle$.

Based on whether players are able to form binding commitments, games can be divided into two categories: cooperative and noncooperative games. Cooperative game theory investigates the utilities that each potential *group of players* can obtain with respect to the relative influencing power held by each player if the group members cooperate with one another. Here, the basic assumption is that users are willing to cooperate with one another if cooperation can help improve their utilities – for example, when there is a binding contract that "enforces" cooperation. Cooperative game theory is most frequently applied to situations in which the concept of relative strength of influence is the most important factor in the game. For instance, the Nash bargaining solution is for situations in which two parties bargain to retain some gains received from reaching agreements and cooperating with each other, and the solution depends only on the relative bargaining strengths of the two parties. The bargaining power is an implicit concept, but it can usually be determined by the inefficient outcome that each individual results in when negotiations break down.

Unlike cooperative games, noncooperative game theory focuses on the analysis of strategic decisions of *individual players*. Players make decisions independently and no enforcement of cooperation can be done by anyone other than the player himself or herself. Noncooperative games are able to model situations to the finest details and produce accurate results. Hence, noncooperative game modeling can help analyze the decision of each user in the media sharing social network, whereas cooperative game modeling focuses more on the outcome of the whole social network – or, the social welfare.

In this chapter, we introduce the basic game theory concepts and tools that will be used later in this book. We start with the noncooperative games and Nash equilibrium, followed by the most commonly used cooperative game, the bargaining theory.

4.2 Noncooperative games

4.2.1 Pure-strategy Nash equilibriums

Given any game, analysts are most interested in the outcome of the game. In noncooperative game theory, we have the luxury of getting into the details of each player's decision. Because all players are assumed to be rational, given other players' choices, they make choices that result in the utilities they prefer the most. *Equilibrium* is the strategy outcome that is the best response of each user given the decision of others. The first step of understanding equilibrium is the concept of dominance.

Under some circumstances, a player has two strategies, X and Y, such that given any combination of strategies of all other players, the utility of this user resulting from choosing strategy X is always better than the utility resulting from choosing strategy Y. In such a case, strategy X dominates strategy Y. As a result, a rational player will never choose to play the dominated strategy, which is strategy Y in this example. The strategy Y is called a *strictly dominated strategy* for user i if there exists a strategy X such that

$$\pi_i(X, \mathbf{a}_{-i}) > \pi_i(Y, \mathbf{a}_{-i}) \; \forall \, \mathbf{a}_{-i}, \tag{4.1}$$

where $\mathbf{a}_{-i}$ is the strategy profile chosen by all other players in the game, and π_i is the utility function of user i. Examining strictly dominated strategies can help eliminate strategies that a rational player will never choose and give precise advice to players on how to play the game.

Although considering only dominated strategies can reduce the possible strategies that a player will take, in many games, there are no dominated strategies. *Nash equilibrium* can provide specific advice on how to play the game in general. The central concept of Nash equilibrium is that it recommends a strategy to each player such that the player's utility cannot be increased given that the other players follow the recommendation of Nash equilibrium. Given the fact that other players are also rational, it is reasonable for each player to expect his or her opponents to follow the recommendation of Nash equilibrium as well. In other words, Nash equilibrium defines the best-response strategy of each player. The formal definition of Nash equilibrium is as follows:

Definition 4.1. The strategy profile $a^* \in A$ is the Nash equilibrium of the strategic game $\langle N, A_i, \pi_i \rangle$ if and only if for every a_i^*,

$$\pi_i(a_i^*, a_{-i}^*) \geq \pi(a_i, a_{-i}^*), \quad \forall \, a_i \in A_i, \tag{4.2}$$

where a_i denotes the strategy of user i, and a_{-i} denotes the strategies of all other users.

The most famous example of Nash equilibrium is the prisoner's dilemma. Suppose there are two prisoners who committed the same crime and a district attorney is going to question the prisoners. To induce the prisoners to talk, the attorney designs the following incentive structure. If neither prisoner talks, both prisoners receive mild sentences, which is quantified as payoff 3. If only one prisoner defects and squeals on the other, the speaking prisoner is let off free and receives payoff 5, whereas the one who refuses to

Table 4.1 Matrix form of prisoner's dilemma

	Cooperate	Defect
Cooperate	(3,3)	(0,5)
Defect	(5,0)	(1,1)

speak is subject to a severe sentence, equivalent to payoff 0. Finally, if both prisoners speak about the crime, they share the punishment, which is a little better than the previous case, and can be set as payoff being equal to 1. We can show the game between these two prisoners as a matrix in Table 4.1.

In Table 4.1, the strategies of prisoner 1 and prisoner 2 serve as row and column labels, respectively, and the corresponding payoffs are listed as pairs in matrix cells such that the first number is the payoff of prisoner 1 and the second number is the payoff of prisoner 2. "Defect" denotes the strategy of confessing to the attorney, and "Cooperate" refers to cooperating with each other and keeping quiet. To search for the Nash equilibrium of this game, we should look over the best response of each player (prisoner).

If player 1 plays "cooperate," then the best response of player 2 is playing "defect," because "defect" yields a payoff of 5, whereas "Cooperate" yields only 3. If player 1 plays "defect," then player 2's best response is again playing "defect," because "defect" yields a payoff of 1, whereas "cooperate" yields only 0. Therefore, regardless of the strategy of player 1, a rational player 2 will always choose to play "defect." By a symmetric argument, a rational player 1 should also always play "defect." As a result, the unique Nash equilibrium of the game, assuming the players are rational, is (defect,defect). This game is known as the prisoner's dilemma because the only rational outcome is (1,1), which is obviously suboptimal, as playing (cooperate,cooperate) yields payoff (3,3) and benefits both prisoners. From this example, we can see that although the Nash equilibrium is the best response of each player to other players' decisions, it does not always give the optimal payoff to all players. We will discuss how to push the Nash equilibrium to the optimal outcome in Section 4.2.3.

4.2.2   Mixed-strategy equilibriums

A game in strategic form does not always have a pure-strategy Nash equilibrium, in which each player deterministically chooses one strategy that gives the best response to others' strategies. Under such circumstances, players may tend to select randomly from their pure strategies with certain probabilities to maximize their expected payoffs. Such a randomized strategy is called a *mixed strategy*.

When mixed strategies are allowed, the Nash equilibrium defined for strategic games in which players take pure strategies can be extended easily. As before, an equilibrium is defined by a (possibly mixed) strategy for each player, in which no player can increase his or her expected payoff by unilateral deviation. As proved by Osborne and Rubinstein [96], any strategic game with finite strategy space has at least one mixed-strategy equilibrium.

Table 4.2 Matrix form of the tennis game

	Back receive	Front receive
Back serve	(0.4,0.6)	(0.7,0.3)
Front serve	(0.8,0.2)	(0.1,0.9)

We use the following example to explain how to derive the mixed-strategy Nash equilibrium of a game. In a tennis game, the server can serve to either the backhand or forehand of the receiver. Given that the tennis ball travels fast and there is only a very short time to make decisions, we assume that the server and the receiver make decisions simultaneously. The receiver also anticipates that the ball will come to either the forehand or backhand side. A receiver who anticipates correctly is more likely to return the ball. In addition, the server has a stronger backhand serve, which makes the receiver more difficult to return a backhand serve. As a result, the receiver will return a correctly anticipated backhand serve with 60 percent probability and a correctly anticipated forehand serve with 90 percent probability. A receiver who wrongly anticipates a forehand serve hits a good return 20 percent of the time, whereas a receiver who wrongly anticipates a backhand serve hits a good return 30 percent of the time. Let the utility of the receiver be the probability of returning the serve and the utility of the server be the probability of the receiver misses the serve; the matrix form of the game can be shown as in Table 4.2. The strategies of the receiver and the server serve as row and column labels, respectively. The corresponding payoffs are listed as pairs in matrix cells such that the first number is the payoff of the server and the second number is the payoff of the receiver.

It is clear that there is no pure-strategy Nash equilibrium for this game: the best response to "front serve" is "front receive," whereas the best response to "front receive" is "back serve." Similar behavior shows in "back serve" as well; hence there is no pure-strategy Nash equilibrium for this game. Instead, the players may look for the mixed-strategy equilibrium. Assume that at the mixed-strategy equilibrium, the server plays "back serve" with probability p, and the receiver plays "back receive" with probability q. Then the utilities of the server π_s and the receiver π_r are

$$\pi_s = pq * 0.4 + p(1-q) * 0.7 + (1-p)q * 0.8 + (1-p)(1-q) * 0.1$$
$$\text{and} \quad \pi_r = pq * 0.6 + p(1-q) * 0.3 + (1-p)q * 0.2 + (1-p)(1-q) * 0.9, \quad (4.3)$$

respectively. According to the definition, at equilibrium, the server's and receiver's utility should satisfy $\frac{\partial \pi_s}{\partial p} = 0$ and $\frac{\partial \pi_r}{\partial q} = 0$. By solving the equations, we obtain $p = 0.7$ and $q = 0.6$.

One may notice that at a mixed-strategy equilibrium, for each player, choosing either strategy gives the same payoff. For example, given the fact that the receiver plays back return with probability 0.6, the server receives payoff 0.48 for playing either back serve or front serve. Such an interpretation is another way to calculate the mixed-strategy equilibrium, which makes every strategy the best response to other players.

4.2.3   Equilibrium refinement

If a game has more than one Nash equilibrium, a theory of strategic interaction should guide players toward the most reasonable equilibrium on which they should focus. Indeed, a large number of works in game theory have been concerned with equilibrium refinement, which attempts to derive conditions that make one equilibrium more plausible or convincing than others.

The first question to ask is whether there exists an optimal equilibrium. Because game theory is the tool to solve multiobjective optimization problems, it is not easy to define optimality in most of the cases because the players may have conflicting objectives. The most popular alternative of optimality is Pareto optimality, which is the payoff profile that no player can increase his or her own payoff without making any other player worse off.

4.2.3.1   Pareto optimality

Definition 4.2. The utility profile π is Pareto optimal if there does not exist another utility profile π' such that $\pi_i' > \pi_i$ for all $i \in N$.

If the utility profile is not Pareto optimal, then at least one user can increase his or her payoff without degrading others'. As we assume all players are rational, there is no reason for players to refuse to help a player to increase his or her payoff while maintaining their own utilities. Therefore, a game with rational users will always go to the equilibriums that are Pareto optimal. Utility profiles that are Pareto optimal form the Pareto optimal set, which usually contains many elements. Hence, more specific refinement of Nash equilibriums is needed to guide the players to desirable outcomes. We introduce the subgame perfection and stability concepts to further refine the Nash equilibriums.

4.2.3.2   Subgame perfection

Definition 4.3. A subgame is a subset of a game that contains an initial node that is the only node in the information set and all nodes that are successors of any node it contains, where an information set is the set of nodes that the player cannot differentiate.

In other words, subgames are the smaller games embedded in the original game. Take the tennis game in Table 4.2 as an example. We first express the game into a tree (extensive form) as in Figure 4.1.

Given the game in Figure 4.1, if the two players make decisions simultaneously, then the only subgame of this game is the game itself. This is because node 1 (the server's move) is the only information set that contains only one node. Node 2 and node 3 (receiver's move after back serve and front serve, respectively) belong to the same information set, as the receiver does not know the server's decision in advance. That is, when the receiver makes a decision, he or she does not know whether the situation is node 2 or node 3, so node 2 and node 3 are in the same information set.

However, the situation will change if the receiver actually makes a decision *after* the server – that is, when there is enough time for the receiver to observe whether it is a

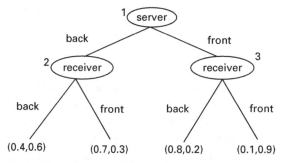

Fig. 4.1 Extensive form of the tennis game

Fig. 4.2 Remaining game of the tennis game after one iteration of backward induction

front or a back serve. As a result, node 2 and node 3 are in different information sets, as the receiver knows where he or she is in the game tree. In such a scenario, the game with initial node 2 or initial node 3 is a subgame of the tennis game. Also, this game has perfect information, as every information set of the game contains only one node.

Definition 4.4. A Nash equilibrium is subgame-perfect if and only if it is a Nash equilibrium in every subgame of the game.

The subgame-perfect Nash equilibrium of a finite-stage perfect-information game can be found by *backward induction*. Backward induction first considers any node that comes just before terminal nodes – that is, after each move stemming from this node, the game ends. If the player who moves at this node acts rationally, he or she will choose the best move for himself or herself. Hence, we select one of the moves that give this player the highest payoff. Assigning the payoff vector associated with this move to the node at hand, we delete all the moves stemming from this node so we have a shorter game, in which our node is a terminal node. Repeat this procedure until we reach the origin.

Take the tennis game with perfect information in Figure 4.1 as an example. First, find the equilibrium of the subgame, starting with node 2. There are two terminal nodes in this game, and the player who makes the decision is the receiver, so apparently the receiver will choose "back receive" to maximize the utility; hence, the utility profile of the equilibrium is (0.4,0.6) in this subgame. The same analysis can be applied to the subgame starting with node 3, in which the utility profile of the equilibrium is (0.1,0.9). Because we have already examined all terminal nodes in the game, we finished one iteration of the backward induction, and the remaining game is as in Figure 4.2. Apparently there is only one subgame in Figure 4.2, which is the game itself. Now the player who makes decision is the server, and to maximize the utility, the server will apparently choose the utility profile (0.4,0.6), which is the subgame-perfect Nash equilibrium of the game.

4.2.3.3   Stable equilibrium

Although subgame perfection can refine Nash equilibriums, it usually needs perfect information among players. However, players often have only limited information about the other players' strategies, or they are even unaware of the game being played. Under such circumstances, subgame perfection is very difficult to meet. Also, what if some players deviate from the equilibrium a little bit? Deviation is very likely to happen as a result of lack of information. If deviation occurs, players may seek *stable* equilibriums that push the outcome back to the stable equilibriums themselves after several rounds.

The *evolutionarily stable strategy* (ESS), inspired by biology mutations, guarantees the stability of the outcome of the game. Different from traditional game theory that emphasizes on the properties of strategy equilibrium, evolutionary game theory focuses on the dynamics of strategy change. ESS provides guidance for a rational player to approach the best strategy against a small number of players who deviate from the best strategy, and thus achieve stability.

Take the prisoner's dilemma in Table 4.1 as an example. Because evolutionary game theory focuses on the behavior of populations, we assume that the game is played by a set of homogeneous players, who have the same form of utility function $\pi_i = \pi$ and action space ("cooperate" and "defect"). To understand how a population of individuals who repeatedly play the prisoner's dilemma evolves, we assume that the population is quite large. In such a case, we can represent the state of the population by simply keeping track of what proportion follows the strategy "cooperate." Let p_c and p_d denote the portions of the population who play cooperate and defect, respectively. Furthermore, the average payoffs of cooperators and defectors are denoted as π_c and π_d, respectively, and the average payoff of the entire population is denoted as $\overline{\pi}$. Then the payoffs can be written as

$$\pi_c = p_c * 3 + p_d * 0,$$
$$\pi_d = p_c * 5 + p_d * 1,$$
$$\text{and}\quad \overline{\pi} = p_c * \pi_c + p_d * \pi_d. \tag{4.4}$$

The key concept of evolution is that the rate of portion change is proportional to the payoff rate; that is,

$$\frac{\partial p_c}{\partial t} = \frac{p_c(\pi_c - \overline{\pi})}{\overline{\pi}}, \quad \text{and} \quad \frac{\partial p_d}{\partial t} = \frac{p_d(\pi_d - \overline{\pi})}{\overline{\pi}}. \tag{4.5}$$

From (4.4), $\pi_d > \pi_c$, which results in both $\pi_c - \overline{\pi} < 0$ and $\partial p_c/\partial t < 0$. Therefore, we see that over time, the proportion of the population choosing the strategy "cooperate" eventually becomes extinct, and "defect" is the only evolutionarily stable equilibrium.

We provide the formal definition of evolutionary stable equilibrium for a two-player game as follows:

Definition 4.5. In a symmetric strategic game with two players $\langle (1, 2), (\mathbf{A}, \mathbf{A}), (\pi_1, \pi_2) \rangle$, an evolutionarily stable strategy is the action $a^* \in \mathbf{A}$ such that $\pi_i(a, a^*) < \pi_i(a^*, a^*)$ and $\pi_i(a, a) < \pi_i(a^*, a)$ for every best response $a \neq a^* \in \mathbf{A}$.

4.2.4 Equilibrium improvement

From the previous analysis, (defect,defect) is the only stable equilibrium of the prisoner's dilemma. However, apparently it is not the best strategy because (cooperate,cooperate) gives higher utilities to both players. In the prisoner's dilemma, equilibriums are inefficient. Remember that the prisoner's dilemma game was designed by the attorney who intended to make both prisoners confess and be punished for the crime. The question is how to change the mechanism of the game so the outcome of the equilibrium can be improved.

The players in the prisoner's dilemma tend to play (defect,defect) because the game is played only once and the players do not build a record of past interactions. What if the game can be played multiple times and the players can learn from their past experience to decide future actions? To model and analyze long-term interactions among players, the repeated game model is used, in which the game is played for multiple rounds. A repeated game is a special form of an extensive-form game, in which each stage is a repetition of the same strategic-form game. The number of rounds may be finite or infinite, but usually the infinite case is more interesting. The reason is that if the game is repeated for finite times, by backward induction, in most cases the equilibriums will be the same as that when the game is played only one time. With the infinitely repeated game, players care about not only the current payoff but also the future payoffs, and a player's current behavior can affect the other players' future behavior. Therefore, cooperation and mutual trust among players, as well as new equilibriums, can be established.

The *Folk theorem* states that in infinitely repeated games, if all players' min–max conditions are satisfied in a strategy profile, such a strategy profile is a feasible equilibrium. Here, the min–max condition means that a player will minimize the maximum possible loss that he or she may face in the game. The Folk theorem tells us that in infinitely repeated games, for any utility profile that gives a nonnegative payoff to every player, there exists at least one equilibrium strategy to reach such a utility profile. Therefore, instead of suffering from inefficient equilibriums, with infinitely repeated games, the problem is to find the desired utility profile and design an interaction mechanism among users to reach such a utility profile.

An example interaction mechanism is as follows. At the beginning of the game, all players know the feasible payoff and try to choose the strategies that will lead to such a desired utility profile. Then, if some players deviate from the desired payoff, such a deviation triggers the the punishment stage, in which the deviating player will be punished by other players such that any gains received by the deviator are at least canceled out. Depending on the mechanism design, such punishment can be kept with the deviator until the game terminates, or until several rounds have been played. In the latter case, if the deviator behaves well later, the deviation behavior will be forgiven and the punishment stopped after a certain period. Either way, there is no incentive for any player to deviate from the strategy that leads to the desired utility profile, and the game will proceed with the suggested strategies that form an equilibrium. With infinitely repeated games, the system designer is able to lead the game to any desired outcome as long as the outcome is feasible and the interaction mechanism is carefully designed.

4.3   Bargaining games

The bargaining game is one of the most interesting cooperative games, in which players have the opportunity to negotiate with one another and reach a mutually beneficial agreement on a division of interest among them. If they fail to reach agreements, the potential benefit will be lost. For instance, both the employer and an employee can gain from reaching an agreement as to more flexible working hours to increase the production rate. Then the question is how the extra benefit that comes from greater flexibility of working hours should be distributed between the laborer and the capital in the form of higher wages or profits. The rearrangement would require an agreement as to how to distribute the resulting burdens and benefits.

In this section, we begin by discussing the Nash axiomatic bargaining model followed by a simple version of the strategic model developed by Rubinstein. We then generalize the simple strategic model, and analyze the relationship in this general context between the Nash and the Rubinstein solutions. Without loss of generality, we consider the two-player bargaining game with $N = \{1, 2\}$. Multiplayer bargaining games can be formulated by simply extending the two-player model. If the players fail to reach any agreement, they receive the disagreement utilities π_1^0 and π_2^0, respectively. The set of all possible utility pairs is the compact and convex feasible set denoted by U. The bargained result is denoted by $\pi^* = (\pi_1^*, \pi_2^*)$. The bargaining solution is a function that maps the bargaining problem to a utility profile, which is $f(U, \pi^0) = \pi^*$.

4.3.1   Axiomatic bargaining model

The basic idea of the axiomatic model is to impose properties that a rational bargaining solution should satisfy, and then look for solutions with these properties other than specifying an explicit bargaining procedure. The axioms imposed on the bargaining solution π' are as follows.

- **Pareto efficiency:** For any utility profile $\pi' \in U$, if there exists another utility profile $\pi^* \in U$ such that $\pi_i^* > \pi_i'$ for $i = 1, 2$, then $f(U, \pi^0) \neq \pi'$.
- **Symmetry:** If the bargaining problem is symmetry – that is, given $(\pi_1, \pi_2) \in U$, (π_2, π_1) is also in U – then $\pi_1^* = \pi_2^*$.
- **Independence of irrelevant alternatives:** For any utility set $U' \subset U$, if $f(U', \pi^0) \in U'$, then $f(U', \pi^0) = f(U, \pi^0)$.

- *Invariance to equivalent utility representations:* If we transform a bargaining problem $\langle U, \pi^0 \rangle$ into $\langle U', \pi'^0 \rangle$ by making $U' = aU + b$ and $\pi'^0 = a\pi^0 + b$, then $f(U', \pi'^0) = af(U, \pi^0) + b$.

It was shown by Nash that there exists a unique bargaining solution satisfying these axioms [97], which is

$$f(U, \pi^0) = \arg\max_{\pi \in U}(\pi_1 - \pi_1^0)(\pi_2 - \pi_2^0). \tag{4.6}$$

The function $(\pi_1 - \pi_1^0)(\pi_2 - \pi_2^0)$ is called the *Nash product*, and the solution f is called the *Nash bargaining solution*. When $\pi_i^0 = 0$ for all $i \in N$, the Nash bargaining solution is equivalent to the proportional fairness solution. In other words, the Nash bargaining solution achieves some degree of fairness among cooperative players through bargaining.

Furthermore, if we drop the symmetry axiom, then there exists the generalized Nash bargaining solution f_θ, which satisfies other three axioms and can be written as

$$f_\theta(U, \pi^0) = \arg\max_{\pi \in U}(\pi_1 - \pi_1^0)^\theta (\pi_2 - \pi_2^0)^{1-\theta}, \tag{4.7}$$

where θ is the bargaining power of player 1. Clearly, the Nash bargaining solution is a special case with $\theta = 0.5$. When $\theta \to 1$, then f_θ maximizes player 1's utility, and vice versa.

4.3.2 Strategic bargaining model

The Nash bargaining model is timeless in the sense that it investigates only the bargained outcome but not the bargaining procedure. Such a property has some advantages; for example, it is very easy to apply the Nash bargaining model to scenarios without explicit bargaining. However, there are circumstances in which the bargaining process is an important issue. For instance, it would be useful to know how bargaining solutions are influenced by the changes in the bargaining procedure to find a more robust bargaining solution. Under such circumstances, studying explicit models of strategic bargaining can provide guidance on these issues. Here we present the strategic bargaining model in which the procedure is modeled explicitly as a game in real time.

We take an example that the sum of the utilities of the two players equals to 1 – that is, $\pi_1 + \pi_2 = 1$ – and the players want to bargain to reach the agreement of how to distribute the utilities. Time proceeds without end as $t = 0, 1, 2, \ldots$; assume that an agreement q gives q to player 1 and $1 - q$ to player 2, where $q \in [0, 1]$. The bargaining procedure is as follows. At $t = 0$, player 1 offers a value of q_1 that player 2 can either *accept* or *reject*. If player 2 accepts, the solution q_1 is implemented and the game ends. If player 2 rejects, both players wait until the next time slot, $t = 1$, at which point player 2 makes an offer q_2, then player 1 decides to accept or reject, and so on. If q_t is accepted at date t, the payoff to player 1 is $\delta_1^t q$ and the payoff to player 2 is $\delta_2^t(1 - q)$, where $\delta_1, \delta_2 \in [0, 1]$ are the discount factors of player 1 and player 2, respectively. If no agreement is ever accepted by both players, both users' payoffs are 0. A strategy for a player is generally a function that specifies the value of q_t if it is his or her turn to make an offer, and an element of {accept, reject} if it is his or her turn to respond.

The subgame perfect bargaining solution (π_1^*, π_2^*) is the reservation values of the players, such that in any subgame, player i will accept any offer that makes $\pi_i \geq \pi_i^*$. To find the subgame perfect bargaining solution, first suppose that at $t = 2$, the lowest payoff that player 1 would offer is $q_2 = M$; then at $t = 1$, player 1 must accept the offer that gives him or her a payoff larger than or equal to $\delta_1 M$, and player 2's offer must be at least $q_1 \geq 1 - \delta_1 M$. Therefore, at $t = 0$, player 2 would accept the offer if it is larger than $\delta_2(1 - \delta_1 M)$. As a result, the lowest payoff that player 1 would offer at $t = 0$ is $1 - \delta_2(1 - \delta_1 M)$. Given the fact that the subgames starting from $t = 0$ and $t = 2$ are exactly the same, $M = 1 - \delta_2(1 - \delta_1 M)$ and $\pi_1^* = M = 1 - \delta_1\delta_2/(\delta_1 + \delta_2)$.

Now let us generalize the above bargaining model. First, instead of letting player 1 always make the offer first, now player i offers first with probability p_i. Second, the utility function of player i is a function of the offer q, which is $\pi_i(q)$. For example, in the previous model, $\pi_1(q) = q$ and $\pi_2(q) = 1 - q$. Also, let the length of time that each bargaining stage takes be ϵ, and the discount factor $\delta_i = 1/(1 + r_i\epsilon)$, where r_i is the discount rate of player i.

The main result of the previous example, that there exists a unique subgame-perfect equilibrium that can be characterized in terms of reservation values q_i^*, continues to hold in this generalized model. It is easy to show that the reservation values satisfy the following recursive relations [98]:

$$\pi_1(q_1^*) = \frac{1}{1 + r_1\epsilon}(p_1\pi_1(q_2^*) + p_2\pi_2(q_1^*))$$

$$\text{and} \quad \pi_2(q_2^*) = \frac{1}{1 + r_2\epsilon}(p_1\pi_2(q_2^*) + p_2\pi_1(q_1^*)). \quad (4.8)$$

Let $q = p_1q_1^* + p_2q_2^*$ be the averaged offer. Following Myerson's analysis [98], we can approximate (4.8) using the first-order Taylor series expansion with respect to q, and calculate the weighted sum of these the two equations in (4.8) to cancel out the common terms. Then, we can get

$$(1 + r_1\epsilon)p_1r_2\pi_2(q)\pi_1'(q) + (1 + r_2\epsilon)p_2r_1\pi_1(q)\pi_2'(q) = \frac{h(\epsilon)}{\epsilon}, \quad (4.9)$$

where $h(\epsilon)$ includes all higher-order terms and $h(\epsilon)/\epsilon \to 0$ when $\epsilon \to 0$.

The generalized Nash bargaining problem in (4.7) can be characterized by the first-order condition [98]:

$$\theta(\pi_2(q) - \pi_2^0)\pi_1'(q) + (1 - \theta)(\pi_1(q) - \pi_1^0)\pi_2'(q) = 0. \quad (4.10)$$

Comparing (4.10) with (4.9), we can see that when $\epsilon \to 0$, the averaged payoff q solves the generalized Nash bargaining problem with $\pi_1^0 = \pi_2^0 = 0$ and with the bargaining power

$$\theta = \frac{p_1r_2}{p_1r_2 + p_2r_1}, \quad (4.11)$$

which is a natural choice that reflects how resistant the two players' utilities are against time.

Hence, the Nash bargaining solution can be considered as an approximation to the equilibrium outcome of the strategic bargaining game when the bargaining duration is very small, which is equivalent to infinite repetitions of bargaining. The advantage of using the strategic model, even if we are interested mainly in the case in which the bargaining time of each round is short, is that it explicitly delivers the bargaining power and indicates how they depend on features of the underlying game.

Part II

Behavior forensics in media-sharing social networks

5 Equal-risk fairness in colluder social networks

As introduced in Chapter 2, multimedia fingerprinting systems involve many users with different objectives, and users influence one another's performance. To demonstrate how to analyze human behavior, we take the equal-risk fairness collusion as an example.

During collusion, colluders contribute their fingerprinted copies and collectively generate the colluded copy. As demonstrated in Section 2.3.2, depending on the way attackers collude, different colluders may have different probabilities to be detected by the digital rights enforcer. Each colluder prefers the collusion strategy that favors him or her the most, and they must agree on risk distribution before collusion. A straightforward solution is to let all colluders have the same probability of being detected, which we call *equal-risk fairness*. Depending on the fingerprinting system and the multimedia content, achieving equal-risk collusion may be complicated, especially when colluders receive fingerprinted copies of different resolutions, as shown in Section 2.3.2. In this chapter, we use equal-risk fairness as an example, analyze how colluders select the collusion parameters to achieve fairness in scalable video fingerprinting systems, and provide a simple example on behavior dynamics analysis and the investigation of the impact of human factors on multimedia systems.

In this chapter, we investigate the ways in which colluders distribute the risk evenly among themselves and achieve fairness of collusion when they receive copies of different resolutions as a result of network and device heterogeneity. We also analyze the effectiveness of such fair collusion in defeating the fingerprinting systems. We then analyze the impact of such colluders' behavior on the fingerprint detector and study the collusion resistance of the scalable fingerprinting system. We evaluate the maximum number of colluders that the embedded fingerprints can withstand in various scenarios with different requirements.

5.1 Equal-risk collusion

In this section, we consider the scalable multimedia fingerprinting system in Chapter 2 and use orthogonal fingerprint modulation, in which fingerprints assigned to different users are orthogonal to each other. When colluders receive fingerprinted copies of different resolution, they form a few subgroups, depending on the resolution of their received copies, and use the two-stage collusion model. Colluders in the same subgroup collude first and then average all the copies they have with the same weight. This intragroup collusion ensures that colluders in the same subgroup have the same probability of being detected. Then, colluders in different subgroups apply the intergroup collusion and select the collusion parameters $\{\alpha_l, \beta_k\}$ to ensure that colluders in different subgroups also have the same risk. At the fingerprint detector's side, we use the collective fingerprint detector, by which fingerprints extracted from all layers are used collectively to identify colluders. Given this model, we investigate how colluders select the collusion parameters in the intergroup collusion to achieve equal-risk fairness.

5.1.1   Analysis of the detection statistics

To let all colluders have the same risk and achieve equal-risk fairness, colluders first need to analyze the detection statistics and calculate each attacker's probability of being detected.

From Chapter 2, for each frame $j_1 \in F_b$ in the base layer, the extracted fingerprint $\mathbf{Y}_j$ is

$$\mathbf{Y}_{j_1} = \frac{\beta_1}{K^b} \sum_{i \in SC^b} \mathbf{W}_{j_1}^{(i)} + \frac{\beta_2}{K^{b,el}} \sum_{i \in SC^{b,el}} \mathbf{W}_{j_1}^{(i)} + \frac{\beta_3}{K^{all}} \sum_{i \in SC^{all}} \mathbf{W}_{j_1}^{(i)} + \mathbf{d}_{j_1}, \quad (5.1)$$

where K^b, $K^{b,el}$, and K^{all} are the number of colluders who receive copies of low, medium, and high resolution, respectively, and $\mathbf{d}_{j_1} = \mathbf{n}_{j_1}/JND_{j_1}$ is the detection noise. If the colluded copy contains frames in the enhancement layers, for each frame $j_2 \in F_{e1}$ in the enhancement layer 1,

$$\mathbf{Y}_{j_2} = \frac{\alpha_1}{K^{b,el}} \sum_{i \in SC^{b,el}} \mathbf{W}_{j_2}^{(i)} + \frac{\alpha_2}{K^{all}} \sum_{i \in S^{all}} \mathbf{W}_{j_2}^{(i)} + \mathbf{d}_{j_2}, \quad (5.2)$$

where $\mathbf{d}_{j_2} = \mathbf{n}_{j_2}/JND_{j_2}$ is the detection noise. If the colluded copy contains all three layers, for each frame $j_3 \in F_{e2}$ in the enhancement layer 2,

$$\mathbf{Y}_{j_3} = \frac{1}{K^{all}} \sum_{i \in SC^{all}} \mathbf{W}_{j_3}^{(i)} + \mathbf{d}_{j_3}, \quad (5.3)$$

where $\mathbf{d}_{j_3} = \mathbf{n}_{j_3}/JND_{j_3}$ is the detection noise.

With orthogonal fingerprint modulation as in Section 2.3.1, because the M originally embedded fingerprints are considered as known signals during fingerprint detection, under the assumption that colluders have reasonably good estimates of JND_j and $\{\mathbf{d}_j\}_{j \in F^c}$ are independent identically distributed (i.i.d.) Gaussian $\mathcal{N}(0, \sigma_n^2)$, it follows that given the colluder set SC, the detection statistics follow Gaussian distribution $p\left(TN^{(i)}|SC\right) \sim \mathcal{N}\left(\mu^{(i)}, \sigma_n^2\right)$ [60]. $\mu^{(i)} = 0$ when user i is innocent, and $\mu^{(i)} > 0$ when i is guilty. For a guilty colluder $i \in SC$, $\mu^{(i)}$ depends on the number of frames in the colluded copy and the number frames that colluder i receives; we provide here a detailed analysis of $\mu^{(i)}$ when the colluded copy has different resolutions.

When the colluded copy contains all three layers – that is, $F^c = F_b \cup F_{e1} \cup F_{e2}$ – we can show that

$$
\mu^{(i)} = \begin{cases}
\dfrac{\beta_1}{K^b}\sqrt{\sum_{j\in F^{(i)}} \|\mathbf{W}_j^{(i)}\|^2} & \text{if } i \in SC^b, \\[2ex]
\dfrac{\beta_2 \sum_{j\in F_b} \|\mathbf{W}_j^{(i)}\|^2 + \alpha_1 \sum_{j\in F_{e1}} \|\mathbf{W}_j^{(i)}\|^2}{K^{b,e1}\sqrt{\sum_{j\in F^{(i)}} \|\mathbf{W}_j^{(i)}\|^2}} & \text{if } i \in SC^{b,e1}, \\[2ex]
\dfrac{\beta_3 \sum_{j\in F_b} \|\mathbf{W}_j^{(i)}\|^2 + \alpha_2 \sum_{j\in F_{e1}} \|\mathbf{W}_j^{(i)}\|^2 + \sum_{j\in F_{e2}} \|\mathbf{W}_j^{(i)}\|^2}{K^{all}\sqrt{\sum_{j\in F^{(i)}} \|\mathbf{W}_j^{(i)}\|^2}} & \text{if } i \in SC^{all}.
\end{cases}
\tag{5.4}
$$

Define $N_b = \sum_{j\in F_b} N_j$, $N_{e1} = \sum_{j\in F_{e1}} N_j$, and $N_{e2} = \sum_{j\in F_{e2}} N_j$ as the lengths of the fingerprints that are embedded in the base layer, enhancement layer 1, and enhancement layer 2, respectively. With orthogonal fingerprint modulation as discussed in Section 2.3.1, we have $\sum_{j_1\in F_b} \|\mathbf{W}_{j_1}^{(i)}\|^2 = N_b \cdot \sigma_w^2$, $\sum_{j_2\in F_{e1}} \|\mathbf{W}_{j_2}^{(i)}\|^2 = N_{e1} \cdot \sigma_w^2$, and $\sum_{j_3\in F_{e2}} \|\mathbf{W}_{j_3}^{(i)}\|^2 = N_{e2} \cdot \sigma_w^2$, where σ_w^2 is the variance of the watermark $\mathbf{W}^{(i)}$. Therefore,

$$
\mu^{(i)} = \begin{cases}
\dfrac{\beta_1 \sqrt{N_b}}{K^b}\sigma_w & \text{if } i \in SC^b, \\[2ex]
\dfrac{\beta_2 N_b + \alpha_1 N_{e1}}{K^{b,e1}\sqrt{N_b + N_{e1}}}\sigma_w & \text{if } i \in SC^{b,e1}, \\[2ex]
\dfrac{\beta_3 N_b + \alpha_2 N_{e1} + N_{e2}}{K^{all}\sqrt{N_b + N_{e1} + N_{e2}}}\sigma_w & \text{if } i \in SC^{all}.
\end{cases}
\tag{5.5}
$$

When the colluded copy contains frames in the base layer and the enhancement layer 1, that is, $F^c = F_b \cup F_{e1}$, similar to the preceding analysis,

$$
\mu^{(i)} = \begin{cases}
\dfrac{\beta_1 \sqrt{N_b}}{K^b}\sigma_w & \text{if } i \in SC^b, \\[2ex]
\dfrac{\beta_2 N_b + \alpha_1 N_{e1}}{K^{b,e1}\sqrt{N_b + N_{e1}}}\sigma_w & \text{if } i \in SC^{b,e1}, \\[2ex]
\dfrac{\beta_3 N_b + \alpha_2 N_{e1}}{K^{all}\sqrt{N_b + N_{e1}}}\sigma_w & \text{if } i \in SC^{all}.
\end{cases}
\tag{5.6}
$$

When the colluded copy contains frames in the base layer only and $F^c = F_b$, we have

$$
\mu^{(i)} = \begin{cases}
\dfrac{\beta_1 \sqrt{N_b}}{K^b}\sigma_w & \text{if } i \in SC^b, \\[2ex]
\dfrac{\beta_2 \sqrt{N_b}}{K^{b,e1}}\sigma_w & \text{if } i \in SC^{b,e1}, \\[2ex]
\dfrac{\beta_3 \sqrt{N_b}}{K^{all}}\sigma_w & \text{if } i \in SC^{all}.
\end{cases}
\tag{5.7}
$$

5.1.2 Selection of the collusion parameters

With the preceding analysis of the detection statistics, given a threshold h, for colluder $\mathbf{u}^{(i)}$ whose detection statistics follow distribution $\mathcal{N}(\mu^{(i)}, \sigma_n^2)$, the probability that colluder i is captured is

$$
P^{(i)} = P\left[TN^{(i)} > h\right] = Q\left(\frac{h - \mu^{(i)}}{\sigma_n}\right),
\tag{5.8}
$$

where $Q(x) = \int_x^\infty \frac{1}{\sqrt{2\pi}} e^{-\frac{t^2}{2}} dt$ is the Gaussian tail function. Therefore, all colluders share the same risk and are equally likely to be detected if and only if their detection statistics have the same mean.

5.1.2.1 Highest-quality colluded copy

When the colluded copy contains frames in all three layers, from (5.5), colluders seek $\{0 \le \beta_k \le 1\}_{k=1,2,3}$ and $\{0 \le \alpha_l \le 1\}_{l=1,2}$ to satisfy

$$\frac{\beta_1 \sqrt{N_b}}{K^b} \sigma_w = \frac{\beta_2 N_b + \alpha_1 N_{e1}}{K^{b,e1} \sqrt{N_b + N_{e1}}} \sigma_w = \frac{\beta_3 N_b + \alpha_2 N_{e1} + N_{e2}}{K^{all} \sqrt{N_b + N_{e1} + N_{e2}}} \sigma_w,$$

subject to $\beta_1 + \beta_2 + \beta_3 = 1$, $\alpha_1 + \alpha_2 = 1$. $\quad\quad$ (5.9)

Note that

$$\frac{\beta_1 \sqrt{N_b}}{K^b} \sigma_w = \frac{\beta_2 N_b + \alpha_1 N_{e1}}{K^{b,e1} \sqrt{N_b + N_{e1}}} \sigma_w \Longleftrightarrow \frac{K^{b,e1} \sqrt{N_b + N_{e1}}}{K^b \sqrt{N_b}} = \frac{\beta_2 N_b + \alpha_1 N_{e1}}{\beta_1 N_b}.$$
$$(5.10)$$

In addition, let $\beta_3 = 1 - \beta_1 - \beta_2$ and $\alpha_2 = 1 - \alpha_1$, we have

$$\frac{\beta_1 \sqrt{N_b}}{K^b} \sigma_w = \frac{\beta_3 N_b + \alpha_2 N_{e1} + N_{e2}}{K^{all} \sqrt{N_b + N_{e1} + N_{e2}}} \sigma_w \Longleftrightarrow \frac{K^{all} \sqrt{N_b + N_{e1} + N_{e2}}}{K^b \sqrt{N_b}}$$
$$= \frac{N_b + N_{e1} + N_{e2}}{\beta_1 N_b} - 1 - \frac{\beta_2 N_b + \alpha_1 N_{e1}}{\beta_1 N_b}. \quad (5.11)$$

Plugging (5.10) into (5.11), we have

$$\frac{N_b + N_{e1} + N_{e2}}{\beta_1 N_b} \underline{\underline{}} \frac{K^{all} \sqrt{N_b + N_{e1} + N_{e2}} + K^{b,e1} \sqrt{N_b + N_{e1}}}{K^b \sqrt{N_b}} + 1. \quad (5.12)$$

Therefore, from (5.10) and (5.12), to ensure equal risk, colluders should choose

$$\beta_1 = \frac{N_b + N_{e1} + N_{e2}}{N_b} \frac{K^b \sqrt{N_b}}{K^b \sqrt{N_b} + K^{b,e1} \sqrt{N_b + N_{e1}} + K^{all} \sqrt{N_b + N_{e1} + N_{e2}}},$$

and $\quad \beta_2 N_b + \alpha_1 N_{e1} = \frac{(N_b + N_{e1} + N_{e2}) K^{b,e1} \sqrt{N_b + N_{e1}}}{K^b \sqrt{N_b} + K^{b,e1} \sqrt{N_b + N_{e1}} + K^{all} \sqrt{N_b + N_{e1} + N_{e2}}}.$

$$(5.13)$$

From Section 2.3.2, the collusion parameters are required to be in the range of $[0, 1]$. From (5.13), $0 \le \beta_1 \le 1$ if and only if

$$\frac{K^b \sqrt{N_b}}{K^b \sqrt{N_b} + K^{b,e1} \sqrt{N_b + N_{e1}} + K^{all} \sqrt{N_b + N_{e1} + N_{e2}}} \le \frac{N_b}{N_b + N_{e1} + N_{e2}}. \quad (5.14)$$

Furthermore, from (5.13),

$$\alpha_1 = \frac{N_b + N_{e1} + N_{e2}}{N_{e1}} \frac{K^{b,e1} \sqrt{N_b + N_{e1}}}{K^b \sqrt{N_b} + K^{b,e1} \sqrt{N_b + N_{e1}} + K^{all} \sqrt{N_b + N_{e1} + N_{e2}}}$$
$$- \beta_2 \frac{N_b}{N_{e1}}. \quad (5.15)$$

Given β_1 as in (5.13), $0 \le \beta_2 \le 1 - \beta_1$. Consequently, from (5.15), we have $\underline{\alpha} \le \alpha_1 \le \overline{\alpha}$, where

$$\underline{\alpha} = \frac{N_b + N_{e1} + N_{e2}}{N_{e1}} \frac{K^b\sqrt{N_b} + K^{b,e1}\sqrt{N_b + N_{e1}}}{K^b\sqrt{N_b} + K^{b,e1}\sqrt{N_b + N_{e1}} + K^{all}\sqrt{N_b + N_{e1} + N_{e2}}}$$

$$- \frac{N_b}{N_e 1}, \quad \text{and}$$

$$\overline{\alpha} = \frac{N_b + N_{e1} + N_{e2}}{N_{e1}} \frac{K^{b,e1}\sqrt{N_b + N_{e1}}}{K^b\sqrt{N_b} + K^{b,e1}\sqrt{N_b + N_{e1}} + K^{all}\sqrt{N_b + N_{e1} + N_{e2}}}.$$

$$(5.16)$$

If $[0, 1] \cap [\underline{\alpha}, \overline{\alpha}]$ is not empty, then there exists at least one α_1^* such that $0 \le \alpha_1^* \le 1$ and $\underline{\alpha} \le \alpha_1^* \le \overline{\alpha}$. Note that $\overline{\alpha} > 0$, so $[0, 1] \cap [\underline{\alpha}, \overline{\alpha}] \ne \emptyset$ if and only if $\underline{\alpha} \le 1$, which is equivalent to

$$\frac{K^{all}\sqrt{N_b + N_{e1} + N_{e2}}}{K^b\sqrt{N_b} + K^{b,e1}\sqrt{N_b + N_{e1}} + K^{all}\sqrt{N_b + N_{e1} + N_{e2}}} \ge \frac{N_{e2}}{N_b + N_{e1} + N_{e2}}.$$

$$(5.17)$$

To summarize, to generate a colluded copy with the highest temporal resolution under the equal-risk fairness constraints, $(K^b, K^{b,e1}, K^{all})$ and (N_b, N_{e1}, N_{e2}) must satisfy (5.14) and (5.17), and colluders should choose the collusion parameters as in (5.13).

5.1.2.2 Medium-quality colluded copy

In this scenario, the colluded copy has medium resolution and contains frames in the base layer and enhancement layer 1. For colluder $i_1 \in SC^{all}$ and colluder $i_2 \in SC^{b,e1}$ who receive copies of the highest and the medium resolution, respectively, the overall lengths of their fingerprints in the colluded copy are the same and equal to $N_b + N_{e1}$. In this scenario, during the intergroup collusion shown in Figure 2.4, colluder i_1 and i_2 let $\alpha_1/K^{b,e1} = \alpha_2/K^{all}$ and $\beta_2/K^{b,e1} = \beta_3/K^{all}$. Such a parameter selection not only guarantees $\mu^{(i_1)} = \mu^{(i_2)}$, but also ensures that for each frame j in the colluded copy, the energies of these two colluders' fingerprints $\mathbf{W}_j^{(i_1)}$ and $\mathbf{W}_j^{(i_2)}$ are reduced by the same ratio. For a given $0 \le \beta_1 \le 1$, it is equivalent to

$$\alpha_1 = \frac{K^{b,e1}}{K^{b,e1} + K^{all}}, \quad \alpha_2 = 1 - \alpha_1,$$

$$\beta_2 = \frac{K^{b,e1}}{K^{b,e1} + K^{all}}(1 - \beta_1), \quad \text{and} \quad \beta_3 = 1 - \beta_1 - \beta_2. \quad (5.18)$$

With the previously selected parameters,

$$\mu^{(i_1)} = \mu^{(i_2)} = \frac{(1 - \beta_1)N_b + N_{e1}}{(K^{b,e1} + K^{all})\sqrt{N_b + N_{e1}}}\sigma_w. \quad (5.19)$$

Colluders seek $0 \le \beta_1 \le 1$ such that

$$\frac{\beta_1\sqrt{N_b}}{K^b}\sigma_w = \frac{(1 - \beta_1)N_b + N_{e1}}{(K^{b,e1} + K^{all})\sqrt{N_b + N_{e1}}}\sigma_w, \quad (5.20)$$

and the solution is

$$\beta_1 = \frac{N_b + N_{e1}}{N_b} \frac{K_b \sqrt{N_b}}{K^b \sqrt{N_b} + (K^{b,e1} + K^{all})\sqrt{N_b + N_{e1}}}. \quad (5.21)$$

With β_1 as in (5.21), $0 \le \beta_1 \le 1$ if and only if

$$\frac{K^b \sqrt{N_b}}{K^b \sqrt{N_b} + (K^{b,e1} + K^{all})\sqrt{N_b + N_{e1}}} \le \frac{N_b}{N_b + N_{e1}}. \quad (5.22)$$

Given $0 \le \beta_1 \le 1$, from (5.18), it is straightforward to show that $0 \le \beta_2, \beta_3, \alpha_1, \alpha_2 \le 1$.

To summarize, to achieve equal-risk fairness, $(K^b, K^{b,e1}, K^{all})$ and (N_b, N_{e1}, N_{e2}) must satisfy (5.22) if colluders wish to generate a colluded copy of medium temporal resolution. Colluders should choose the collusion parameters as in (5.18) and (5.21).

5.1.2.3 Lowest-quality colluded copy

When the colluded copy contains frames in the base layer only, colluders choose $\{0 \le \beta_k \le 1\}_{k=1,2,3}$ with $\beta_1 + \beta_2 + \beta_3 = 1$ to satisfy

$$\frac{\beta_1 \sqrt{N_b}}{K^b}\sigma_w = \frac{\beta_2 \sqrt{N_b}}{K^{b,e1}}\sigma_w = \frac{\beta_3 \sqrt{N_b}}{K^{all}}\sigma_w, \quad (5.23)$$

and the solution is

$$\beta_1 = \frac{K^b}{K^b + K^{b,e1} + K^{all}}, \quad \beta_2 = \frac{K^{b,e1}}{K^b + K^{b,e1} + K^{all}},$$

$$\text{and} \quad \beta_3 = \frac{K^{all}}{K^b + K^{b,e1} + K^{all}}. \quad (5.24)$$

In this scenario, there are no constraints on $(K^b, K^{b,e1}, K^{all})$ and (N_b, N_{e1}, N_{e2}), and colluders can always generate a colluded copy containing frames in the base layer only.

5.1.3 Summary of parameter selection to achieve fairness during collusion

Table 5.1 summarizes the constraints and the parameter selection during collusion to ensure equal-risk fairness in three scenarios, in which the colluded copy has the highest, medium, and lowest temporal resolutions, respectively. From Table 5.1, if colluders want to generate a colluded copy of higher resolution, the constraints are more severe to distribute the risk of being detected evenly among all attackers.

To select the collusion parameters, colluders need to estimate $N_b : N_{e1} : N_{e2}$, the ratio of the lengths of the fingerprints embedded in different layers. Because adjacent frames in a video sequence are similar to each other and have approximately the same number of embeddable coefficients, colluders can use the following approximation $N_b : N_{e1} : N_{e2} \approx |F_b| : |F_{e1}| : |F_{e2}|$.

Table 5.1 Constraints and selection of collusion parameters during collusion to achieve fairness

$F^c = F_b \cup F_{e1} \cup F_{e2}$ (highest resolution)	Fairness constraints: $$\begin{cases} \dfrac{K^b \sqrt{N_b}}{K^b\sqrt{N_b}+K^{b,e1}\sqrt{N_b+N_{e1}}+K^{all}\sqrt{N_b+N_{e1}+N_{e2}}} \leq \dfrac{N_b}{N_b+N_{e1}+N_{e2}}, \\[2em] \dfrac{K^{all}\sqrt{N_b+N_{e1}+N_{e2}}}{K^b\sqrt{N_b}+K^{b,e1}\sqrt{N_b+N_{e1}}+K^{all}\sqrt{N_b+N_{e1}+N_{e2}}} \geq \dfrac{N_{e2}}{N_b+N_{e1}+N_{e2}}. \end{cases}$$ Parameter selection: $$\begin{cases} \beta_1 = \dfrac{N_b+N_{e1}+N_{e2}}{N_b} \dfrac{K^b\sqrt{N_b}}{K^b\sqrt{N_b}+K^{b,e1}\sqrt{N_b+N_{e1}}+K^{all}\sqrt{N_b+N_{e1}+N_{e2}}}, \\[1.5em] \beta_2 N_b + \alpha_1 N_{e1} = \dfrac{(N_b+N_{e1}+N_{e2})K^{b,e1}\sqrt{N_b+N_{e1}}}{K^b\sqrt{N_b}+K^{b,e1}\sqrt{N_b+N_{e1}}+K^{all}\sqrt{N_b+N_{e1}+N_{e2}}}, \\[1.5em] \beta_3 = 1 - \beta_1 - \beta_2, \ \alpha_2 = 1 - \alpha_1. \end{cases}$$
$F^c = F_b \cup F_{e1}$ (medium resolution)	Fairness constraints: $$\dfrac{K^b\sqrt{N_b}}{K^b\sqrt{N_b}+(K^{b,e1}+K^{all})\sqrt{N_b+N_{e1}}} \leq \dfrac{N_b}{N_b+N_{e1}}.$$ Parameter selection: $$\begin{cases} \beta_1 = \dfrac{N_b+N_{e1}}{N_b} \dfrac{K_b\sqrt{N_b}}{K^b\sqrt{N_b}+(K^{b,e1}+K^{all})\sqrt{N_b+N_{e1}}}, \\[1.5em] \beta_2 = \dfrac{K^{b,e1}}{K^{b,e1}+K^{all}}(1-\beta_1), \ \beta_3 = 1 - \beta_1 - \beta_2, \\[1.5em] \alpha_1 = \dfrac{K^{b,e1}}{K^{b,e1}+K^{all}}, \ \alpha_2 = 1 - \alpha_1. \end{cases}$$
$F^c = F_b$ (lowest resolution)	Fairness constraints: No constraints on $(K^b, K^{b,e1}, K^{all})$ and (N_b, N_{e1}, N_{e2}). Parameter selection: $$\beta_1 = \dfrac{K^b}{K^b+K^{b,e1}+K^{all}}, \ \beta_2 = \dfrac{K^{b,e1}}{K^b+K^{b,e1}+K^{all}}, \ \beta_3 = \dfrac{K^{all}}{K^b+K^{b,e1}+K^{all}}.$$

5.2 Influence on the detector's side: collusion resistance

In this section, we analyze the impact of equal-risk fair collusion on multimedia fingerprinting. We study the effectiveness of collusion attacks and examine the traitor-tracing performance of scalable fingerprinting systems when colluders follow Table 5.1 to select the collusion parameters.

5.2.1 Statistical analysis on traitor-tracing capacity

Assume that there are a total of M users. From the analysis in the previous section, if colluders select the collusion parameters as in Table 5.1, then given a colluder set SC, for each user i,

$$TN^{(i)} \sim \begin{cases} \mathcal{N}(\mu, \sigma_n^2) & \text{if } i \in SC, \\ \mathcal{N}(0, \sigma_n^2) & \text{if } i \notin SC, \end{cases} \tag{5.25}$$

where σ_n^2 is the variance of the detection noise $\mathbf{d}_j$, and the M detection statistics $\{TN^{(i)}\}_{i=1,\cdots,M}$ are independent of each other because of the orthogonality of the fingerprints. In addition, for $i \in SC$,

$$\mu = \frac{\beta_1 \sqrt{N_b}}{K^b} \sigma_w$$

$$= \begin{cases} \dfrac{N_b + N_{e1} + N_{e2}}{K^b \sqrt{N_b} + K^{b,e1}\sqrt{N_b + N_{e1}} + K^{all}\sqrt{N_b + N_{e1} + N_{e2}}} \sigma_w & \text{if } F^c = F_b \cup F_{e1} \cup F_{e2}, \\[2ex] \dfrac{N_b + N_{e1}}{K^b \sqrt{N_b} + (K^{b,e1} + K^{all})\sqrt{N_b + N_{e1}}} \sigma_w & \text{if } F^c = F_b \cup F_{e1}, \\[2ex] \dfrac{\sqrt{N_b}}{K^b + K^{b,e1} + K^{all}} \sigma_w & \text{if } F^c = F_b. \end{cases}$$

$$(5.26)$$

Note that

$$\frac{N_b + N_{e1}}{K^b \sqrt{N_b} + (K^{b,e1} + K^{all})\sqrt{N_b + N_{e1}}} = \frac{\sqrt{N_b + N_{e1}}}{K^b \sqrt{\dfrac{N_b}{N_b + N_{e1}}} + K^{b,e1} + K^{all}}$$

$$\geq \frac{\sqrt{N_b + N_{e1}}}{K^b + K^{b,e1} + K^{all}}$$

$$\geq \frac{\sqrt{N_b}}{K^b + K^{b,e1} + K^{all}}. \quad (5.27)$$

Similarly, we can also show that

$$\frac{N_b + N_{e1} + N_{e2}}{K^b \sqrt{N_b} + K^{b,e1}\sqrt{N_b + N_{e1}} + K^{all}\sqrt{N_b + N_{e1} + N_{e2}}}$$

$$\geq \frac{N_b + N_{e1}}{K^b \sqrt{N_b} + (K^{b,e1} + K^{all})\sqrt{N_b + N_{e1}}}. \quad (5.28)$$

Therefore, under the equal-risk fairness constraints, μ in (5.26) is larger when the colluded copy has higher resolution.

Given a threshold h, from (5.25), we can have

$$P_d = P\left[\max_{i \in SC} TN^{(i)} > h\right] = 1 - \left[1 - Q\left(\frac{h - \mu}{\sigma_n}\right)\right]^K,$$

$$P_{fp} = P\left[\max_{i \notin SC} TN^{(i)} > h\right] = 1 - \left[1 - Q\left(\frac{h}{\sigma_n}\right)\right]^{M-K},$$

$$E[F_d] = \sum_{i \in SC} P\left[TN^{(i)} > h\right]/K = Q\left(\frac{h - \mu}{\sigma_n}\right),$$

$$\text{and} \quad E[F_{fp}] = \sum_{i \notin SC} P\left[TN^{(i)} > h\right]/(M - K) = Q\left(\frac{h}{\sigma_n}\right). \quad (5.29)$$

From (5.26) and (5.29), the effectiveness of equal-risk collusion in defeating the scalable fingerprinting systems depends on the total number of colluders K as well as the temporal resolution of the colluded copy. For a fixed resolution of the colluded copy, when there are more colluders in the systems, they are less likely to be captured and the collusion attack is more effective. For a fixed total number of colluders K, when the colluded copy has a higher resolution, the extracted fingerprint is longer and provides more information on colluders' identities to the detector. Therefore, colluders have a larger probability of being detected. During collusion, colluders must take into consideration the tradeoff between the risk of being detected and the resolution of the colluded copy.

5.2.2 Simulation results with ideal gaussian models

When simulating the scalable fingerprinting systems and collusion attacks using ideal Gaussian models, we test a total of forty frames as an example. We consider a temporally scalable coding system in which frames $F_b = \{1, 5, \ldots, 37\}$ are encoded in the base layer, frames $F_{e1} = \{3, 7, \ldots, 39\}$ are in enhancement layer 1, and enhancement layer 2 consists of frames $F_{e2} = \{2, 4, \ldots, 40\}$. User $i_1 \in \mathbf{U}^b$ receives the base layer only and reconstructs a fingerprinted copy of ten frames, including frame $1, 5, \ldots$, and frame 37. For user $i_2 \in \mathbf{U}^{b,el}$ who receives the base layer and enhancement layer 1, the fingerprinted copy includes all twenty odd-numbered frames. User $i_3 \in \mathbf{U}^{all}$ subscribes to all three layers and receives a fingerprinted copy of all forty frames.

From human visual models [58], not all coefficients are embeddable because of imperceptibility constraints. For real video sequences such as "akiyo" and "carphone," the number of embeddable coefficients in each frame varies from 3000 to 7000, depending on the characteristics of the video sequences. In our simulations, we assume that the length of the fingerprints embedded in each frame is 5000, and the lengths of the fingerprints embedded in the base layer, enhancement layer 1, and enhancement layer 2 are $N_b = 50000$, $N_{e1} = 50000$, and $N_{e2} = 100000$, respectively. We assume that there are a total of $M = 450$ users and $|\mathbf{U}^b| = |\mathbf{U}^{b,el}| = |\mathbf{U}^{all}| = 150$. We first generate independent vectors following Gaussian distribution $\mathcal{N}(0, \sigma_w^2)$ with $\sigma_w^2 = 1/9$, and then apply Gram–Schmidt orthogonalization to produce fingerprints that satisfy (2.1). In each fingerprinted copy, fingerprints embedded in adjacent frames are correlated with each other.

We assume that $0 \leq K^b$, $K^{b,el}$, $K^{all} \leq 150$ are the number of colluders in subgroups SC^b, $SC^{b,el}$, and SC^{all}, respectively. During collusion, colluders apply the intragroup collusion followed by the intergroup collusion, as in Figure 2.4. Furthermore, we assume that the detection noise follows Gaussian distribution, with zero mean and variance $\sigma_n^2 = 2\sigma_w^2$.

In Figure 5.1, we fix the ratio $K^b : K^{b,el} : K^{all} = 1:1:1$, and assume that the colluded copy has medium resolution and includes all twenty odd-numbered frames. In Figure 5.1(a), we select the threshold h to fix the probability of accusing at least one innocent user as 10^{-3} and plot the probability of capturing at least one colluder P_d when the total number of colluders K increases. In Figure 5.1(b), $E[F_{fp}] = 10^{-3}$ and we plot the expected fraction of colluders that are captured when K increases. From Figure 5.1, the collusion is more effective in removing traces of the fingerprints when there are more colluders.

We then fix the total number of colluders $K = 150$, and compare the effectiveness of the collusion attacks when the temporal resolution of the colluded copy changes. Define

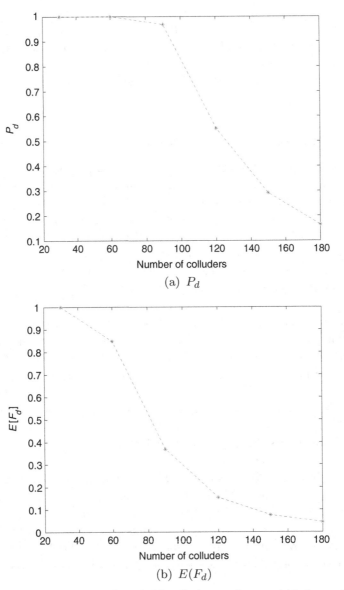

(a) P_d

(b) $E(F_d)$

Fig. 5.1 Effectiveness of equal-risk collusion attacks on scalable fingerprinting systems. $K^b{:}K^{b,e1}{:}K^{all} = 1{:}1{:}1$

the lines $\overline{AB}$ and $\overline{CD}$ as

$$\overline{AB} \triangleq \left\{ \left(K^b, K^{b,e1}, K^{all}\right) : \frac{K^b \sqrt{N_b}}{K^b \sqrt{N_b} + K^{b,e1} \sqrt{N_b + N_{e1}} + K^{all} \sqrt{N_b + N_{e1} + N_{e2}}} \right.$$

$$= \frac{N_b}{N_b + N_{e1} + N_{e2}}, 0 \leq K^b \leq |\mathbf{U}^b|, 0 \leq K^{b,e1} \leq |\mathbf{U}^{b,e1}|, 0 \leq K^{all} \leq |\mathbf{U}^{all}|,$$

$$\left. K^b + K^{b,e1} + K^{all} = K \right\}, \quad \text{and} \tag{5.30}$$

$$\overline{CD} \triangleq \left\{ \left(K^b, K^{b,e1}, K^{all} \right) : \frac{K^{all} \sqrt{N_b + N_{e1} + N_{e2}}}{K^b \sqrt{N_b} + K^{b,e1} \sqrt{N_b + N_{e1}} + K^{all} \sqrt{N_b + N_{e1} + N_{e2}}} \right.$$

$$= \frac{N_{e2}}{N_b + N_{e1} + N_{e2}}, 0 \le K^b \le |\mathbf{U}^b|, 0 \le K^{b,e1} \le |\mathbf{U}^{b,e1}|, 0 \le K^{all} \le |\mathbf{U}^{all}|,$$

$$\left. K^b + K^{b,e1} + K^{all} = K \right\}, \tag{5.31}$$

respectively, as shown in Figure 5.2(a). Line $\overline{AB}$ and line $\overline{CD}$ are the boundaries of the two constraints to achieve fairness, respectively, when generating an attacked copy of the highest resolution. For a fixed $K = 150$, we study the effectiveness of collusion when $\left(K^b, K^{b,e1}, K^{all} \right)$ takes different values on line $\overline{AB}$ and line $\overline{CD}$, respectively. In our simulations, we assume that colluders generate a colluded copy of the highest possible resolution under the constraints in Table 5.1. Figure 5.2(b) plots the regions in which colluders can generate a colluded copy of high resolution and regions where colluders can generate a medium resolution copy under the fairness constraints in Table 5.1.

Figures 5.3 and 5.4 show the simulation results when $K = 150$ is fixed and $\left(K^b, K^{b,e1}, K^{all} \right)$ takes different values on line $\overline{AB}$ (5.30). In Figures 5.3 and 5.4, a given value of K^{all} corresponds to a unique point on line $\overline{AB}$, and therefore, a unique triplet $\left(K^b, K^{b,e1}, K^{all} \right)$. Figure 5.3(a) shows the number of frames in the colluded copy $L^c = |F^c|$. $L^c = 20$ when the attacked copy has medium resolution and $L^c = 40$ when attackers generate a copy including all three layers. Figure 5.3(b) shows the means of the detection statistics of the guilty colluders. In Figure 5.4(a), we select the threshold used to fix $P_{fp} = 10^{-3}$ and we compare P_d of the collusion attacks when the triplet $\left(K^b, K^{b,e1}, K^{all} \right)$ takes different values on line $\overline{AB}$. In Figure 5.4(b), $E[F_{fp}] = 10^{-3}$ by selecting the threshold in the simulation runs and we compare $E[F_d]$ of the fair collusion for different triplets $\left(K^b, K^{b,e1}, K^{all} \right)$ on line $\overline{AB}$.

Similarly, Figures 5.5 and 5.6 show the simulation results when K is fixed as 150 and $\left(K^b, K^{b,e1}, K^{all} \right)$ moves on line $\overline{CD}$ (5.31). In these two figures, each K^b represents one point on line $\overline{CD}$ and a unique $\left(K^b, K^{b,e1}, K^{all} \right)$. Figure 5.5(a) plots the total number of frames in the colluded copy. $L^c = 10$, $L^c = 20$, and $L^c = 40$ correspond to the scenarios in which the colluded copy has low, medium, and high resolution, respectively. Figure 5.5(b) shows the mean of the guilty colluders' detection statistics. In Figure 5.6(a), P_{fp} is fixed as 10^{-3} and we compare P_d when $\left(K^b, K^{b,e1}, K^{all} \right)$ moves from left to right on line $\overline{CD}$. Figure 5.6(b) fixes $E[F_{fp}] = 10^{-3}$ and plots $E[F_d]$ for different $\left(K^b, K^{b,e1}, K^{all} \right)$ on Line $\overline{CD}$.

From Figures 5.3 to 5.6, we see that when the colluded copy has higher temporal resolution, the attacked copy contains more information on the attackers' fingerprints, and colluders have a larger probability to be captured. This is in agreement with our statistical analysis in Section 5.2.1. Colluders must consider the tradeoff between the probability of being detected and the resolution of the attacked copy during collusion.

From the preceding figures, if we fix the total number of colluders K and the resolution of the colluded copy $L^c = |F^c|$, P_d and $E[F_d]$ have larger values when K^b is smaller (or equivalently, when K^{all} is larger). This is because, with fixed $K = K^b + K^{b,e1} + K^{all}$

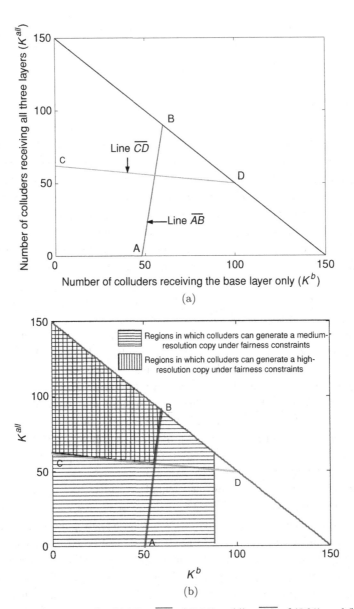

Fig. 5.2 An example of (a) line $\overline{AB}$ of (5.30) and line $\overline{CD}$ of (5.31), and (b) regions in which colluders can generate a medium-resolution or a high-resolution copy while still ensuring equal-risk fairness. $K = 150$

and fixed $F^c = F_b \cup F_{el}$, from (5.26),

$$\mu = \frac{N_b + N_{el}}{K^b\sqrt{N_b} + (K^{b,el} + K^{all})\sqrt{N_b + N_{el}}}\sigma_w$$

$$= \frac{N_b + N_{el}}{K^b\sqrt{N_b} + (K - K^b)\sqrt{N_b + N_{el}}}\sigma_w$$

$$= \frac{N_b + N_{el}}{\sqrt{N_b + N_{el}} + K^b(\sqrt{N_b} - \sqrt{N_b + N_{el}})}\sigma_w \qquad (5.32)$$

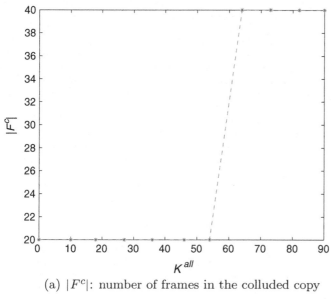

(a) $|F^c|$: number of frames in the colluded copy

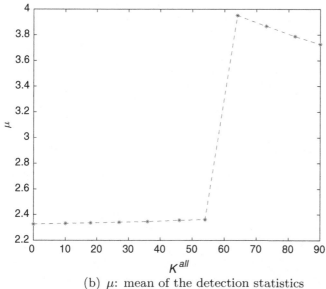

(b) μ: mean of the detection statistics

Fig. 5.3 Fair collusion when $(K^b, K^{b,e1}, K^{all})$ takes different values on line $\overline{AB}$ (5.30). $K = 150$

is an increasing function of K^b. Therefore, μ takes larger values when K^b increases, and the equal-risk collusion attacks are less effective. The analysis is similar with fixed $F^c = F_b \cup F_{e1} \cup F_{e2}$ and fixed K.

5.2.3　Simulation results on video sequences

In our simulations on real videos, we test the first forty frames of sequence "carphone" as an example. We choose $F_b = \{1, 5, \ldots, 37\}$, $F_{e1} = \{3, 7, \ldots, 39\}$, and

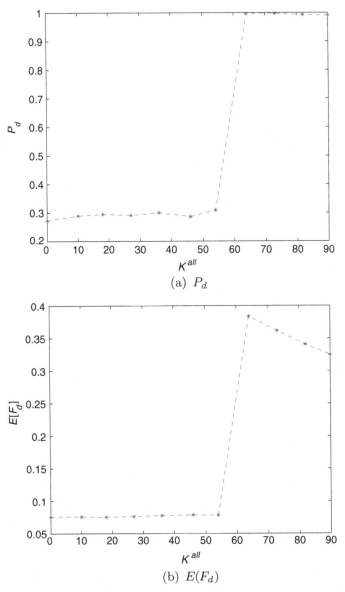

Fig. 5.4 Effectiveness of fair collusion when $(K^b, K^{b,e1}, K^{all})$ takes different values on line $\overline{AB}$ (5.30). $K = 150$

$F_{e2} = \{2, 4, \ldots, 40\}$ as an example of the temporal scalability. Assume that there are a total of $M = 450$ users and $|\mathbf{U}^b| = |\mathbf{U}^{b,e1}| = |\mathbf{U}^{all}| = 150$. We adopt the human visual model-based spread spectrum embedding presented by Podilchuk and Zeng [58], and embed the fingerprints in the discrete cosine transform (DCT) domain. The lengths of the embedded fingerprints in the base layer, enhancement layer 1, and enhancement layer 2 are $N_b = 39988$, $N_{e1} = 39934$, and $N_{e2} = 79686$, respectively. We first generate independent vectors following Gaussian distribution $\mathcal{N}(0, 1/9)$ and apply Gram–Schmidt

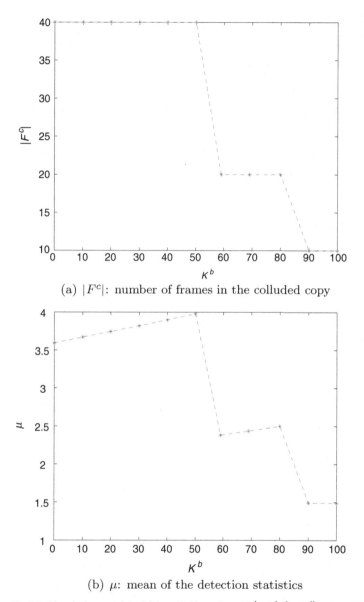

(a) $|F^c|$: number of frames in the colluded copy

(b) μ: mean of the detection statistics

Fig. 5.5 Simulation results of fair collusion when $(K^b, K^{b,e1}, K^{all})$ takes different values on line $\overline{CD}$ (5.31). $K = 150$

orthogonalization to produce fingerprints satisfying the strict orthogonality and equal energy requirements in (2.1). In each fingerprinted copy, the fingerprints embedded in different frames are correlated with one another, depending on the similarity between the host frames.

During collusion, we fix the total number of colluders as $K = 150$ and assume that the collusion attack is also in the DCT domain. In our simulations, colluders apply the intragroup collusion attacks followed by the intergroup attacks, as in Section 2.3.2. We

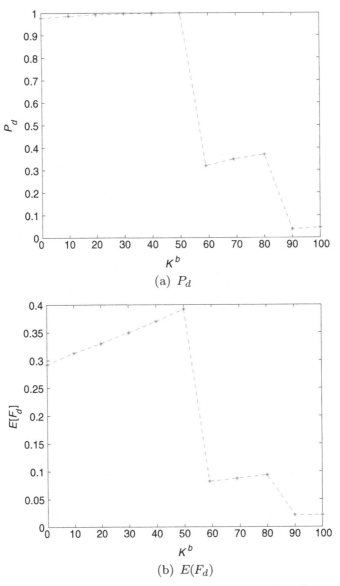

(a) P_d

(b) $E(F_d)$

Fig. 5.6 Simulation results of fair collusion when $(K^b, K^{b,e1}, K^{all})$ takes different values on line $\overline{CD}$ (5.31). $K = 150$

adjust the power of the additive noise such that $||\mathbf{n}_j||^2/||JND_j\mathbf{W}_j^{(i)}||^2 = 2$ for every frame $j \in F^c$ in the colluded copy. In our simulations, we assume that colluders generate a colluded copy of the highest possible resolution under the equal-risk fairness constraint.

At the detector's side, we consider a nonblind detection scenario in which the host signal is removed from the colluded copy before the colluder identification process. The detector follows the detection process in Section 2.3.3 and estimates the indices of the colluders $\widehat{SC}$.

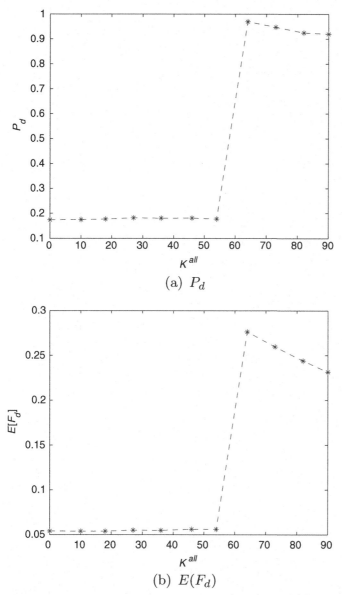

(a) P_d

(b) $E(F_d)$

Fig. 5.7 Simulation results on sequence "carphone." (K^b, K^{be}, K^{all}) are on line $\overline{AB}$ (5.30)

Figure 5.7 and 5.8 show the simulation results. In Figure 5.7, the x axis is the number of colluders who receive all three layers K^{all}, and each value of K^{all} represents a unique triplet $(K^b, K^{b,e1}, K^{all})$ on line $\overline{AB}$ (5.30). In Figure 5.8, the x axis is the number of colluders who receive the base layer only, and a given K^b corresponds to a triplet $(K^b, K^{b,e1}, K^{all})$ on line $\overline{CD}$ (5.31). In Figures 5.7(a) and 5.8(a), we select the threshold to fix $P_{fp} = 10^{-3}$ and compare P_d when $(K^b, K^{b,e1}, K^{all})$ takes different values. In Figures 5.7(b) and 5.8(b), $E[F_{fp}]$ is fixed as 10^{-3}, and we compare $E[F_d]$ of the collusion attacks with different $(K^b, K^{b,e1}, K^{all})$.

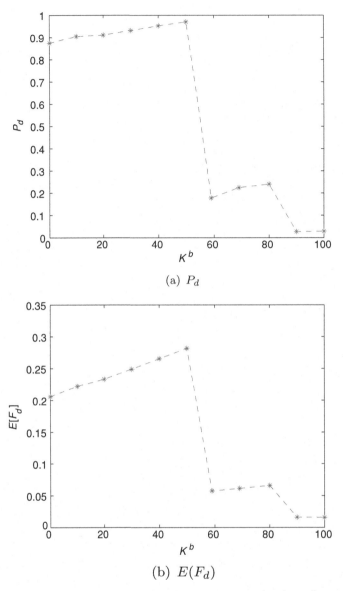

(a) P_d

(b) $E(F_d)$

Fig. 5.8 Simulation results on sequence "carphone." (K^b, K^{be}, K^{all}) are on line $\overline{CD}$ (5.31)

From Figures 5.7 and 5.8, we see that the effectiveness of collusion in defeating the scalable fingerprinting systems depends on the resolution of the colluded copy. When the colluded copy has higher resolution, the extracted fingerprint gives the detector more information about the colluders' identities, and attackers take a greater risk of being detected. The simulation results on real videos agree with our analytical results and are comparable with the simulation results in Section 5.2.2.

5.3   Traitor-tracing capability of scalable fingerprints

In this section, we quantify the traitor-tracing capacity of scalable fingerprinting systems under the equal-risk collusion attack, and examine K_{max}, the maximum number of colluders that the fingerprinting system can successfully resist under the system requirements. As done by Wang et al. [61], we consider three different scenarios with different collusion resistance requirements, and analyze the tracing-tracing capability one by one.

5.3.1   Catch-one

In the *catch-one* scenario, the fingerprinting systems wish to maximize the chance to capture one colluder while minimizing the probability of falsely accusing an innocent user. An example of such a scenario is to provide trustworthy digital evidence in a court of law. The performance criteria in this scenario are the probability of capturing at least one colluder P_d and the probability of accusing at least one innocent user P_{fp}. From the detector's point of view, the detector fails if either it does not capture any of the colluders or it falsely accuses an innocent user as a colluder. Consequently, the system requirements are

$$P_d \geq \gamma_d \quad \text{and} \quad P_{fp} \leq \gamma_{fp}. \tag{5.33}$$

5.3.1.1   Upper and lower bounds of K_{max}

To quantify the collusion resistance of the scalable fingerprinting system in Section 2.3 and analyze K_{max}, we first need to analyze P_d and P_{fp}. From (5.26) and (5.29), if we fix the probability of accusing at least one innocent user $P_{fp} = \gamma_{fp}$, given the system parameters ($|\mathbf{U}^b|$, $|\mathbf{U}^{b,e1}|$, $|\mathbf{U}^{all}|$) and (N_b, N_{e1}, N_{e2}), the performance of the detector in Section 2.3.3 depends on the number of colluders in different subgroups (K^b, $K^{b,e1}$, K^{all}) and the temporal resolution of the colluded copy L^c. For a fixed total number of colluders K, we define

$$P_d^U(K) \overset{\triangle}{=} \max_{L^c, (K^b, K^{b,e1}, K^{all})} P_d,$$

$$s.t. \quad K^b + K^{b,e1} + K^{all} = K, \ 0 \leq K^b \leq |\mathbf{U}^b|,$$

$$0 \leq K^{b,e1} \leq |\mathbf{U}^{b,e1}|, \ 0 \leq K^{all} \leq |\mathbf{U}^{all}|,$$

and the fairness constraints in Table 5.1 are satisfied; $\qquad$ (5.34)

and $\quad P_d^L(K) \overset{\triangle}{=} \min_{L^c, (K^b, K^{b,e1}, K^{all})} P_d,$

$$s.t. \quad K^b + K^{b,e1} + K^{all} = K, \ 0 \leq K^b \leq |\mathbf{U}^b|,$$

$$0 \leq K^{b,e1} \leq |\mathbf{U}^{b,e1}|, \ 0 \leq K^{all} \leq |\mathbf{U}^{all}|,$$

and the fairness constraints in Table 5.1 are satisfied. $\qquad$ (5.35)

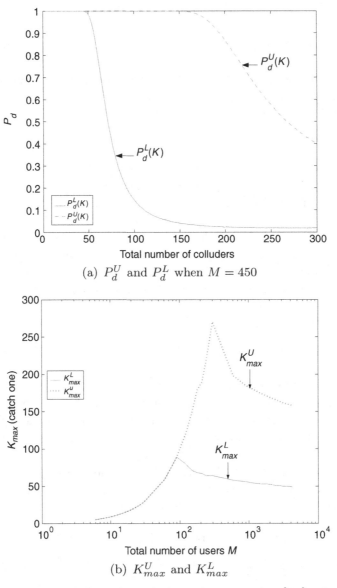

(a) P_d^U and P_d^L when $M = 450$

(b) K_{max}^U and K_{max}^L

Fig. 5.9 The collusion resistance in the catch-one scenario; $\sigma_n^2/\sigma_w^2 = 2$. $\gamma_d = 0.8$ and $\gamma_{fp} = 10^{-3}$. In (a), there are a total of $M = 450$ users in the system and $|\mathbf{U}^b| = |\mathbf{U}^{b,e1}| = |\mathbf{U}^{all}| = 150$

P_d reaches the upper bound $P_d^U(k)$ when colluders generate a colluded copy of the highest resolution, whereas P_d equals to $P_d^L(K)$ when the colluded copy contains the base layer only. Figure 5.9(a) shows an example of $P_d^U(K)$ and $P_d^L(K)$ when there are a total of $M = 450$ users and $\gamma_{fp} = 10^{-3}$. In Figure 5.9, $|\mathbf{U}^b|:|\mathbf{U}^{b,e1}|:|\mathbf{U}^{all}| = 1:1:1$ and $(N_b, N_{e1}, N_{e2}) = (50000, 50000, 100000)$. From Figure 5.9(a), the fingerprinting system's performance degrades when K becomes larger. Under the requirements $P_d \geq 0.8$

and $P_{fp} \leq 10^{-3}$, we can see from Figure 5.9(a) that when the total number of colluders is larger than 210, $P_d^U(K) < 0.8$ and the fingerprinting systems will always fail no matter which resolution the colluded copy has. When there are fewer than sixty attackers, $P_d^L(k) \geq 0.8$ and colluders can never bypass the detector without being detected, even if they generate only a colluded copy of low resolution.

In the catch-one scenario, given the system parameters $(|\mathbf{U}^b|, |\mathbf{U}^{b,e1}|, |\mathbf{U}^{all}|)$ and the total number of users M, we further define

$$K_{max}^U \triangleq \arg\max_K\{P_d^U(K) \geq \gamma_d\},$$

$$\text{and} \quad K_{max}^L \triangleq \arg\max_K\{P_d^I(K) \geq \gamma_d\}. \tag{5.36}$$

Given the parameters $(|\mathbf{U}^b|, |\mathbf{U}^{b,e1}|, |\mathbf{U}^{all}|)$ and (N_b, N_{e1}, N_{e2}), when the total number of colluders K is smaller than K_{max}^L, no matter what values L^c and $(K^b, K^{b,e1}, K^{all})$ take, the system requirements of (5.33) are always satisfied. On the contrary, if the total number of colluders K is larger than K_{max}^U, for all possible values of L^c and $(K^b, K^{b,e1}, K^{all})$, the detector will always fail under the system requirements. Therefore, K_{max}^U and K_{max}^L provide the upper and lower bounds of K_{max}, respectively.

From the colluders' point of view, if they can collect no more than K_{max}^L independent copies, no matter how they collude, the collusion attack will always fail. However, if they manage to collect more than K_{max}^U copies, they can be guaranteed success even if they generate a colluded copy of the highest resolution. From the content owner's point of view, if he or she can ensure that potential colluders cannot collect more than K_{max}^L independent copies, the fingerprinting system is essentially collusion-resistant.

Figure 5.9(b) shows K_{max}^U and K_{max}^L as functions of the total number of users M under the system requirements $\gamma_{fp} = 10^{-3}$ and $\gamma_d = 0.8$. In Figure 5.9(b), $|\mathbf{U}^b|:|\mathbf{U}^{b,e1}|:|\mathbf{U}^{all}| = 1:1:1$ and $(N_b, N_{e1}, N_{e2}) = (50000, 50000, 100000)$. As shown in Figure 5.9(b), with thousands of users, the fingerprinting system can withstand 50 colluders if the colluded copy has low resolution, and it can resist attacks with up to 150 colluders if the colluded copy has high resolution. Furthermore, if the content owner distributes no more than 100 copies, the detection performance will always satisfy the requirement (5.33) even if all users participate in collusion. Consequently, the fingerprinting system is also collusion-secure if $M \leq 100$.

In Figure 5.9(b), K_{max} first increases and then decreases, as the total number of users M increases. The intuitive explanation of this behavior is the same as that given by Wang et al. [61]. When the total number of users is small (e.g., $M \leq 20$), even if all users participate in collusion, the fingerprinting system can still successfully capture them with $P_d = 1$, as shown in Figure 5.9(a). Therefore, when M is small, $K_{max} = M$ and it increases as M increases. When M continues to increase, owing to the energy reduction of the embedded fingerprints during collusion, P_d starts to drop when there are more colluders, and it is more likely to make errors when identifying colluders: it either fails to detect any colluders or falsely accuses innocents. Thus, K_{max} drops as M increases when the total number of users is sufficiently large.

5.3.1.2　Calculation of K_{max}^U and K_{max}^L

To calculate K_{max}^U and K_{max}^L, we need to first find $P_d^U(K)$ and $P_d^L(K)$. From the analysis in Section 5.2.1, the detector has the worst performance when the colluded copy contains frames in the base layer only and $F^c = F_b$. In this scenario, for a guilty colluder $i \in SC$, the mean of his or her detection statistics is $\mu = \sqrt{N_b} \cdot \sigma_w / K$, where N_b is the length of the fingerprints embedded in the base layer and σ_w^2 is the variance of the fingerprint. Therefore, from (5.29), for a given K, the lower bound of P_d is

$$P_d^L(K) = 1 - \left[1 - Q\left(\frac{h - \sqrt{N_b} \cdot \sigma_w / K}{\sigma_n} \right) \right]^K, \tag{5.37}$$

where σ_n^2 is the variance of the detection noise and the detection threshold h is chosen to satisfy $P_{fp} = \gamma_{fp}$.

To calculate the upper bound of P_d, given (N_b, N_{e1}, N_{e2}) and K, we define

$$\mathbb{RC}^3 \triangleq \Big\{ (K^b, K^{b,e1}, K^{all}) : \frac{K^b \sqrt{N_b}}{K^b \sqrt{N_b} + K^{b,e1} \sqrt{N_b + N_{e1}} + K^{all} \sqrt{N_b + N_{e1} + N_{e2}}}$$

$$\leq \frac{N_b}{N_b + N_{e1} + N_{e2}}, \frac{K^{all} \sqrt{N_b + N_{e1} + N_{e2}}}{K^b \sqrt{N_b} + K^{b,e1} \sqrt{N_b + N_{e1}} + K^{all} \sqrt{N_b + N_{e1} + N_{e2}}}$$

$$\geq \frac{N_{e2}}{N_b + N_{e1} + N_{e2}}, 0 \leq K^b \leq |\mathbf{U}^b|, 0 \leq K^{b,e1} \leq |\mathbf{U}^{all}|, 0 \leq K^{all} \leq |\mathbf{U}^{all}|,$$

$$K^b + K^{b,e1} + K^{all} = K, \Big\} \quad \text{and} \tag{5.38}$$

$$\mathbb{RC}^2 \triangleq \Big\{ (K^b, K^{b,e1}, K^{all}) : \frac{K^b \sqrt{N_b}}{K^b \sqrt{N_b} + (K^{b,e1} + K^{all}) \sqrt{N_b + N_{e1}}}$$

$$\leq \frac{N_b}{N_b + N_{e1}}, 0 \leq K^b \leq |\mathbf{U}^b|, 0 \leq K^{b,e1} \leq |\mathbf{U}^{all}|, 0 \leq K^{all} \leq |\mathbf{U}^{all}|,$$

$$K^b + K^{b,e1} + K^{all} = K. \Big\} \tag{5.39}$$

From Section 5.2.1, for a given K, P_d is maximized when the colluded copy has the highest possible temporal resolution under the equal-risk fairness constraint. If $\mathbb{RC}^3 \neq \emptyset$, then there exists at least one triplet $(K^b, K^{b,e1}, K^{all})$ that satisfies the fairness constraint in Table 5.1 for generating an attacked copy of the highest resolution with $F^c = F_b \cup F_{e1} \cup F_{e2}$. Therefore,

$$P_d^U(K) = \max_{F^c = F_b \cup F_{e1} \cup F_{e2}, (K^b, K^{b,e1}, K^{all})} P_d,$$

such that $\quad K^b + K^{b,e1} + K^{all} = K, \ 0 \leq K^b \leq |\mathbf{U}^b|,$

$$0 \leq K^{b,e1} \leq |\mathbf{U}^{b,e1}|, \ 0 \leq K^{all} \leq |\mathbf{U}^{all}|,$$

$$\frac{K^b \sqrt{N_b}}{K^b \sqrt{N_b} + K^{b,e1} \sqrt{N_b + N_{e1}} + K^{all} \sqrt{N_b + N_{e1} + N_{e2}}} \leq \frac{N_b}{N_b + N_{e1} + N_{e2}},$$

$$\frac{K^{all} \sqrt{N_b + N_{e1} + N_{e2}}}{K^b \sqrt{N_b} + K^{b,e1} \sqrt{N_b + N_{e1}} + K^{all} \sqrt{N_b + N_{e1} + N_{e2}}} \geq \frac{N_{e2}}{N_b + N_{e1} + N_{e2}}. \tag{5.40}$$

From (5.29), maximizing P_d when $F^c = F_b \cup F_{e1} \cup F_{e2}$ is equivalent to maximizing the corresponding mean of the detection statistics

$$\mu = \frac{N_b + N_{e1} + N_{e2}}{K^b \sqrt{N_b} + K^{b,e1}\sqrt{N_b + N_{e1}} + K^{all}\sqrt{N_b + N_{e1} + N_{e2}}} \sigma_w. \qquad (5.41)$$

It is also equivalent to minimizing the denominator of μ, which is $\vartheta(K) \triangleq K^b\sqrt{N_b} + K^{b,e1}\sqrt{N_b + N_{e1}} + K^{all}\sqrt{N_b + N_{e1} + N_{e2}}$. Consequently, the optimization problem of (5.40) can be simplified to

$$\vartheta^L(K) \triangleq \min_{(K^b, K^{b,e1}, K^{all})} K^b\sqrt{N_b} + K^{b,e1}\sqrt{N_b + N_{e1}} + K^{all}\sqrt{N_b + N_{e1} + N_{e2}}$$

$$(5.42)$$

with the same constraints as in (5.40). We can use linear programming [99] to solve the optimization problem of (5.42), and then calculate

$$P_d^U(K) = 1 - \left[1 - Q\left(\frac{h - (N_b + N_{e1} + N_{e2})\sigma_w/\vartheta^L(K)}{\sigma_n}\right)\right]^K. \qquad (5.43)$$

If $\mathbb{RC}^3 = \emptyset$ and $\mathbb{RC}^2 \neq \emptyset$, no matter what value the triplet $(K^b, K^{b,e1}, K^{all})$ takes, colluders cannot generate a colluded copy of the highest resolution while still achieving fairness of collusion. However, there exists at least one $(K^b, K^{b,e1}, K^{all})$ with which colluders can generate an attacked copy of medium resolution with $F^c = F_b \cup F_{e1}$ and still guarantee the equal risk of all colluders. In this scenario, the calculation of $P_d^U(K)$ is similar to that when $\mathbb{RC}^3 \neq \emptyset$ and is not repeated here.

If $\mathbb{RC}^3 = \emptyset$ and $\mathbb{RC}^2 = \emptyset$, to ensure that all attackers have the same risk, colluders can generate only a colluded copy of the lowest resolution with $F^c = F_b$. In this scenario, $P_d^U(K) = P_d^L(K)$.

Once we obtain $P_d^U(K)$ and $P_d^L(K)$, the analysis of K_{max}^U and K_{max}^L is the same as that of Wang *et al.* [61] and is omitted.

5.3.2 Catch-more

In the *catch-more* scenario, the goal of the fingerprinting system is to capture as many colluders as possible, though possibly at a cost of accusing more innocent users. For this scenario, the set of performance criteria consists of the expected fraction of colluders that are successfully captured, $E[F_d]$, and the expected fraction of innocent users that are falsely placed under suspicion, $E[F_{fp}]$. The system requirements for such applications are $E[F_d] \geq \lambda_d$ and $E[F_{fp}] \leq \lambda_{fp}$.

Similar to the catch-one scenario, if we fix $E[F_{fp}]$ as λ_{fp}, given $(|\mathbf{U}^b|, |\mathbf{U}^{b,e1}|, |\mathbf{U}^{all}|)$, (N_b, N_{e1}, N_{e2}), and the total number of colluders K, we define

$$F_d^U(K) \triangleq \max_{L^c, (K^b, K^{b,e1}, K^{all})} E[F_d],$$

$$s.t. \quad K^b + K^{b,e1} + K^{all} = K, \ 0 \leq K^b \leq |\mathbf{U}^b|,$$

$$0 \leq K^{b,e1} \leq |\mathbf{U}^{b,e1}|, \ 0 \leq K^{all} \leq |\mathbf{U}^{all}|,$$

and the fairness constraints in Table 5.1 are satisfied; $\qquad (5.44)$

$$\text{and} \quad F_d^L(K) \stackrel{\triangle}{=} \min_{L^c, (K^b, K^{b,e1}, K^{all})} E[F_d],$$

$$s.t. \quad K^b + K^{b,e1} + K^{all} = K, \ 0 \le K^b \le |\mathbf{U}^b|,$$

$$0 \le K^{b,e1} \le |\mathbf{U}^{b,e1}|, 0 \le K^{all} \le |\mathbf{U}^{all}|,$$

and the fairness constraints in Table 5.1 are satisfied, (5.45)

which are the upper and lower bounds of $E[F_d]$, respectively. $F_d^U(K)$ and $F_d^L(K)$ are decreasing functions of K, as the collusion attack is more effective in undermining the tracing capacity with larger number of attackers. Then, we define

$$K_{max}^U \stackrel{\triangle}{=} \arg\max_K \{F_d^U(K) \ge \lambda_d\}$$

$$\text{and} \quad K_{max}^L \stackrel{\triangle}{=} \arg\max_K \{F_d^L(K) \ge \lambda_d\}, \quad (5.46)$$

which are the upper and lower bounds of K_{max} in the catch-more scenario, respectively. The analysis of $(F_d^U(K), F_d^L(K))$ and (K_{max}^U, K_{max}^L) in the catch-more scenario is similar to that in the catch-one scenario and is thus omitted here. Similar to the scenario in which users receive copies of the same resolution [100], in scalable fingerprinting systems, the detection threshold h is determined only by λ_{fp}, and K_{max} is not affected by the total number of users in the catch-more scenario.

Figure 5.10 shows the simulation results on the collusion resistance of the fingerprinting systems in the catch-more scenario. In our simulation, $(N_b, N_{e1}, N_{e2}) = (50000, 50000, 100000)$ and $\sigma_n^2 = 2\sigma_w^2$. Figure 5.10(a) plots $F_d^U(K)$ and $F_d^L(K)$ versus the total number of colluders K when $|\mathbf{U}^b| = |\mathbf{U}^{b,e1}| = |\mathbf{U}^{all}| = 300$ and $\lambda_{fp} = 0.01$. Under the requirements that $E[F_d] \ge 0.5$ and $E[F_{fp}] \le 0.01$, from Figure 5.10(a), K_{max}^U is approximately 180 and K_{max}^L is around 70. Figure 5.10(b) plots K_{max}^U and K_{max}^L versus λ_{fp} with fixed $\lambda_d = 0.5$. From Figure 5.10(b), the fingerprinting system can resist a few dozen to hundreds of colluders, depending on the resolution of the colluded copy, as well as the system requirements. If the fingerprinting system can afford to put a large fraction of innocents under suspicion, it can withstand more colluders.

5.3.3 Catch-all

In this scenario, the fingerprints are designed to maximize the probability of capturing all colluders, while maintaining an acceptable number of innocents being falsely accused. This goal arises when data security is of great concern and any information leakage could result in serious damage. Assume that there are a total of M users and a total of K colluders in the system. This set of performance criteria consists of measuring the probability of capturing all colluders $P_{d,all} = P\left[\min_{i \in SC} T_N^{(i)} > h\right]$, and the efficiency rate $R = \frac{(M-K) \cdot E[F_{fp}]}{K \cdot E[F_d]}$, which describes the number of innocents falsely accused per colluder successfully captured. The system requirements for these applications are $R \le \theta_r$ and $P_{d,all} \ge \theta_d$.

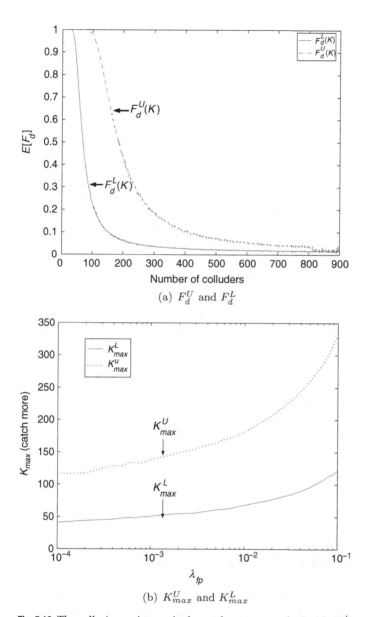

(a) F_d^U and F_d^L

(b) K_{max}^U and K_{max}^L

Fig. 5.10 The collusion resistance in the catch-more scenario. In (a), $|\mathbf{U}^b| = |\mathbf{U}^{b,e1}| = |\mathbf{U}^{all}| = 300$ and $\lambda_{fp} = 0.01$. In (b), $\lambda_d = 0.5$

Similar to the catch-one scenario, given $(|\mathbf{U}^b|, |\mathbf{U}^{b,e1}|, |\mathbf{U}^{all}|)$ and (N_b, N_{e1}, N_{e2}), for a fixed total number of colluders K and fixed $P_{d,all} = \theta_d$, define

$$R^U(K) \stackrel{\triangle}{=} \max_{L^c, (K^b, K^{b,e1}, K^{all})} R,$$

$$s.t. \quad K^b + K^{b,e1} + K^{all} = K, \; 0 \le K^b \le |\mathbf{U}^b|,$$

$$0 \le K^{b,e1} \le |\mathbf{U}^{b,e1}|, \; 0 \le K^{all} \le |\mathbf{U}^{all}|,$$

and the fairness constraints in Table 5.1 are satisfied,   (5.47) and $R^L(K) \stackrel{\triangle}{=} \min_{L^c,(K^b,K^{b,e1},K^{all})} R,$

s.t. $K^b + K^{b,e1} + K^{all} = K,\ 0 \leq K^b \leq |\mathbf{U}^b|,$

$0 \leq K^{b,e1} \leq |\mathbf{U}^{b,e1}|,\ 0 \leq K^{all} \leq |\mathbf{U}^{all}|,$

and the fairness constraints in Table 5.1 are satisfied,   (5.48)

which are the upper and lower bounds of R, respectively. We further define

$$K_{max}^U \stackrel{\triangle}{=} \arg\max_K\{R^L(K) \leq \theta_r\}$$

and $$K_{max}^L \stackrel{\triangle}{=} \arg\max_K\{R^U(K) \leq \theta_r\}$$ (5.49)

as the upper and lower bounds of K_{max}, respectively. The analysis of K_{max}^U and K_{max}^L in the catch-all scenario is similar to that in the catch-one scenario and is not repeated.

In our simulations of the catch-one scenario, we let $|\mathbf{U}^b|:|\mathbf{U}^{b,e1}|:|\mathbf{U}^{all}| = 1{:}1{:}1$ and $(N_b, N_{e1}, N_{e2}) = (50000, 50000, 100000)$. Figure 5.11(a) plots $R^U(K)$ and $R^L(K)$ versus the total number of colluders K when there are $M = 450$ users and $\theta_d = 0.99$. We consider a scenario that is required to catch all colluders with probability larger than 0.99 ($P_{d,all} \geq 0.99$) and accuse no more than one innocent for every 100 colluders captured ($R \leq 0.01$). Under these requirements, from Figure 5.11(a), attackers should collect more than $K_{max}^U = 65$ different copies to ensure the success of collusion, and the scalable fingerprinting system is collusion-free when there are fewer than $K_{max}^L = 25$ colluders. Figure 5.11(b) shows K_{max}^U and K_{max}^L versus the total number of users M when $\theta_d = 0.99$ and $\theta_r = 0.01$. From Figure 5.11(b), in the catch-all scenario with thousands of users, the scalable fingerprinting systems can survive collusion by twenty to sixty attackers, depending on the resolution of the colluded copy. It is collusion-secure if the content owner distributes no more than thirty different copies. The non-monotonic behavior in Figure 5.11 can be explained in the same way as in the catch one scenario.

5.4   Chapter summary and bibliographical notes

In this chapter, we use multimedia fingerprinting system as an example, and analyze how colluders achieve equal-risk fairness and how such a decision influences the fingerprint detector's performance.

We first investigate how to distribute the risk of being detected evenly to all colluders when they receive copies of different resolutions because of network and device heterogeneity. We show that a higher resolution of the colluded copy puts more severe constraints on achieving fairness of collusion. We then analyze the effectiveness of such fair collusion attacks. Both our analytical and simulation results show that colluders are more likely to be captured when the colluded copy has higher resolution. Colluders must take into consideration the tradeoff between the probability of being detected and the resolution of the colluded copy during collusion.

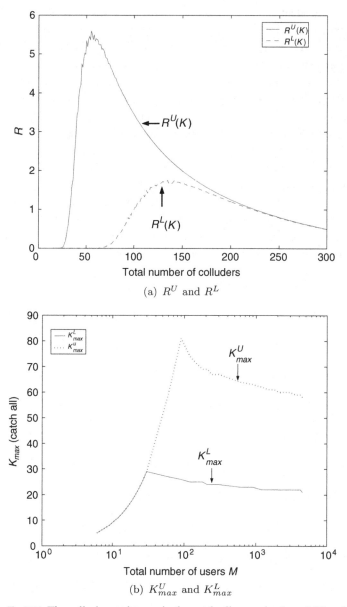

(a) R^U and R^L

(b) K^U_{max} and K^L_{max}

Fig. 5.11 The collusion resistance in the catch-all scenario. $\theta_d = 0.99$ and $\theta_r = 0.01$. In (a), $M = 450$ and $|\mathbf{U}^b| = |\mathbf{U}^{b,e1}| = |\mathbf{U}^{all}| = 150$

We also analyze the collusion resistance of the scalable fingerprinting system for various fingerprinting scenarios with different requirements. We evaluate the maximum number of colluders that the system can resist, and show that scalable fingerprinting system can withstand dozens to hundreds of colluders, depending on the resolution of the colluded copy as well as the system requirements. We also provide the lower and upper bounds of K_{max}. From the colluders' point of view, K^U_{max} tells attackers how many independent copies are required to guarantee the success of collusion under all circumstances. From the content owner's point of view, to achieve a collusion-free condition, a desired security requirement is to make the potential colluders very unlikely to collect more than K_{max}^L copies.

Recent research on collusion attacks against multimedia fingerprinting systems has focused on many different topics. Intracontent collusion attacks [101–104] are specific for video content. Multiuser collusion was modeled by Ergun *et al.* [63] as averaging different copies with equal weights followed by an additive noise, and such a model was generalized to multiple-input–single-output linear shift invariant filtering followed by an additive Gaussian noise by Su *et al.* [65]. The collusion attack model of nonlinear collusion attacks were examined and analyzed by Stone and by Zhao *et al.* [66,67]. Furthermore, to support multimedia forensics, there have been a number of works on anticollusion multimedia fingerprint design [62,69,72,73], which can resist such multiuser collusion as well as common signal processing and attacks on a single copy [105,106]. Interested readers can also refer to reference [76] for scalable video coding techniques.

6 Leveraging side information in colluder social networks

In general, *side information* is the information other than the target signal that can help improve system performance. For instance, in digital communications, side information about channel conditions at the transmitter's side can help reduce the bit error rate, and in learning theory, the side information map can also improve the classification accuracy [107]. In this chapter, we use multimedia fingerprinting as an example and discuss how side information affects user behavior in media-sharing social networks.

In the scalable fingerprinting system in Chapter 5, given a test copy, the fingerprint detector simply uses fingerprints extracted from all layers collectively to identify colluders, and does not use any other information in the detection process. Intuitively, if some information about collusion can be made available during the colluder identification process, using such side information can help improve the traitor-tracing performance [108]. In this chapter, we investigate two important issues in multimedia fingerprinting social networks that are related to side information: which side information can help improve the traitor-tracing performance, and how it affects user behavior in multimedia fingerprinting systems.

In this chapter, we first examine which side information can help improve the traitor-tracing performance; our analysis shows that information about the statistical means of the detection statistics can significantly improve the detection performance. We then explore possible techniques for the fingerprint detector to probe and use such side information, and analyze its performance. Side information not only improves the traitor-tracing performance, but also affects each colluder's probability of being detected and colluders' strategy to achieve equal-risk fairness. In this chapter, we build a game-theoretic framework to model the colluder-detector dynamics, analyze how colluders minimize their probability of being detected under the equal-risk constraint, and examine the impact of side information on the Nash equilibrium of this game.

6.1   Probing and using side information

This section analyzes which side information about collusion can help improve collusion resistance, and studies how to probe such side information from the colluded copy. As in Chapter 5, the discussion is under the framework of three-layer scalable video coding and fingerprinting. Here, we consider the scenario in which the colluded video contains all three layers, and thus has the highest quality. The analysis for other scenarios is similar and is thus omitted. Without loss of generality, users in U^{all} are used as an example to demonstrate the detection process and analyze the performance. For users in $U^{b,e1}$ and U^b, the colluder identification process and the performance analysis are similar.

6.1.1   Different fingerprint detection strategies

A nonblind detection scenario is considered, in which the host signal is first removed from the test copy before colluder identification. The detector then extracts the fingerprint Y_j from the jth frame V_j in the colluded copy. Then the detector calculates the similarity between the extracted fingerprint Y and each of the original fingerprints $\{W^{(i)}\}$, compares it with a predetermined threshold h, and outputs the estimated identities of the colluders $\widehat{SC}$. However, because the fingerprint in the colluded copy Y contains three layers, when detecting fingerprints, there are many different ways to measure the similarity between the extracted fingerprint Y and the originally embedded one $W^{(i)}$.

6.1.1.1   Collective fingerprint detector

The previous chapter considered a simple fingerprint detector that uses fingerprints extracted from all layers collectively to identify colluders. For user i, the detector first calculates $\check{F}^{(i)} = F^{(i)} \cap F^c$, where $F^{(i)}$ contains the indices of the frames received by user i and $F^c = F_b \cup F_{e1} \cup F_{e2}$. Then, the detector calculates

$$TN_c^{(i)} = \left(\sum_{j \in \check{F}^{(i)}} \langle Y_j, W_j^{(i)} \rangle \right) \Big/ \sqrt{\sum_{j \in \check{F}^{(i)}} ||W_j^{(i)}||^2}, \qquad (6.1)$$

where $||W_j^{(i)}||$ is the Euclidean norm of $W_j^{(i)}$. Given a predetermined threshold h, $\widehat{SC}_c = \{i : TN_c^{(i)} > h\}$.

With orthogonal fingerprint modulation, for a user i who receives a high-resolution fingerprinted copy, under the assumption that the detection noises are i.i.d. Gaussian $\mathcal{N}(0, \sigma_n^2)$, the detection statistics $\{TN_c^{(i)}\}$ in (6.1) are independent Gaussian with marginal distribution

$$TN_c^{(i)} \sim \begin{cases} \mathcal{N}\left(\mu_c^{(i)}, \sigma_n^2\right), & \text{if } i \in SC, \\ \mathcal{N}\left(0, \sigma_n^2\right), & \text{if } i \notin SC, \end{cases}$$

$$\text{where} \quad \mu_c^{(i)} = \frac{(1 - \beta_1 - \beta_2)N_b + (1 - \alpha_1)N_{e1} + N_{e2}}{K^{all}\sqrt{N_b + N_{e1} + N_{e2}}} \sigma_w^2. \qquad (6.2)$$

Here, N_b, N_{e1}, and N_{e2} are the lengths of the fingerprints embedded in the base layer, enhancement layer 1, and enhancement layer 2, respectively. If user i is a guilty colluder, his or her chance of being detected is

$$P_s^{(i)} = Q\left(\frac{h - \mu_c^{(i)}}{\sigma_n}\right), \qquad (6.3)$$

and the probability of falsely accusing user i if he or she is innocent is

$$P_{fa}^{(i)} = Q\left(\frac{h}{\sigma_n}\right). \tag{6.4}$$

With the collective detector, if colluders follow Table 5.1 to select the collusion parameters, $P_s^{(i)}$ is the same for all colluders and they have the same probability of being detected.

6.1.1.2 Fingerprint detection at each layer

Given $\mathbf{Y}_{e2}$, $\mathbf{Y}_{e1}$, and $\mathbf{Y}_b$, which are the fingerprints extracted from enhancement layer 2, enhancement layer 1, and the base layer, respectively, in addition to the collective detector (6.1), the digital rights enforcer can also examine $\mathbf{Y}_{e2}$, $\mathbf{Y}_{e1}$, and $\mathbf{Y}_b$ independently and use the detection results at each individual layer to estimate the colluders' identities. Therefore, in addition to the collective detector, the digital rights enforcer can also use detectors at base layer, enhancement layer 1, and enhancement layer 2. To demonstrate this colluder identification process and analyze its performance, we look at users in $\mathbf{U}^{all}$ who receive all three layers as an example. The analysis for users in $\mathbf{U}^{b,e1}$ and $\mathbf{U}^b$ is similar and is thus omitted.

Let F_t be the set of indices of the frames in layer t in which $t = b, e1, e2$ represents the base layer, enhancement layer 1, and enhancement layer 2, respectively. For users $i \in \mathbf{U}^{all}$ who receive all three layers from the content owner, given $\{\mathbf{Y}_j\}_{j\in F_t}$, the fingerprints from layer t of the colluded copy, the detector at layer t calculates the detection statistics

$$TN_t^{(i)} = \left(\sum_{j\in F_t}\langle \mathbf{Y}_j, \mathbf{W}_j^{(i)}\rangle\right) \Big/ \sqrt{\sum_{j\in F_t}||\mathbf{W}_j^{(i)}||^2} \tag{6.5}$$

to measure the similarity between the extracted fingerprint and the originally embedded fingerprint. The detector at layer t accuses user i as a colluder if $TN_t^{(i)} > h$, and sets $i \in \widehat{SC}$, which is the suspicious-colluder set. Here, h is a predetermined threshold.

The analysis of the detection statistics $TN_t^{(i)}$ in (6.6) below is similar to that of $TN^{(i)}$ in (6.1). If the detection noises are i.i.d. and follow Gaussian distribution $\mathcal{N}(0, \sigma_n^2)$, for user $i \in \mathbf{U}^{all}$, $TN_t^{(i)}$ are independent Gaussian with marginal distribution

$$TN_t^{(i)} \sim \begin{cases} \mathcal{N}(\mu_t^{(i)}, \sigma_n^2) & \text{if } i \in SC, \\ \mathcal{N}(0, \sigma_n^2) & \text{if } i \notin SC, \end{cases}$$

$$\text{where} \quad \mu_b^{(i)} = (1 - \beta_1 - \beta_2)\frac{\sqrt{N_b}}{K^{all}}\sigma_w,$$

$$\mu_{e1}^{(i)} = (1 - \alpha_1)\frac{\sqrt{N_{e1}}}{K^{all}}\sigma_w, \quad \text{and} \quad \mu_{e2}^{(i)} = \frac{\sqrt{N_{e2}}}{K^{all}}\sigma_w. \tag{6.6}$$

Therefore, the probability of successfully capturing user $i \in \mathbf{U}^{all}$ if he or she is guilty is

$$P_s^{(i)} = Q\left(\frac{h - \mu_t^{(i)}}{\sigma_n}\right), \tag{6.7}$$

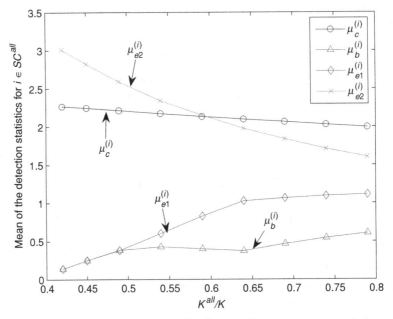

Fig. 6.1 Comparison of μ_c in (6.2), $\mu_{e2}^{(i)}$, $\mu_{e1}^{(i)}$, and $\mu_b^{(i)}$ in (6.7) for $i \in SC^{all}$

and the probability of falsely accusing this user if he or she is innocent is

$$P_{fa}^{(i)} = Q\left(\frac{h}{\sigma_n}\right). \tag{6.8}$$

It is clear from (6.6) and (6.7) that the higher the $\mu_t^{(i)}$, the better the traitor-tracing performance.

6.1.2   Performance comparison of detection types

This section compares the performance of the four detection statistics (6.1) and (6.6) when identifying colluders in SC^{all}. From the preceding analysis, for a given h and a fixed P_{fp}, comparing P_d of different detection statistics is equivalent to comparing their means.

For a colluder $i \in SC^{all}$, Figure 6.1 shows an example of the means of the detection statistics in (6.1) and (6.6). In Figure 6.1, we first generate independent vectors following Gaussian distribution $\mathcal{N}(0, 1)$, and then apply Gram–Schmidt orthogonalization to generate orthogonal fingerprints for different users. The lengths of the fingerprints embedded in the base layer, enhancement layer 1, and enhancement layer 2 are $N_b = 50000$, $N_{e1} = 50000$, and $N_{e2} = 100000$, respectively. In Figure 6.1, we fix the total number of colluders as $K = 250$, and $K^b = 50$ of them receive the fingerprinted base layer only. Each point on the x axis corresponds to a unique triplet (K^b, K^{e1}, K^{e2}). Colluders follow discussion in Chapter 5 to select the collusion parameters and generate a colluded copy with all three layers under the fairness constraints.

From Figure 6.1, $TN_c^{(i)}$ in (6.1) has the best performance when more than 60 percent of the colluders receive a high-quality copy with all three layers. This is because in this scenario, colluder i's fingerprint spreads all over the entire colluded copy $\mathbf{V}$, and $\mathbf{W}^{(i)}$'s energy is evenly distributed in the three layers of $\mathbf{V}$. Therefore, from detection theory [60], fingerprints extracted from all layers should be used during detection to improve performance. When $K^{all}/K < 0.6$, owing to the selection of the collusion parameters, a significant portion of $\mathbf{W}^{(i)}$'s energy is in enhancement layer 2, whereas the other two layers of the colluded copy contain little information on colluder i's identity. Thus, in this scenario, $TN_{e2}^{(i)}$ in (6.6) gives the best detection performance.

6.1.3   Colluder identification with side information

For the four detection statistics in Section 6.1.1, the traitor-tracing capability is determined by their *statistical means*. The larger the statistical mean is, the better the performance. From the preceding analysis, the collusion parameters ($\{\alpha_l\}$ and $\{\beta_k\}$ in the two-stage collusion model) determine the statistical means of the detection statistics. Thus, if side information about the statistical means of different detection statistics (or equivalently, the collusion parameters) is available to the fingerprint detector, the detector should select the detection statistics with the largest statistical mean to improve the traitor-tracing capability.

During the fingerprint detection and colluder identification process, the fingerprint detector should first examine the colluded copy and probe such side information, then select the best detection statistics and identify colluders. As an example, to identify colluders who receive all three layers, the key steps in probing the means of the detection statistics and selecting the optimum detection statistics are as follows:

- For every user $i \in \mathbf{U}^{all}$, the detector first calculates $TN_c^{(i)}$, $TN_{e2}^{(i)}$, $TN_{e1}^{(i)}$ and $TN_b^{(i)}$ as in Section 6.1.1, and obtains

$$\widehat{SC}_c^{all} = \{i : TN_c^{(i)} > h_t\}, \quad \widehat{SC}_{e2}^{all} = \{i : TN_{e2}^{(i)} > h_t\},$$
$$\widehat{SC}_{e1}^{all} = \{i : TN_{e1}^{(i)} > h_t\}, \quad \text{and} \quad \widehat{SC}_b^{all} = \{i : TN_b^{(i)} > h_t\} \tag{6.9}$$

  for a given h_t.
- The detector combines these four sets of estimated colluders in $\mathbf{U}^{all}$ and lets $\widehat{SC}^{all} = \widehat{SC}_c^{all} \cup \widehat{SC}_{e2}^{all} \cup \widehat{SC}_{e1}^{all} \cup \widehat{SC}_b^{all}$.
- Given $\widehat{SC}^{all}$, the detector estimates the means of the four detection statistics in Section 6.1.1

$$\hat{\mu}_c = \sum_{k \in \widehat{SC}^{all}} \frac{TN_c^{(k)}}{|\widehat{SC}^{all}|}, \quad \hat{\mu}_{e2} = \sum_{k \in \widehat{SC}^{all}} \frac{TN_{e2}^{(k)}}{|\widehat{SC}^{all}|},$$
$$\hat{\mu}_{e1} = \sum_{k \in \widehat{SC}^{all}} \frac{TN_{e1}^{(k)}}{|\widehat{SC}^{all}|}, \quad \text{and} \quad \hat{\mu}_b = \sum_{k \in \widehat{SC}^{all}} \frac{TN_b^{(k)}}{|\widehat{SC}^{all}|}. \tag{6.10}$$

- The detector compares $\hat{\mu}_c$, $\hat{\mu}_{e2}$, $\hat{\mu}_{e1}$ and $\hat{\mu}_b$ and selects the detection statistics with the largest estimated mean. For example, the collective detector in (6.1) is chosen if $\hat{\mu}_c$ has the largest value.

When identifying colluders in $SC^{b,e1}$, the side information probing process is similar and is not repeated. Then, the fingerprint detector follows Section 6.1.1 and estimates the identities of the colluders.

6.1.4   Performance analysis and simulation results

In our simulations, we simulate three different fingerprint detectors: the simple collective detector in (6.1); the optimum detector with perfect knowledge of the statistical means of the four detection statistics; and the self-probing detector, which first uses the algorithm in Section 6.1.3 to select the best detection statistics and then follows Section 6.1.1 to identify colluders.

The simulation setup is the same as that in Figure 6.1, and the simulation results shown in Figure 6.2 are based on 10000 simulation runs. There are a total of $K = 250$ colluders, and $K^b = 50$ of them receive the fingerprinted base layer only. Each point on the x axis in Figure 6.2 corresponds to a unique triplet $(K^b, K^{b,e1}, K^{all})$. Colluders select $\{\alpha_l, \beta_k\}$ in the same way as in Chapter 5 and generate a colluded copy with all three layers. For each frame j in the colluded copy, we adjust the power of the additive noise such that $||\mathbf{n}_j||^2 = ||\mathbf{W}_j^{(i)}||^2$. Other values give the same trend. When estimating the statistical means of different detection statistics, h_t is chosen to let $P_{fa}^{(i)} = 10^{-2}$ for an innocent user $i \notin SC$.

Figure 6.2(a) plots the probability that the probing algorithm in Section 6.1.3 selects the optimum detection statistics when identifying colluders in $\mathbf{U}^{all}$. In the example in Figure 6.1, we choose only between $TN^{(i)}$ and $TN_{e2}^{(i)}$, as $TN_{e1}^{(i)}$ and $TN_b^{(i)}$ never outperform the other two. From Figure 6.2(a), the probing algorithm selects the optimum detection statistics with probability 0.6 when $K^{all}/K \approx 0.6$, whereas in other scenarios, the detector always picks the best detection statistics. From Figure 6.1, when $K^{all}/K \approx 0.6$, μ_c and $\mu_{e2}^{(i)}$ have similar values and, therefore, $TN^{(i)}$ and $TN_{e2}^{(i)}$ have approximately the same performance. Consequently, in this scenario, choosing the suboptimal detection statistics does not significantly deteriorate the detection performance. When μ_c and $\mu_{e2}^{(i)}$ differ significantly from each other, the self-probing detector always chooses the optimal detection statistics when identifying colluders in $\mathbf{U}^{all}$.

To evaluate the traitor-tracing performance of the colluder identification algorithm with side information, we consider the catch-one scenario, in which the fingerprint detector aims to capture at least one colluder without falsely accusing any innocents. In this scenario, the criterion used to measure the performance is P_d and P_{fp}. The analysis for other scenarios using other performance criteria is similar and gives the same trend. For a fixed $P_{fp} = 10^{-3}$, Figure 6.2(b) shows P_d of the three detectors. From Figure 6.2(b), using side information about the means of different detection statistics can help the fingerprint detector significantly improve its performance, especially when K^{all}/K is small and the colluders' fingerprints are not evenly distributed in the three layers of the colluded copy. Furthermore, from Figure 6.2(b), when the difference between μ_c and

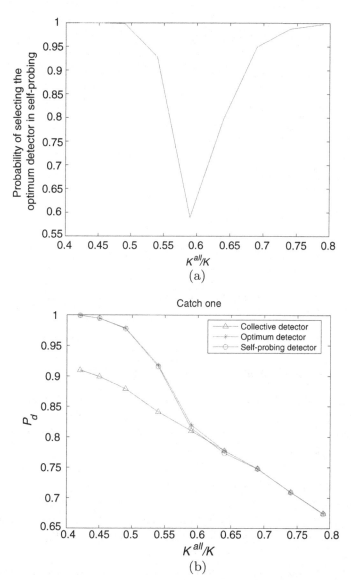

Fig. 6.2 Performance of the self-probing fingerprint detector for the example in Figure 6.1. (a) Probability of selecting the optimum detection statistics when identifying colluders in $\mathbf{U}^{all}$. (b) P_d of the collective detector, the optimum detector with perfect knowledge of the detection statistics' means, and the self-probing detector that probes the side information itself

$\mu_{e2}^{(i)}$ is large, the side information probing algorithm in Section 6.1.3 helps the detector choose the best detection statistics; achieve the optimal performance. When μ_c and $\mu_{e2}^{(i)}$ are approximately the same, the performance of the self-probing fingerprint detector is almost the same as that of the optimal detector with perfect knowledge of the means of the detection statistics; the difference between these two is no larger than 0.005 and can be ignored.

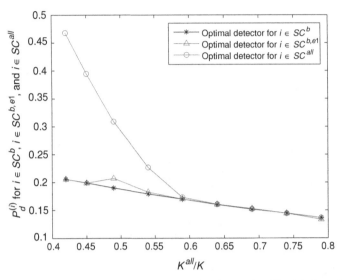

Fig. 6.3 Each colluder's probability of being detected ($P_s^{(i)}$) with the self-probing fingerprint detector

6.1.5   Impact of side information on fairness of multiuser collusion

Without probing side information, the detector will always use all frames collectively to identify colluders, hoping that more frames will give the detector more information about colluders' identities. On the other side, colluders adjust the collusion parameters to seek the collective equal-risk fairness. Under such circumstances, colluders and the fingerprint detector reach the *collective fairness equilibrium*. However, side information about collusion not only improves the fingerprint detector's performance, but it also affects each colluder's probability of being detected and influences how they collude. Thus, side information breaks the collective fairness equilibrium between colluders and the fingerprint detector, and both sides need to search for a new equilibrium.

To demonstrate how side information breaks the collective fairness equilibrium, Figure 6.3 shows each colluder's probability of being detected with the self-probing fingerprint detector. The simulation setup is the same as that in Figure 6.2. From Figure 6.3, when $K^{all}/K < 0.6$, colluders who receive all three layers have a much larger probability of being detected than the others. In this example, during collusion, attackers consider only the collective detector in (6.2), and they select the parameters $\{\alpha_j\}$ and $\{\beta_l\}$ such that $\{TN_c^{(i)}\}$ in (6.2) have the same statistical mean for all attackers. However, during the colluder identification process, the fingerprint detector considers all possible detection strategies in Section 6.1.1, probes side information about detection statistics, and uses the one that gives the best collusion resistance. Therefore, with the self-probing fingerprint detector in Section 6.1.3, colluders must find a new set of collusion parameters to ensure the equal risk of all attackers.

6.2   Game-theoretic analysis of colluder detector dynamics

In this section, we study the impact of side information on user behavior in multimedia fingerprinting social networks, build a game-theoretic framework to analyze the dynamics between colluders and the fingerprint detector, and investigate how they find the new equilibrium.

In multimedia fingerprinting social networks, different users have different goals and utilities. Colluders wish to generate a high-quality colluded copy for illegal redistribution without being detected, whereas the fingerprint detector aims to accurately identify colluders without falsely accusing others. We formulate their interaction as a game with two players with conflicting objectives, and in this game, one's loss is another's gain.

- **Players:** There are two players in this game. Colluders make their decisions first and act as the *leader*, followed by the fingerprint detector, which detects colluders and acts as a *follower*.
- **Payoff function definition:** To analyze the dynamics between colluders and the forensic detector, we consider the scenario in which all colluders agree to share the same risk and reward. Therefore, every colluder has the same goal and aims to minimize his or her risk of being detected $P_s^{(i)}$ under the constraint that $\{P_s^{(i)}\}$ are the same for all colluders. Thus, a natural definition of colluder i's payoff function is $\pi^C = 1 - P_s^{(i)}$, the probability that colluder i successfully removes traces of his or her fingerprint during collusion. The fingerprint detector aims to catch as many colluders as possible without falsely accusing others; our analysis in Section 5.2.1 shows that this is equivalent to maximizing $P_s^{(i)}$. Therefore, we define the detector's payoff as $\pi^D = P_s^{(i)} = 1 - \pi^C$.
- **Colluders' strategies:** Colluders' strategies are the sets of collusion parameters $(\alpha_1, \beta_1, \beta_2)$ that achieve equal-risk fairness.
- **Detector's strategies:** The fingerprint detector's strategies include the collective detector in (6.1) and the single-layer detector in (6.6).

In this game, colluders are the leader and the fingerprint detector is the follower. For a given colluded copy – that is, a given set of collusion parameters $(\alpha_1, \beta_1, \beta_2)$ – side information about the statistical means of the detection statistics can help improve the detection performance; the fingerprint detector should always select the detection statistics with the largest chance of successfully capturing colluders. Furthermore, the self-probing detector in Section 6.1.3 achieves approximately the same performance as the optimal detector with perfect information of the statistical means, and thus, the self-probing detector is the fingerprint detector's best response and maximizes the digital rights enforcer's payoff. In addition, from a game theory perspective, probing side information at the detector's side is equivalent to observing colluders' action. That is, in this game, the fingerprint detector (follower) can observe colluders' actions, and colluders (leader) have perfect knowledge of the fingerprint detector's response. Therefore, the colluder-detector game is a Stackelberg game with perfect information.

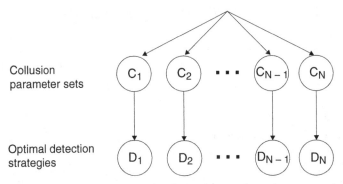

Fig. 6.4 Game tree illustration of colluder-detector dynamics

6.3   Equilibrium analysis

In the Stackelberg game, because the follower (detector) can observe the leader's (colluders') strategy, the game model can be solved by backward induction. It starts from the last stage of the game, and finds the optimal detection strategy for each possible set of collusion parameters $(\alpha_1, \beta_1, \beta_2)$. From our previous discussion, the self-probing detector is the fingerprint detector's best response and maximizes the detector's payoff. Then, given the self-probing detector, it moves one stage up and decides the optimal strategy for colluders. Here, with the self-probing fingerprint detector, colluders should consider the worst-case scenario, in which the fingerprint detector always selects the detection statistics with the best detection performance. They select the collusion parameters that minimize their risk under the constraint that all colluders have the same probability of being detected. This is illustrated in Figure 6.4, in which $C_1, C_2, \ldots, C_N$ are the possible sets of collusion parameters that achieve absolute fairness when the fingerprint detector uses the optimal detection statistics to identify colluders; and $D_1, D_2, \ldots, D_N$ are the corresponding optimal fingerprint detection strategies. Colluders select from among $\{C_1, C_2, \ldots, C_N\}$ the one that gives them the smallest probability of being detected. Thus, with side information, the equilibrium of the colluder-detector game can be modeled as a *min–max problem*.

6.3.1   Min–max formulation

For each user i, define $\mathfrak{D}^{(i)}$ as the set including all possible detection statistics that can be used to measure the similarity between the extracted fingerprint $\mathbf{Y}$ and user i's fingerprint $\mathbf{W}^{(i)}$. For example, $\mathfrak{D}^{(i)} = \{TN_c^{(i)}, TN_{e2}^{(i)}, TN_{e1}^{(i)}, TN_b^{(i)}\}$ for a colluder $i \in SC^{all}$ who receives all three layers, whereas $\mathfrak{D}^{(i)} = \{TN_c^{(i)}, TN_{e1}^{(i)}, TN_b^{(i)}\}$ for user $i \in SC^{b,e1}$ who receives a medium-resolution copy.

Define $P_s^{(i)}\left(\mathfrak{D}^{(i)}, \{\alpha_k, \beta_l\}\right)$ as the probability that colluder i is captured by the digital rights enforcer. Consequently, we can model the colluder-detector dynamics as a min–max problem:

$$\min_{\{\alpha_l,\beta_k\}} \max_{\mathfrak{D}^{(i)}} P_s^{(i)}\left(\mathfrak{D}^{(i)}, \{\alpha_l, \beta_k\}\right)$$

such that $\max_{\mathfrak{D}^{(i_1)}} P_s^{(i_1)}\left(\mathfrak{D}^{(i_1)}, \{\alpha_l, \beta_k\}\right) = \max_{\mathfrak{D}^{(i_2)}} P_s^{(i_2)}\left(\mathfrak{D}^{(i_2)}, \{\alpha_l, \beta_k\}\right), \forall i_1, i_2 \in SC.$

$$(6.11)$$

From the analysis in the previous section, for a given threshold h and fixed σ_n^2, $P_s^{(i)}$ is determined by the mean of the detection statistics that are used. Therefore, for colluder $i_1 \in SC^b$, $i_2 \in SC^{b,e1}$ and $i_3 \in SC^{all}$, (6.11) can be simplified to

$$\min_{\{\alpha_l,\beta_k\}} \mu = \mu_{max}^{(i_1)} = \mu_{max}^{(i_2)} = \mu_{max}^{(i_3)},$$

$$\text{s.t. } 0 \le \alpha_l \le 1, \ 0 \le \beta_k \le 1,$$

$$\text{where } \mu_{max}^{(i_1)} = \mu_c^{(i_1)}, \mu_{max}^{(i_2)} = \max\{\mu_b^{(i_2)}, \mu_{e1}^{(i_2)}, \mu_c^{(i_2)}\},$$

$$\mu_{max}^{(i_3)} = \max\{\mu_b^{(i_3)}, \mu_{e1}^{(i_3)}, \mu_{e2}^{(i_3)}, \mu_c^{(i_3)}\}. \tag{6.12}$$

In (6.12),

$$\mu_c^{(i_1)} = \frac{\beta_1 \sqrt{N_b}}{K^b} \sigma_w,$$

$$\mu_b^{(i_2)} = \frac{\beta_2 \sqrt{N_b}}{K^{b,e1}} \sigma_w, \qquad \mu_{e1}^{(i_2)} = \frac{\alpha_1 \sqrt{N_{e1}}}{K^{b,e1}} \sigma_w,$$

$$\mu_c^{(i_2)} = \frac{\beta_2 N_b + \alpha_1 N_{e1}}{K^{b,e1} \sqrt{N_b + N_{e1}}} \sigma_w,$$

$$\mu_b^{(i_3)} = \frac{(1 - \beta_1 - \beta_2)\sqrt{N_b}}{K^{all}} \sigma_w, \qquad \mu_{e1}^{(i_3)} = \frac{(1 - \alpha_1)\sqrt{N_{e1}}}{K^{all}} \sigma_w,$$

$$\mu_{e2}^{(i_3)} = \frac{\sqrt{N_{e2}}}{K^{all}} \sigma_w,$$

$$\text{and} \quad \mu_c^{(i_3)} = \frac{(1 - \beta_1 - \beta_2)N_b + (1 - \alpha_1)N_{e1} + N_{e2}}{K^{all}\sqrt{N_b + N_{e1} + N_{e2}}} \sigma_w \tag{6.13}$$

from the analysis in Section 6.1.1.

Given $(K^b, K^{b,e1}, K^{all})$ and (N_b, N_{e1}, N_{e2}), for colluder $i_1 \in SC^b$, $i_2 \in SC^{b,e1}$ and $i_3 \in SC^{all}$ who receive fingerprinted copies of different resolutions, they first find all possible sets of collusion parameters $\{\alpha_l, \beta_k\}$ that satisfy $\mu_{max}^{(i_1)} = \mu_{max}^{(i_2)} = \mu_{max}^{(i_3)}$. Then, they select the one that gives them the minimum risk of being detected.

To summarize, without side information, colluders and the detector achieve the collective fairness equilibrium: the fingerprint detector uses the collective detection statistics in (6.1), and colluders select the collusion parameter as in Chapter 5 to ensure the same risk under the collective detector. Probing and using side information moves the equilibrium of the colluder-detector game from the collective one to the min–max solution.

6.3.2 Analysis of the maximum mean

To solve the min–max problem in (6.12), we first need to analyze $\mu_{max}^{(i)}$ for each colluder i and study which detection statistic gives the maximal mean under which condition.

Note that for a colluder i_1 who receives the base layer only, there is only one detection statistic $\mu_c^{(i_1)}$. Thus, we will analyze only $\mu_{max}^{(i)}$ for colluders who receive at least two layers.

6.3.2.1   For colluder $i \in SC^{b,e1}$

For colluder $i \in SC^{b,e1}$ who receives a medium-resolution copy, there are three possibilities: $\mu_{max}^{(i)} = \mu_b^{(i)}$, $\mu_{max}^{(i)} = \mu_{e1}^{(i)}$, and $\mu_{max}^{(i)} = \mu_c^{(i)}$.

- $\mu_{max}^{(i)} = \mu_b^{(i)}$: If $\mu_{max}^{(i)} = \mu_b^{(i)}$, then $\mu_b^{(i)} \geq \mu_{e1}^{(i)}$ and $\mu_b^{(i)} \geq \mu_c^{(i)}$. From (6.13), we have

$$\mu_b^{(i)} \geq \mu_{e1}^{(i)} \Leftrightarrow \frac{\beta_2 \sqrt{N_b}}{K^{b,e1}} \sigma_w \geq \frac{\alpha_1 \sqrt{N_{e1}}}{K^{b,e1}} \sigma_w \Leftrightarrow \beta_2 \geq \frac{\alpha_1 \sqrt{N_{e1}}}{\sqrt{N_b}}. \tag{6.14}$$

Similarly, we have

$$\mu_b^{(i)} \geq \mu_c^{(i)} \Leftrightarrow \frac{\beta_2 \sqrt{N_b}}{K^{b,e1}} \sigma_w \geq \frac{\beta_2 N_b + \alpha_1 N_{e1}}{K^{b,e1} \sqrt{N_b + N_{e1}}} \sigma_w$$

$$\Leftrightarrow \beta_2 \geq \frac{\alpha_1 N_{e1}}{\sqrt{N_b}(\sqrt{N_b + N_{e1}} - \sqrt{N_b})}. \tag{6.15}$$

Note that $\sqrt{N_b} + \sqrt{N_{e1}} \geq \sqrt{N_b + N_{e1}}$ — that is, $\sqrt{N_{e1}} \geq \sqrt{N_b + N_{e1}} - \sqrt{N_b}$. Therefore, we have $\frac{\alpha_1 N_{e1}}{\sqrt{N_b}(\sqrt{N_b + N_{e1}} - \sqrt{N_b})} \geq \frac{\alpha_1 \sqrt{N_{e1}}}{\sqrt{N_b}}$, and combining (6.14) and (6.15), for colluder $i \in SC^{b,e1}$, we have

$$\mu_{max}^{(i)} = \mu_b^{(i)} \quad \text{if and only if} \quad \beta_2 \geq \frac{\alpha_1 N_{e1}}{\sqrt{N_b}(\sqrt{N_b + N_{e1}} - \sqrt{N_b})}. \tag{6.16}$$

- $\mu_{max}^{(i)} = \mu_{e1}^{(i)}$: In this case, $\mu_{e1}^{(i)} \geq \mu_b^{(i)}$ and $\mu_{e1}^{(i)} \geq \mu_c^{(i)}$. We have

$$\mu_{e1}^{(i)} \geq \mu_b^{(i)} \Leftrightarrow \frac{\alpha_1 \sqrt{N_{e1}}}{K^{b,e1}} \sigma_w \geq \frac{\beta_2 \sqrt{N_b}}{K^{b,e1}} \sigma_w \Leftrightarrow \beta_2 \leq \frac{\alpha_1 \sqrt{N_{e1}}}{\sqrt{N_b}},$$

$$\text{and} \quad \mu_{e1}^{(i)} \geq \mu_c^{(i)} \Leftrightarrow \frac{\alpha_1 \sqrt{N_{e1}}}{K^{b,e1}} \sigma_w \geq \frac{\beta_2 N_b + \alpha_1 N_{e1}}{K^{b,e1} \sqrt{N_b + N_{e1}}} \sigma_w$$

$$\Leftrightarrow \beta_2 \leq \frac{\alpha_1 \sqrt{N_{e1}}(\sqrt{N_b + N_{e1}} - \sqrt{N_{e1}})}{N_b}. \tag{6.17}$$

Because $\sqrt{N_b + N_{e1}} - \sqrt{N_{e1}} \leq \sqrt{N_b}$ and $\frac{\alpha_1 \sqrt{N_{e1}}(\sqrt{N_b + N_{e1}} - \sqrt{N_{e1}})}{N_b} \geq \frac{\alpha_1 \sqrt{N_{e1}}}{\sqrt{N_b}}$, combining the results in (6.17), we have

$$\mu_{max}^{(i)} = \mu_{e1}^{(i)} \quad \text{if and only if} \quad \beta_2 \leq \frac{\alpha_1 \sqrt{N_{e1}}(\sqrt{(N_b + N_{e1})} - \sqrt{N_{e1}})}{N_b}. \tag{6.18}$$

- $\mu_{max}^{(i)} = \mu_c^{(i)}$: This scenario happens if $\mu_c^{(i)} \geq \mu_b^{(i)}$ and $\mu_c^{(i)} \geq \mu_{e1}^{(i)}$. Following the same analysis as in the previous two scenarios,

$$\mu_c^{(i)} \geq \mu_b^{(i)} \Leftrightarrow \beta_2 \leq \frac{\alpha_1 N_{e1}}{\sqrt{N_b}(\sqrt{N_b + N_{e1}} - \sqrt{N_b})},$$

$$\text{and} \quad \mu_c^{(i)} \geq \mu_{e1}^{(i)} \Leftrightarrow \beta_2 \geq \frac{\alpha_1 \sqrt{N_{e1}}(\sqrt{N_b + N_{e1}} - \sqrt{N_{e1}})}{N_b}. \tag{6.19}$$

Note that $\sqrt{N_b + N_{e1}} \leq \sqrt{N_b} + \sqrt{N_{e1}}$. Therefore, we have

$$\sqrt{N_b + N_{e1}} - \sqrt{N_b} \leq \sqrt{N_{e1}} \quad \text{and} \quad \sqrt{N_b + N_{e1}} - \sqrt{N_{e1}} \leq \sqrt{N_b}$$

$$\Leftrightarrow \frac{\sqrt{N_b + N_{e1}} - \sqrt{N_{e1}}}{\sqrt{N_b}} \leq \frac{\sqrt{N_{e1}}}{\sqrt{N_b + N_{e1}} - \sqrt{N_b}}$$

$$\Leftrightarrow \frac{\alpha_1 \sqrt{N_{e1}}(\sqrt{N_b + N_{e1}} - \sqrt{N_{e1}})}{N_b} \leq \frac{\alpha_1 N_{e1}}{\sqrt{N_b}(\sqrt{N_b + N_{e1}} - \sqrt{N_b})}. \tag{6.20}$$

Consequently, $\mu_{max}^{(i)} = \mu_c^{(i)}$ if and only if

$$\frac{\alpha_1 \sqrt{N_{e1}}(\sqrt{(N_b + N_{e1})} - \sqrt{N_{e1}})}{N_b} \leq \beta_2 \leq \frac{\alpha_1 N_{e1}}{\sqrt{N_b}(\sqrt{N_b + N_{e1}} - \sqrt{N_b})}. \tag{6.21}$$

6.3.2.2 For colluder $i \in SC^{all}$

For colluder $i \in SC^{all}$, if the colluded copy includes all three layers, there are four possibilities for $\mu_{max}^{(i)}$: $\mu_{max}^{(i)} = \mu_b^{(i)}$, $\mu_{max}^{(i)} = \mu_{e1}^{(i)}$, $\mu_{max}^{(i)} = \mu_{e2}^{(i)}$, and $\mu_{max}^{(i)} = \mu_c^{(i)}$.

- $\mu_{max}^{(i)} = \mu_b^{(i)}$: Following the same analysis as in the previous section,

$$\mu_{max}^{(i)} = \mu_b^{(i)} \Leftrightarrow \mu_b^{(i)} \geq \mu_{e1}^{(i)}, \quad \mu_b^{(i)} \geq \mu_{e2}^{(i)}, \quad \text{and} \quad \mu_b^{(i)} \geq \mu_c^{(i)},$$

$$\mu_b^{(i)} \geq \mu_{e1}^{(i)} \Leftrightarrow \beta_1 + \beta_2 \leq 1 - (1 - \alpha_1)\frac{\sqrt{N_{e1}}}{\sqrt{N_b}},$$

$$\mu_b^{(i)} \geq \mu_{e2}^{(i)} \Leftrightarrow \beta_1 + \beta_2 \leq 1 - \frac{\sqrt{N_{e2}}}{\sqrt{N_b}}, \quad \text{and}$$

$$\mu_b^{(i)} \geq \mu_c^{(i)} \Leftrightarrow \beta_1 + \beta_2 \leq 1 - \frac{(1 - \alpha_1)N_{e1} + N_{e2}}{\sqrt{N_b}(\sqrt{N_b + N_{e1} + N_{e2}} - \sqrt{N_b})}. \tag{6.22}$$

We have the constraint $0 \leq \beta_1, \beta_2 \leq \beta_1 + \beta_2 \leq 1$ in (6.12) when selecting the collusion parameters. Therefore, from (6.22), to satisfy $\mu_b^{(i)} \geq \mu_{e2}^{(i)}$ and let $\mu_b^{(i)} = \max\{\mu_b^{(i)}, \mu_{e1}^{(i)}, \mu_{e2}^{(i)}, \mu_c^{(i)}\}$, $N_{e2} \leq N_b$ must be true. This observation explains why, in the example shown in Figure 6.1 in which $N_{e2} = 2N_b$, among the four detection statistics, $TN_b^{(i)}$ never achieves the best performance.

- $\mu_{max}^{(i)} = \mu_{e1}^{(i)}$: In this scenario,

$$\mu_{max}^{(i)} = \mu_{e1}^{(i)} \Leftrightarrow \mu_{e1}^{(i)} > \mu_b^{(i)}, \quad \mu_{e1}^{(i)} \geq \mu_{e2}^{(i)}, \quad \text{and} \quad \mu_{e1}^{(i)} \geq \mu_c^{(i)},$$

$$\mu_{e1}^{(i)} \geq \mu_b^{(i)} \Leftrightarrow \alpha_1 \geq 1 - (1 - \beta_1 - \beta_2)\frac{\sqrt{N_{e2}}}{\sqrt{N_{e1}}},$$

$$\mu_{e1}^{(i)} \geq \mu_{e2}^{(i)} \Leftrightarrow \alpha_1 \leq 1 - \frac{\sqrt{N_{e2}}}{\sqrt{N_{e1}}}, \quad \text{and}$$

$$\mu_{e1}^{(i)} \geq \mu_c^{(i)} \Leftrightarrow \alpha_1 \leq 1 - \frac{(1 - \beta_1 - \beta_2)N_b + N_{e2}}{\sqrt{N_{e1}}(\sqrt{N_b + N_{e1} + N_{e2}} - \sqrt{N_{e1}})}. \tag{6.23}$$

From (6.23), $N_{e2} \leq N_{e1}$ must hold to let $\mu_{e1}^{(i)} = \max\{\mu_b^{(i)}, \mu_{e1}^{(i)}, \mu_{e2}^{(i)}, \mu_c^{(i)}\}$, which is the reason that in Figure 6.1 with $N_{e2} = 2N_{e1}$, $TN_{e1}^{(i)}$ never gives the best traitor-tracing performance.

- $\mu_{max}^{(i)} = \mu_{e2}^{(i)}$: Here, $\mu_{max}^{(i)} = \mu_{e2}^{(i)}$ if and only if

$$\beta_1 + \beta_2 \geq 1 + \frac{(1-\alpha_1)N_{e1} - \sqrt{N_{e2}}(\sqrt{N_b + N_{e1} + N_{e2}} - \sqrt{N_{e2}})}{N_b},$$

and $\quad \alpha_1 \geq 1 - \dfrac{\sqrt{N_{e2}}}{\sqrt{N_{e1}}}.$ (6.24)

- $\mu_{max}^{(i)} = \mu_c^{(i)}$: Following the same analysis as in the previous section, $\mu_{max}^{(i)} = \mu_c^{(i)}$ if and only if

$$\beta_1 + \beta_2 \leq 1 - \frac{(1-\alpha_1)\sqrt{N_{e1}}(\sqrt{N_b + N_{e1} + N_{e2}} - \sqrt{N_{e1}}) - N_{e2}}{N_b},$$

$$\beta_1 + \beta_2 \leq 1 + \frac{(1-\alpha_1)N_{e1} - \sqrt{N_{e2}}(\sqrt{N_b + N_{e1} + N_{e2}} - \sqrt{N_{e2}})}{N_b},$$

and $\quad \beta_1 + \beta_2 \geq 1 - \dfrac{(1-\alpha_1)N_{e1} + N_{e2}}{\sqrt{N_b}(\sqrt{N_b + N_{e1} + N_{e2}} - \sqrt{N_b})}.$ (6.25)

6.3.3   Analysis of the feasible set

Given the parameters (N_b, N_{e1}, N_{e2}) and $(K^b, K^{b,e1}, K^{all})$ and the previous analysis of $\mu_{max}^{(i)}$, the next step is to study how attackers achieve fairness of collusion and let $\mu_{max}^{(i)}$ be the same for all colluders. This section investigates the constraints on collusion to ensure the fair play of the attack.

Without loss of generality, in this section, we use $N_b : N_{e1} : N_{e2} = 1:1:2$ as an example to illustrate how colluders achieve the equal-risk fairness of the attack and analyze the constraints on collusion. We consider the scenario in which colluders generate a high-resolution colluded copy including all three layers. In this scenario, from the earlier analysis, for a colluder $i_2 \in SC^{b,e1}$ who receives a medium-resolution copy, $\mu_{max}^{(i_2)}$ has three possible values: $\mu_{max}^{(i_2)} = \mu_b^{(i_2)}$, $\mu_{max}^{(i_2)} = \mu_{e1}^{(i_2)}$, and $\mu_{max}^{(i_2)} = \mu_c^{(i_2)}$. Furthermore, for a colluder $i_3 \in SC^{all}$ who receives all three layers, $\mu_{max}^{(i_3)}$ is either $\mu_{e1}^{(i_3)}$ or $\mu_c^{(i_3)}$, and neither $\mu_b^{(i_3)}$ nor $\mu_{e1}^{(i_3)}$ can be the maximum. Thus, there are a total of six possible scenarios, which are:

(1) $\mu_{max}^{(i_2)} = \mu_b^{(i_2)}$ for $i_2 \in SC^{b,e1}$ and $\mu_{max}^{(i_3)} = \mu_{e2}^{(i_3)}$ for $i_3 \in SC^{all}$,

(2) $\mu_{max}^{(i_2)} = \mu_{e1}^{(i_2)}$ for $i_2 \in SC^{b,e1}$ and $\mu_{max}^{(i_3)} = \mu_{e2}^{(i_3)}$ for $i_3 \in SC^{all}$,

(3) $\mu_{max}^{(i_2)} = \mu_c^{(i_2)}$ for $i_2 \in SC^{b,e1}$ and $\mu_{max}^{(i_3)} = \mu_{e2}^{(i_3)}$ for $i_3 \in SC^{all}$,

(4) $\mu_{max}^{(i_2)} = \mu_b^{(i_2)}$ for $i_2 \in SC^{b,e1}$ and $\mu_{max}^{(i_3)} = \mu_c^{(i_3)}$ for $i_3 \in SC^{all}$,

(5) $\mu_{max}^{(i_2)} = \mu_{e1}^{(i_2)}$ for $i_2 \in SC^{b,e1}$ and $\mu_{max}^{(i_3)} = \mu_c^{(i_3)}$ for $i_3 \in SC^{all}$, and

(6) $\mu_{max}^{(i_2)} = \mu_c^{(i_2)}$ for $i_2 \in SC^{b,e1}$ and $\mu_{max}^{(i_3)} = \mu_c^{(i_3)}$ for $i_3 \in SC^{all}$.

This section analyzes the six scenarios one by one.

6.3.3.1   Scenario 1

In this scenario, for three colluders $i_1 \in SC^b$, $i_2 \in SC^{b,e1}$ and $i_3 \in SC^{all}$, $\mu_{max}^{(i_2)} = \mu_b^{(i_2)}$ for $i_2 \in SC^{b,e1}$ and $\mu_{max}^{(i_3)} = \mu_{e2}^{(i_3)}$ for $i_3 \in SC^{all}$ – that is, (6.13),

$$\mu^{(i_1)} = \frac{\beta_1 \sqrt{N_b}}{K^b}\sigma_w, \quad \mu_{max}^{(i_2)} = \frac{\beta_2 \sqrt{N_b}}{K^{b,e1}}\sigma_w, \quad \text{and} \quad \mu_{max}^{(i_3)} = \frac{\sqrt{N_{e2}}}{K^{all}}\sigma_w. \quad (6.26)$$

To achieve the equal-risk fairness, colluders select the collusion parameters $\{\alpha_l, \beta_k\}$ such that $\mu^{(i_1)} = \mu_{max}^{(i_2)} = \mu_{max}^{(i_3)}$. Therefore, we have

$$\beta_1 = \frac{\sqrt{N_{e2}}}{\sqrt{N_b}}\frac{K^b}{K^{all}} = \sqrt{2}\frac{K^b}{K^{all}}, \quad \text{and} \quad \beta_2 = \frac{K^{b,el}}{K^b}\beta_1 = \sqrt{2}\frac{K^{b,el}}{K^{all}}. \quad (6.27)$$

For colluder i_2 who receives the medium resolution, because $\mu_b^{(i_2)}$ is the largest among $\{\mu_b^{(i_2)}, \mu_{el}^{(i_2)}, \mu_c^{(i_2)}\}$, from (6.16), the selected collusion parameters must satisfy

$$\alpha_1 \leq \beta_2 \frac{\sqrt{N_b}(\sqrt{N_b + N_{e1}} - \sqrt{N_b})}{N_{e1}}$$

$$= A \triangleq \frac{\sqrt{2N_b}(\sqrt{N_b + N_{e1}} - \sqrt{N_b})}{N_{e1}}\frac{K^{b,el}}{K^{all}}. \quad (6.28)$$

In our example with $N_b:N_{e1}:N_{e2} = 1:1:2$, $A = (2 - \sqrt{2})K^{b,el}/K^{all}$. Similarly, for colluder $i_3 \in SC^{all}$, to ensure $\mu_{max}^{(i_3)} = \mu_{e2}^{(i_3)}$, from (6.24) and (6.27), α_1, β_1 and β_2 must satisfy

$$\alpha_1 \geq 1 + \frac{(1 - \beta_1 - \beta_2)N_b - \sqrt{N_{e2}}(\sqrt{N_b + N_{e1} + N_{e2}} - \sqrt{N_{e2}})}{N_{e1}}$$

$$= B \triangleq 1 + \frac{N_b}{N_{e1}} - \frac{\sqrt{N_{e2}}(\sqrt{N_b + N_{e1} + N_{e2}} - \sqrt{N_{e2}})}{N_{e1}}$$

$$- \frac{\sqrt{2}N_b}{N_{e1}} \cdot \frac{K^b + K^{b,el}}{K^{all}}. \quad (6.29)$$

$B = 4 - 2\sqrt{2} - \sqrt{2}(K^b + K^{b,el})/K^{all}$ if $N_b:N_{e1}:N_{e2} = 1:1:2$.

Define $R^b = K^b/K$, $R^{b,el} = K^{b,el}/K$ and $R^{all} = K^{all}/K$ as the percentages of colluders who are in SC^b, $SC^{b,el}$, and SC^{all}, respectively. By combining (6.28) and (6.29), scenario 1 will happen if and only if

$$\frac{\sqrt{2}}{1 + \sqrt{2}} \leq R^{all} \leq \min\left\{\frac{2 - (2 - \sqrt{2})R^b}{6 - 2\sqrt{2}}, 1 - R^b\right\}. \quad (6.30)$$

To summarize, if $(R^b, R^{b,el}, R^{all})$ satisfies (6.30), colluders can achieve equal risk for all colluders by following (6.27) through (6.29) when selecting the collusion parameters.

6.3.3.2 Scenario 2

In this scenario, $\mu_{max}^{(i_2)} = \mu_{el}^{(i_2)}$ for $i_2 \in SC^{b,el}$ and $\mu_{max}^{(i_3)} = \mu_{e2}^{(i_3)}$ for $i_3 \in SC^{all}$. Following the same analysis as in Section 6.3.3.1, for the example of $N_b:N_{e1}:N_{e2} = 1:1:2$, if $(R^b, R^{b,el}, R^{all})$ satisfies

$$\max\left\{\sqrt{2}R^b, (2 - \sqrt{2})(1 - R^b)\right\} \leq R^{all}$$

$$\leq \min\left\{\frac{2 - (2 - \sqrt{2})R^b}{6 - 2\sqrt{2}}, \frac{\sqrt{2} - \sqrt{2}R^b}{3 - \sqrt{2}}, 1 - R^b\right\}, \quad (6.31)$$

colluders can guarantee the equal risk of all attackers by selecting

$$\alpha_1 = \sqrt{2}\frac{K^{b,e1}}{K^{all}}, \quad \beta_1 = \sqrt{2}\frac{K^b}{K^{all}},$$

and $\quad 4 - \sqrt{2} - \sqrt{2}\frac{K}{K^{all}} \le \beta_2 \le \min\left\{1 - \sqrt{2}\frac{K^b}{K^{all}}, (2 - \sqrt{2})\frac{K^{b,e1}}{K^{all}}\right\}. \quad (6.32)$

6.3.3.3 Scenario 3

Given $N_b:N_{e1}:N_{e2} = 1:1:2$, if $(R^b, R^{b,e1}, R^{all})$ satisfies

$$\max\left\{\frac{2 - (2 - \sqrt{2})R^b}{4}, \frac{(2 - \sqrt{2}) + (2\sqrt{2} - 2)R^b}{3 - \sqrt{2}}\right\}$$

$$\le R^{all} \le \left\{\frac{2 - (2 - \sqrt{2})R^b}{6 - 2\sqrt{2}}, 1 - R^b\right\}, \quad (6.33)$$

and colluders select

$$\beta_1 = \sqrt{2}\frac{K^b}{K^{all}},$$

$$\max\left\{2\frac{K^{b,e1}}{K^{all}} - 1, (2 - \sqrt{2})\frac{K^{b,e1}}{K^{all}}\right\} \le \beta_2 \le \min\left\{\sqrt{2}\frac{K^{b,e1}}{K^{all}}, 1 - \sqrt{2}\frac{K^b}{K^{all}}\right\},$$

and $\quad \alpha_1 = 2\frac{K^{b,e1}}{K^{all}} - \beta_2, \quad (6.34)$

then $\mu^{(i_2)}_{max} = \mu^{(i_2)}_c$ for $i_2 \in SC^{b,e1}$ and $\mu^{(i_3)}_{max} = \mu^{(i_3)}_{e2}$ for $i_3 \in SC^{all}$, and all colluders have the same probability of being detected.

6.3.3.4 Scenario 4

Given $N_b:N_{e1}:N_{e2} = 1:1:2$, if

$$\max\left\{2 - \sqrt{2}, \frac{4 - \sqrt{2} + (\sqrt{2} - 1)R^b}{6 - \sqrt{2}}\right\} \le R^{all} \le 1 - R^b, \quad (6.35)$$

by choosing the collusion parameters as

$$\beta_1 \ge \max\left\{\frac{4K^b}{\sqrt{2}K - (\sqrt{2} - 1)K^b + (2 - \sqrt{2})K^{all}}, \frac{3K^b}{K + K^{all}}, \frac{\sqrt{2}K^b}{K^{all}}\right\},$$

$$\beta_1 \le \min\left\{\frac{K^b}{K - K^{all}}, \frac{4K^b}{K + K^{all}}\right\},$$

$$\beta_2 = \frac{K^{b,e1}}{K^b}\beta_1, \quad \text{and} \quad \alpha_1 = 4 - \frac{K + K^{all}}{K^b}\beta_1, \quad (6.36)$$

colluders achieve equal-risk fairness. In this scenario, $\mu^{(i_2)}_{max} = \mu^{(i_2)}_b$ for $i_2 \in SC^{b,e1}$ and $\mu^{(i_3)}_{max} = \mu^{(i_3)}_c$ for $i_3 \in SC^{all}$.

6.3.3.5   Scenario 5

Here, under the constraint that $(R^b, R^{b,el}, R^{all})$ satisfies

$$\max\left\{4R^b - 1, \sqrt{2}R^b, \frac{4 - \sqrt{2} - (5 - \sqrt{2})R^b}{6 - \sqrt{2}}, \frac{\sqrt{2}}{4 - \sqrt{2}}\right\} \le R^{all}$$

$$\le 1 - R^b, \tag{6.37}$$

all colluders have the same probability of being detected if they select

$$\beta_1 \ge \max\left\{\frac{4K^b}{\sqrt{2}K - (\sqrt{2} - 1)K^b + (2 - \sqrt{2})K^{all}}, \frac{3K^b}{K + K^{all} - K^b}, \frac{\sqrt{2}K^b}{K^{all}}\right\},$$

$$\beta_1 \le \min\left\{\frac{K^b}{K^{b,el}}, \frac{4K^b}{K + K^{all}}\right\},$$

$$\beta_2 = 4 - \frac{K + K^{all}}{K^b}\beta_1, \quad \text{and} \quad \alpha_1 = \frac{K^{b,el}}{K^b}\beta_1 \tag{6.38}$$

during collusion. In this scenario, $\mu_{max}^{(i_2)} = \mu_{el}^{(i_2)}$ for $i_2 \in SC^{b,el}$ and $\mu_{max}^{(i_3)} = \mu_c^{(i_3)}$ for $i_3 \in SC^{all}$.

6.3.3.6   Scenario 6

If $(R^b, R^{b,el}, R^{all})$ satisfies the constraint

$$\max\left\{\frac{3\sqrt{2} - 4 - (3\sqrt{2} - 7)R^b}{3\sqrt{2} - 2}, \frac{\sqrt{2} - (\sqrt{2} - 1)R^b}{3\sqrt{2} - 2}\right\} \le R^{all} \le 1 - R^b \tag{6.39}$$

and if the selected parameters are

$$\beta_1 = \frac{4K^b}{\sqrt{2}K - (\sqrt{2} - 1)K^b + (2 - \sqrt{2})K^{all}},$$

$$\alpha_1 \ge \max\left\{\frac{3\sqrt{2}K^{b,el} + 3K^b - 2K^{all}}{\sqrt{2}K - (\sqrt{2} - 1)K^b + (2 - \sqrt{2})K^{all}}, \right.$$

$$\left. \frac{4(\sqrt{2} - 1)K^{b,el}}{\sqrt{2}K - (\sqrt{2} - 1)K^b + (2 - \sqrt{2})K^{all}}\right\},$$

$$\alpha_1 \le \frac{4K^{b,el}}{\sqrt{2}K - (\sqrt{2} - 1)K^b + (2 - \sqrt{2})K^{all}},$$

$$\text{and} \quad \beta_2 + \alpha_1 = \frac{4\sqrt{2}K^{b,el}}{\sqrt{2}K - (\sqrt{2} - 1)K^b + (2 - \sqrt{2})K^{all}}, \tag{6.40}$$

then all colluders have the same risk and $\mu_{max}^{(i_2)} = \mu_c^{(i_2)}$ for $i_2 \in SC^{b,el}$ and $\mu_{max}^{(i_3)} = \mu_c^{(i_3)}$ for $i_3 \in SC^{all}$.

6.3.4   Min–max solution

Given the analysis in Section 6.3.3, for three colluders $i_1 \in SC^b$, $i_2 \in SC^{b,el}$ and $i_3 \in SC^{all}$, they first identify all possible collusion parameters $\{\alpha_l, \beta_k\}$ that satisfy $\mu_{max}^{(i_1)} = \mu_{max}^{b,el} = \mu_{max}^{all}$, and then select the one that gives them the minimum probability of being detected.

To demonstrate this process, we use the system setup in Figure 6.2 as an example, in which the lengths of the fingerprints embedded in the base layer, enhancement layer 1, and enhancement layer 2 are $N_b = 5000$, $N_{e1} = 5000$, and $N_{e2} = 10000$, respectively. When generating fingerprints, we first generate independent Gaussian vectors following distribution $\mathcal{N}(0, 1)$ and then apply Gram–Schmidt orthogonalization to produce fingerprints that have equal energies and are strictly orthogonal to each other.

Assume that there are a total of $K = 250$ colluders. Among the 250 colluders, we use $K^b = 50$, $K^{b,el} = 25$, and $K^{all} = 175$ as an example – that is, $(R^b, R^{b,el}, R^{all}) = (0.2, 0.1, 0.7)$. From Section 6.3.3, $(R^b, R^{b,el}, R^{all})$ satisfies constraint (6.35) in scenario 4, constraint (6.37) in scenario 5, and constraint (6.39) in scenario 6.

- Because $(R^b, R^{b,el}, R^{all})$ satisfies constraint (6.35) in scenario 4, colluders can guarantee the equal risk of all colluders if they choose

$$\beta_1 \geq \max \left\{ \frac{4K^b}{\sqrt{2}K - (\sqrt{2}-1)K^b + (2-\sqrt{2})K^{all}}, \frac{3K^b}{K + K^{all}}, \frac{\sqrt{2}K^b}{K^{all}} \right\} = 0.4594,$$

$$\beta_1 \leq \min \left\{ \frac{K^b}{K - K^{all}}, \frac{4K^b}{K + K^{all}} \right\} = 0.4706,$$

$$\beta_2 = \frac{K^{b,el}}{K^b}\beta_1, \quad \text{and} \quad \alpha_1 = 4 - \frac{K + K^{all}}{K^b}\beta_1. \tag{6.41}$$

Here, $\mu_{max}^{(i_2)} = \mu_b^{(i_2)}$ for colluder $i_2 \in SC^{b,el}$ and $\mu_{max}^{(i_3)} = \mu_c^{(i_3)}$ for colluder $i_3 \in SC^{all}$. For any colluder $i \in SC$, $\mu_{max}^{(i)}$ has the smallest possible value of 2.0545 when $\beta_1 = 0.4594$, $\beta_2 = 0.2297$, and $\alpha_1 = 0.0951$.

- Following (6.38), when colluders select parameters

$$\beta_1 \geq \max \left\{ \frac{4K^b}{\sqrt{2}K - (\sqrt{2}-1)K^b + (2-\sqrt{2})K^{all}}, \right.$$

$$\left. \frac{3K^b}{K + K^{all} - K^b}, \frac{\sqrt{2}K^b}{K^{all}} \right\} = 0.4594,$$

$$\beta_1 \leq \min \left\{ \frac{K^b}{K^{b,el}}, \frac{4K^b}{K + K^{all}} \right\} = 0.4706,$$

$$\beta_2 = 4 - \frac{K + K^{all}}{K^b}\beta_1, \quad \text{and} \quad \alpha_1 = \frac{K^{b,el}}{K^b}\beta_1, \tag{6.42}$$

they have the same probability of being detected. Here, $\mu_{max}^{(i_2)} = \mu_{e1}^{(i_2)}$ for colluder $i_2 \in SC^{b,el}$ and $\mu_{max}^{(i_3)} = \mu_c^{(i_3)}$ for colluder $i_3 \in SC^{all}$. For any colluder $i \in SC$, $\mu_{max}^{(i)}$

reaches its minimum value of 2.0545 when $\beta_1 = 0.4594$, $\beta_2 = 0.0951$, and $\alpha_1 = 0.2297$.

- Following (6.40), colluders can also achieve fairness of collusion by selecting

$$\beta_1 = \frac{4K^b}{\sqrt{2}K - (\sqrt{2} - 1)K^b + (2 - \sqrt{2})K^{all}} = 0.4594,$$

$$\alpha_1 \geq \max \left\{ \frac{3\sqrt{2}K^{b,el} + 3K^b - 2K^{all}}{\sqrt{2}K - (\sqrt{2} - 1)K^b + (2 - \sqrt{2})K^{all}}, \right.$$

$$\left. \frac{4(\sqrt{2} - 1)K^{b,el}}{\sqrt{2}K - (\sqrt{2} - 1)K^b + (2 - \sqrt{2})K^{all}} \right\} = 0.2297,$$

$$\alpha_1 \leq \frac{4K^{b,el}}{\sqrt{2}K - (\sqrt{2} - 1)K^b + (2 - \sqrt{2})K^{all}} = 0.0951, \quad \text{and}$$

$$\beta_2 = \frac{4\sqrt{2}K^{b,el}}{\sqrt{2}K - (\sqrt{2} - 1)K^b + (2 - \sqrt{2})K^{all}} - \alpha_1 = 0.3248 - \alpha_1 \qquad (6.43)$$

during collusion. In this scenario, $\mu_{max}^{(i_2)} = \mu_c^{(i_2)}$ for colluder $i_2 \in SC^{b,el}$ and $\mu_{max}^{(i_3)} = \mu_c^{(i_3)}$ for colluder $i_3 \in SC^{all}$, and $\mu_{max}^{(i)} = 2.0545$ for all colluders.

The means of the detection statistics in these three scenarios are the same; therefore, colluders can choose either (6.41), (6.42), or (6.43) during collusion. In fact, (6.41) and (6.42) are the two boundaries of (6.43).

In the example of $(K^b, K^{b,el}, K^{all}) = (50, 75, 125)$, the constraints (6.31) in scenario 2 and (6.33) in scenario 3 are satisfied, and the minimum value of $\mu_{max}^{(i)}$ is 2.5298, when colluders select $(\beta_1 = 0.5657, \beta_2 = 0.0544, \alpha_1 = 0.4485)$ or use $(\beta_1 = 0.5657, \beta_2 = 0.3929, \alpha_1 = 0.4071)$ during collusion.

If $(K^b, K^{b,el}, K^{all}) = (50, 125, 75)$, none of the six constraints in Section 6.3.3 is satisfied, and colluders cannot generate a high-resolution colluded copy while still achieving equal-risk fairness. They must lower the resolution of the attacked copy to medium to guarantee the equal risk of all colluders.

6.4 Simulation results

In our simulations, we test over the first forty frames of "carphone," and use $F_b = \{1, 5, \ldots, 37\}$, $F_{el} = \{3, 7, \ldots, 39\}$, and $F_{e2} = \{2, 4, \ldots, 40\}$ as an example of the temporal scalability. The lengths of the fingerprints embedded in the base layer, enhancement layer 1, and enhancement layer 2 are $N_b = 42987$, $N_{el} = 42951$ and $N_{e2} = 85670$, respectively. We assume that there are a total of $M = 750$ users and $|U^b| = |U^{b,el}| = |U^{all}| = 250$. We first generate independent vectors following Gaussian distribution $\mathcal{N}(0, 1/9)$, and then apply Gram–Schmidt orthogonalization to generate orthogonal fingerprints for different users.

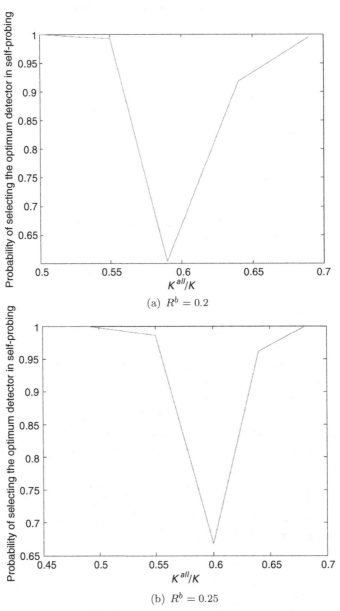

Fig. 6.5 Probability that the self-probing detector selects the optimum detection statistics with the largest mean. The results are based on 1000 simulation runs on the first 40 frames of sequence "carphone"

We assume that $0 \leq K^b, K^{b,e1}, K^{all} \leq 250$ are the number of colluders in subgroups SC^b, $SC^{b,e1}$, and SC^{all}, respectively, and the total number of colluders is fixed at 250. During collusion, colluders apply the intragroup collusion followed by the inter-group collusion, and follow Section 6.3 when choosing the collusion parameters. In our

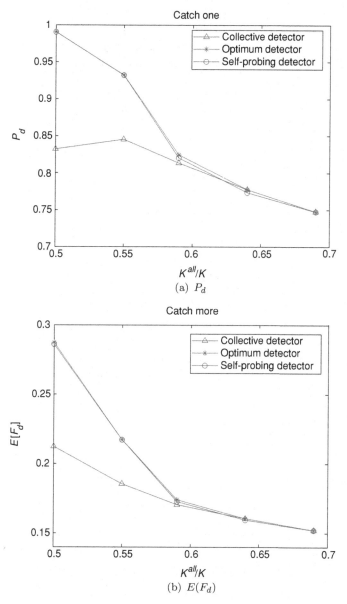

Fig. 6.6 Simulation results on the first 40 frames of sequence "carphone" with $R^b = 0.2$

simulations, we adjust the power of the additive noise such that $||\mathbf{n}_j||^2 = ||JND_j\mathbf{W}_j^{(i)}||^2$ for every frame j in the video sequence.

We compare the performance of three detectors: the simple collective detector in (6.1); the optimum detector, which always selects the detection statistics with the largest mean; and the self-probing detector in Section 6.1.3. The self-probing fingerprint detector follows Section 6.1.3 when identifying selfish colluders. The detector first estimates the means of different detection statistics, selects the detection statistics with the largest

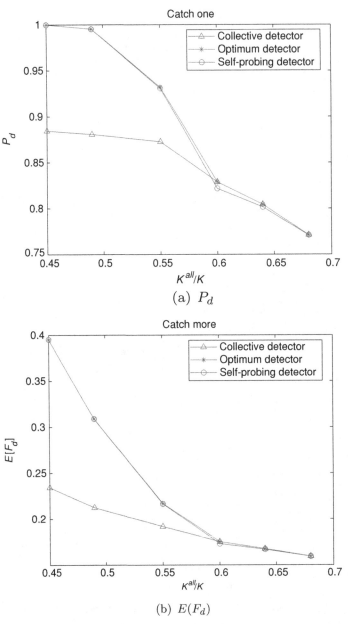

(a) P_d

(b) $E(F_d)$

Fig. 6.7 Simulation results on the first 40 frames of sequence "carphone" with $R^b = 0.25$

estimated mean, and then identifies the colluders. In our simulations, we fix $P_{fp} = 10^{-3}$ and $E[F_{fp}] = 10^{-3}$.

Figure 6.5 plots the probability that the self-probing detector selects the optimum detection statistics with the largest mean. Figures 6.6 and 6.7 show the performance of the fingerprint detector when $R^b = 0.2$ and $R^b = 0.25$, respectively. The results

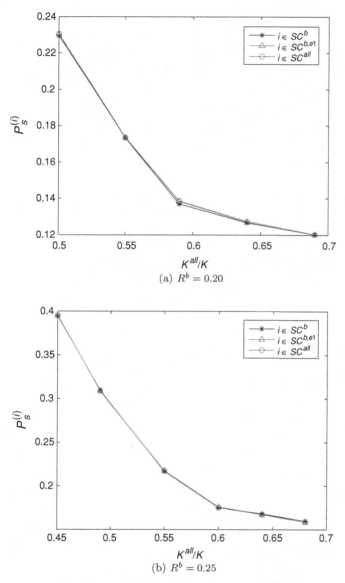

Fig. 6.8 Each colluder's probability of being detected ($P_s^{(i)}$) when they follow Section 6.2 to select the collusion parameters. $K = 250$

are based on 10000 simulation runs. In Figures 6.5 through 6.7, the x axis is R^b, the percentage of colluders who receive the low-resolution copy, and each point on the x axis corresponds to a unique triplet (K^b, $K^{b,e1}$, K^{all}) where $K^b = K \cdot R^b$ and $K^{b,e1} = K - K^b - K^{all}$.

From Figure 6.5, when the statistical means of different detection statistics differ significantly from each other, the self-probing detector in Section 6.1.3 always selects the

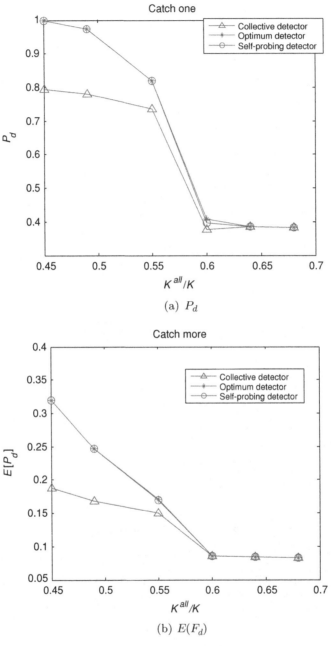

(a) P_d

(b) $E(F_d)$

Fig. 6.9 Simulation results on first 28 frames of "tennis" from 1000 simulation runs. $P_{fp} = 10^{-3}$ in (a), and $E[F_{fp}] = 10^{-3}$ in (b)

optimum detection statistics with the largest mean. When the difference between different means is small, the optimum and the suboptimum detection statistics have approximately the same statistical means, and selecting the suboptimum detection strategy does not significantly deteriorate the detection performance when compared with the optimum detection statistics. In Figures 6.6 and 6.7, the performance gap is smaller than 2×10^{-3} and can be ignored. In addition, by exploring side information about collusion, the probability of catching at least one colluder has been improved by 17 percent when $R^{all} = 0.5$ and $R^b = 0.2$, and by 12 percent when $R^{all} = 0.45$ and $R^b = 0.25$. These simulation results support our conclusions that exploring side information about collusion can significantly help improve the detection performance, and the self-probing detector has approximately the same performance as the optimum detector with perfect knowledge of the detection statistics' means.

Figure 6.8 plots each colluder's probability of being detected when the colluders follow Section 6.3 to select the collusion parameters. The simulation setup is similar to that in Figures 6.5 through 6.7, and the results are based on 10000 simulation runs. It is obvious that in this example, all colluders have the same probability of being detected, and this multiuser collusion achieves equal-risk fairness of the attack with the self-probing detector.

To show that the self-probing detector can be applied to various types of videos, we also run the simulation on "tennis", which is a fast-motion video. We use the first twenty-eight frames of "tennis," and use $F_b = \{1, 5, \ldots, 25\}$, $F_{e1} = \{3, 7, \ldots, 27\}$, and $F_{e2} = \{2, 4, \ldots, 28\}$ as an example of the temporal scalability. The lengths of the fingerprints embedded in the base layer, enhancement layer 1, and enhancement layer 2 are $N_b = 45092$, $N_{e1} = 45103$, and $N_{e2} = 90174$, respectively. Other settings are the same as the earlier ones. Figure 6.9 shows P_d and $E[F_d]$ of the optimal detector, the self-probing detector, and the collective detector when $R^b = 0.25$. It is clear from Figure 6.9 that the self-probing detector achieves almost the same performance as the optimal detector, which has perfect information about the mean value. Such a result shows that the detection performance of the self-probing detector is not influenced by the video characteristics.

6.5 Chapter summary and bibliographical notes

This chapter uses multimedia fingerprinting as an example to illustrate the impact of side information on human behavior. We study how side information about multiuser collusion can help the fingerprint detector increase the traitor-tracing capability, and investigate its influence on colluder and the forensic detector's strategies. We model the dynamics between colluders and the fingerprint detector as a Stackelberg game, analyze its Nash equilibrium, and derive the optimal strategies for both players.

We first investigate multimedia forensics with side information, and show that information about the statistical means of the detection statistics can help significantly improve the collusion resistance. We then propose a self-probing detector for the fingerprint detector to probe such side information from the colluded copy itself. We show that the self-probing detector has approximately the same performance as the optimal fingerprint detector with perfect information about the statistical means of the detection statistics, and the difference between these two detectors can be ignored.

Side information not only improves the fingerprint detector's collusion resistance, but also affects each colluder's probability of being detected and makes some colluders take a larger risk than others. Thus, it breaks the collective fairness equilibrium between the colluders and the fingerprint detector, and they must choose different strategies and look for a new equilibrium. We model the colluder-detector dynamics with side information as a two-player Stackelberg game in which the colluders are the leader and the fingerprint detector is the follower. We show that under the assumption that colluders demand equal-risk fairness, the min–max solution is the Nash equilibrium and is the optimal strategy for both colluders and the fingerprint detector. Neither of them can further increase their payoff and, therefore, they have no incentive to deviate from this equilibrium.

Researchers have already started to study modeling and analysis of collusion [75,103,104]. The work of Wang *et al.* [61] studied the relationship between the maximum allowable colluders by a fingerprinting system and other parameters, such as the fingerprint length, the total number of users, and the system requirements. Techniques from different disciplines, including error-correcting codes, finite-projective geometry, and combinatorial theories, have been used in the literature to design multimedia fingerprints that can resist collusion attacks [69,71–73].

7 Risk–distortion analysis of multiuser collusion

In the previous chapter, we used multimedia fingerprinting social network as an example and examined the impact of side information on users' strategies. We showed that information about the statistical means of detection statistics is useful to improve the detection performance. A straightforward question to ask is whether there is other information that may potentially influence user dynamics. In this chapter, we investigate how side information changes the risk–distortion relationship in linear video collusion attacks against Gaussian fingerprints.

Video data have the unique characteristic that temporally adjacent frames are similar but usually not identical. Video collusion attacks include not only the intercopy attack that combines the same frames from different copies, but also the intracopy attack that combines temporally adjacent frames within the same copy. Because temporally adjacent frames are not exactly the same, an intracopy collusion attack will introduce distortion. Therefore, for a video collusion attack, there exists a tradeoff between the fingerprint remaining in the colluded copy that determines colluders' probability of being detected – that is, their risk – and the quality of the colluded copy – the distortion. It is extremely important for colluders to learn the risk–distortion tradeoff, as knowing this tradeoff would help them choose the best strategy when generating the colluded copy. It is also essential for the fingerprint detector to understand the risk–distortion tradeoff, as it would help predict colluders' behavior and design an anticollusion strategy [109].

To understand how side information influences the video-collusion social network, we first explicitly explore the relationship between risk and distortion by conducting a theoretical analysis of linear video collusion attacks against Gaussian fingerprinting. We write the risk and the distortion as functions of the temporal filter coefficients used in intracopy collusion, and model the collusion attack as an optimization problem that minimizes the distortion subject to a given risk constraint. By varying the risk constraint and solving the corresponding optimization problem, we can derive the optimal risk–distortion curve.

Furthermore, we show that the detector can improve the detection performance given the optimal coefficients that attackers use, and similarly, attackers can improve the attack effectiveness given the optimal coefficients the detector uses. The knowledge about optimal coefficients is the side information of video-collusion social networks. We formulate the collusion attack and the colluder identification problem as a dynamic cat-and-mouse game and study the optimal strategy for each player in the game, given knowledge of the opponent's strategy. In practice, because attackers need to act first, a powerful detector will be able to estimate attackers' strategy. In such a case, we show that attackers' best strategy is the min–max strategy – that is, to minimize the risk under the assumption that the detector has perfect knowledge of their strategy. We also discuss attackers' min–max strategy when they add additive white Gaussian noise (AWGN) to the colluded copy to further hinder the detection performance. Finally, we conduct several experiments to verify the risk–distortion model using real video data.

7.1   Video fingerprinting

In this section, we introduce video fingerprint embedding, detection, and a video-collusion attack model.

7.1.1   Fingerprint embedding

Let S_t be the tth frame of the host video sequence, which can be the pixel values or the DCT coefficients. Let $X_t^{(k)}$ and $W_t^{(k)}$ be the tth frame of the fingerprinted video and fingerprint signal for user k, respectively. Then, the fingerprint embedding process of the tth frame for the kth user can be written as

$$X_t^{(k)} = S_t + W_t^{(k)}. \tag{7.1}$$

Here, we drop the term *JND*, the just-noticeable difference from human visual models, to simplify the notations.

As in the previous chapters, orthogonal fingerprint modulation is used; $\langle W_t^{(i)}, W_t^{(j)} \rangle = \sigma_w^2 \delta_{i,j}$, where $\delta_{i,j} = 1$ if $i = j$ and $\delta_{i,j} = 0$ if $i \neq j$. Moreover, to resist an intracopy collusion attack [110,111], the fingerprints $W^{(k)}$ between neighboring frames for the same user k are correlated with each other, whereas the correlation is determined by the similarity of the host frames and the temporal distance of the indices [110] as given by

$$\rho[W_{t_1}^{(k)}, W_{t_2}^{(k)}] = \gamma^{|t_1 - t_2|} \rho[S_{t_1}, S_{t_2}], \tag{7.2}$$

where $\rho[A, B] = \frac{cov(A,B)}{\sqrt{var(A)var(B)}}$ is the statistical correlation coefficient between random variables A and B, and γ is a scaling parameter ($0 \leq \gamma \leq 1$) that controls the tradeoff between the visual quality and the degree of the resistance. If γ is large, then the degree of the resistance against the intracopy attack is high. However, the visual quality of the fingerprinted video becomes poor owing to the veiling artifacts. On the contrary, if γ is small, then the veiling artifacts are less significant, and the fingerprinted video becomes vulnerable to the intracopy attack.

7.1.2   Fingerprint detection

Without loss of generality, we analyze the frame-based detection. Similar analysis can be easily extended to group-of-picture (GOP)-based or sequence-based detection. For each frame V_t of the colluded copy, the detector extracts the fingerprint using

$$Y_t = V_t - S_t. \tag{7.3}$$

Then, for each user k who receives frame t, we compute the detection statistics using

$$TN_t^{(k)} = \frac{\langle W_t^{(k)}, Y_t \rangle}{||W_t^{(k)}||} = \frac{W_t^{(k)^T} Y_t}{||W_t^{(k)}||}. \tag{7.4}$$

Finally, given a threshold h that is determined by false alarm probability, the estimated attacker set for frame t is $\hat{SC}_t = \{i : TN_t^{(i)} > h\}$.

7.1.3   Video-collusion attack model

With orthogonal fingerprint modulation, nonlinear intercopy collusion attacks can be approximated as a linear intercopy collusion attack followed by AWGN [61], and we consider only linear intercopy collusion attacks. Let M be the total number of colluders. Each attacker first performs an intracopy attack by applying temporal filtering on the temporally adjacent video frames. Then, all attackers collude together to perform an intercopy attack. Because the fingerprint in every frame for each attacker $W_t^{(k)}$ is independent and identically distributed (i.i.d), if we assume that all attackers share the same risk, then the weights allocated to the intracopy and intercopy attacks would be the same for all attackers. Therefore, frame t in the colluded copy is

$$V_t = \sum_{k=1}^{M} \frac{1}{M} \left[\sum_{i=-n}^{n} a_i X_{t+i}^{(k)} \right], \tag{7.5}$$

where $\sum_{i=-n}^{n} a_i = 1$. Attackers choose $\{a_i\}$ to minimize the collusion distortion under a certain risk constraint, and the detector is to estimate the a_i that attackers use and explore side information to improve the detection performance.

7.2   Risk–distortion modeling

In this section, we analyze the analytical model of the relationship between the colluders' risk and the distortion of the colluded copy without side information.

7.2.1   Risk of the colluders

Given the colluded frame V_t, the detector extracts the fingerprint Y_t by

$$Y_t = V_t - S_t = \left(\mathbf{e}_S + \frac{1}{M} \sum_{i=1}^{M} \mathbf{W}^{(i)} \right) \mathbf{a}, \tag{7.6}$$

where $\mathbf{e}_S = [S_{t-n} - S_t, \dots, S_{t+n} - S_t]$, $\mathbf{W}^{(i)} = [W_{t-n}^{(i)}, \dots, W_{t+n}^{(i)}]$, and $\mathbf{a} = [a_{-n}, \dots, a_n]^T$.

Because the linear combination of the Gaussian distribution is also a Gaussian distribution, if we assume that the residue satisfies Gaussian distribution [112,113] – that is,

$[S_{t+i} - S_t] \sim N(0, \sigma_i^2)$ – then $\mathbf{e}_S\mathbf{a} \sim N(0, ||\Lambda\mathbf{a}||_2^2)$, where $\Lambda = diag\{\sigma_{-n}, \ldots, \sigma_n\}$ and $||x||_2$ is L_2-norm of x.

According to (7.4), the detection statistic $TN_t^{(k)}$ can be written as

$$TN_t^{(k)} = \frac{\langle W_t^{(k)}, Y_t \rangle}{||W_t^{(k)}||} = \frac{\langle W_t^{(k)}, \mathbf{e}_S\mathbf{a} + \frac{1}{M}\sum_{i=1}^{M} \mathbf{W}^{(i)}\mathbf{a} \rangle}{||W_t^{(k)}||}. \tag{7.7}$$

From (7.7), we know that the detection statistic of the attacker k, $TN_t^{(k)}$, satisfies Gaussian distribution $N(\mu^{(k)}, ||\Lambda\mathbf{a}||_2^2)$ [60], where the mean $\mu^{(k)}$ is given by

$$\mu^{(k)} = E\left[\frac{\langle Y_t, W_t^{(k)} \rangle}{||W_t^{(k)}||} \right] = \frac{1}{M}\mathbf{p}^T\mathbf{a}, \tag{7.8}$$

where

$$\mathbf{p} = \left[E\left(\frac{\langle W_{t-n}^{(k)}, W_t^{(k)} \rangle}{||W_t^{(k)}||} \right), \ldots, E\left(\frac{\langle W_{t+n}^{(k)}, W_t^{(k)} \rangle}{||W_t^{(k)}||} \right) \right]^T. \tag{7.9}$$

In this chapter, we use R to denote colluders' risk – that is, their probability of being detected. Given a detection threshold h, according to (7.7) and (7.8), the risk R can be computed by

$$R = P[TN_t^{(k)} > h] = Q\left(\frac{h - \frac{1}{M}\mathbf{p}^T\mathbf{a}}{||\Lambda\mathbf{a}||_2} \right), \tag{7.10}$$

where $Q(x)$ is the Gaussian tail function $Q(x) = \int_x^\infty \frac{1}{\sqrt{2\pi}} \exp^{-\frac{t^2}{2}} dt$.

Similarly, the detection statistic of an innocent user satisfies Gaussian distribution $N(0, ||\Lambda\mathbf{a}||_2^2)$ [60]. Therefore, the probability of an innocent user to be falsely detected as an attacker, P_{fa}, is given by

$$P_{fa} = Q\left(\frac{h}{||\Lambda\mathbf{a}||_2} \right). \tag{7.11}$$

From (7.10) and (7.11), we can see that the threshold h controls the tradeoff between the positive detection probability R and the false alarm probability P_{fa}. If the desired false alarm probability P_{fa} is upper-bounded by α, then $h = Q^{-1}(\alpha)||\Lambda\mathbf{a}||_2$ and the risk R becomes

$$R = Q\left(\frac{Q^{-1}(\alpha)||\Lambda\mathbf{a}||_2 - \frac{1}{M}\mathbf{p}^T\mathbf{a}}{||\Lambda\mathbf{a}||_2} \right). \tag{7.12}$$

7.2.2   Distortion of the colluded frame

From (7.6), we can see that the difference between the colluded frame V_t and the original frame S_t is Y_t. Therefore, the distortion D of the colluded copy, which is defined as the mean square of the difference, can be computed by

$$D = E\left[||Y_t||^2 \right] = \mathbf{a}^T\mathbf{K}\mathbf{a}, \tag{7.13}$$

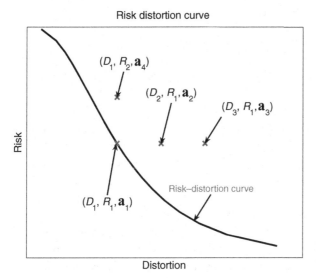

Risk distortion curve

$(D_1, R_2, \mathbf{a}_4)$

$(D_2, R_1, \mathbf{a}_2)$

$(D_3, R_1, \mathbf{a}_3)$

Risk–distortion curve

$(D_1, R_1, \mathbf{a}_1)$

Risk

Distortion

Fig. 7.1 Risk–distortion curve

where $\mathbf{K} = E[||\mathbf{e}_S||^2] + \frac{1}{M}E[||\mathbf{W}^{(1)}||^2]$ and the second equality follows from the independence between S and $W^{(k)}$ and the orthogonal fingerprint modulation with $E[||\mathbf{W}^{(1)}||^2] = E[||\mathbf{W}^{(k)}||^2]$, for all k.

7.2.3 The risk-distortion relationship

From (7.12) and (7.13), we can see that both the distortion and the risk are determined by the coefficients of the temporal filter $\mathbf{a}$. In Figure 7.1, we show the risk–distortion plot using different coefficients. We can see that, for a fixed risk R_1, there are several different coefficients, $\mathbf{a}_1$, $\mathbf{a}_2$, and $\mathbf{a}_3$, which would lead to different amounts of distortions, D_1, D_2, and D_3. A rational attacker will choose the optimal coefficient $\mathbf{a}$ that minimizes the distortion to generate the colluded copy, which leads to the risk distortion curve. Therefore, the attacker's problem can be formulated as

$$\min_{\mathbf{a}} \; \frac{1}{2}D = \frac{1}{2}\mathbf{a}^T\mathbf{K}\mathbf{a}$$

$$s.t. \; R = Q\left(\frac{Q^{-1}(\alpha)||\Lambda\mathbf{a}||_2 - \frac{1}{M}\mathbf{p}^T\mathbf{a}}{||\Lambda\mathbf{a}||_2}\right) \le R_0,$$

$$\mathbf{1}^T\mathbf{a} = 1, \tag{7.14}$$

where the scale factor $\frac{1}{2}$ in the objective function is only for computation convenience.

Obviously, this optimization problem is nonconvex owing to the existence of the quadratic term $||\Lambda\mathbf{a}||_2$ in the denominator of the first constraint. However, because the Gaussian tail function $Q(x)$ is a monotonically decreasing function, we can rewrite the optimization problem as

$$\min_{\mathbf{a}} \frac{1}{2}D = \frac{1}{2}\mathbf{a}^T \mathbf{K} \mathbf{a}$$

$$s.t. \quad \left[Q^{-1}(R_0) - Q^{-1}(\alpha)\right] ||\Lambda\mathbf{a}||_2 + \frac{1}{M}\mathbf{p}^T\mathbf{a} \le 0,$$

$$\mathbf{1}^T\mathbf{a} = 1. \tag{7.15}$$

The optimization problem above is a quadratically constrained quadratic program (QCQP) problem [114]. If $Q^{-1}(R_0) \ge Q^{-1}(\alpha)$ – that is, $R_0 \le \alpha$, the problem is a convex optimization problem. We can find the optimal solution using numerical methods such as the interior point methods [114].

If $Q^{-1}(R_0) < Q^{-1}(\alpha)$, which means $R_0 > \alpha$, the problem is nonconvex. In general, a nonconvex QCQP problem is a nondeterministic polynomial-time (NP)-hard problem [114] and it is very difficult to find the global optimal solution. However, by approximating the concave term with its first-order Taylor expansion, a locally optimal solution can be solved using a constrained concave–convex procedure (CCCP) [115], and the relaxed optimization problem becomes

$$\min_{\mathbf{a}} \frac{1}{2}D = \frac{1}{2}\mathbf{a}^T \mathbf{K} \mathbf{a}$$

$$s.t. \quad \left[Q^{-1}(R_0) - Q^{-1}(\alpha)\right] \frac{\mathbf{a}^T \Lambda^T \Lambda \mathbf{a}}{||\Lambda\mathbf{a}||_2} + \frac{1}{M}\mathbf{p}^T\mathbf{a} \le 0,$$

$$\mathbf{1}^T\mathbf{a} = 1. \tag{7.16}$$

Given an initial $\mathbf{a}^{(0)}$, CCCP computes $\mathbf{a}^{(t+1)}$ from $\mathbf{a}^t$ iteratively, using (7.16). It can be shown that CCCP converges to a locally optimal solution of the original optimization problem (7.14) [116].

According to (7.15) and (7.16), the optimal coefficient $\mathbf{a}$ that minimizes the distortion subject to a predefined risk constraint R_0 can be found using numerical optimization methods. The minimal distortion for the given risk constraint R_0 can then be computed using (7.13). In this way, the optimal risk–distortion relationship for the colluders can be obtained. Now the only question is how to find a good initial $\mathbf{a}^{(0)}$ for the CCCP process to converge to a good local optimum.

7.2.4   Initialization for CCCP

According to (7.15), the reason that we need to use CCCP to find the locally optimal solution is the quadratic term $||\Lambda\mathbf{a}||_2$ in the constraint. When the value of $||\Lambda\mathbf{a}||_2$ is around a constant β, we can relax the optimization problem by approximating $||\Lambda\mathbf{a}||_2$ with β. Then, the relaxed optimization problem becomes

$$\min_{\mathbf{a}} \frac{1}{2}D = \frac{1}{2}\mathbf{a}^T \mathbf{K} \mathbf{a}$$

$$s.t. \quad \frac{1}{M}\mathbf{p}^T\mathbf{a} - \eta \le 0,$$

$$\mathbf{1}^T\mathbf{a} = 1, \tag{7.17}$$

where $\eta = [Q^{-1}(\alpha) - Q^{-1}(R_0)]\beta$.

From (7.17), we can see that the objective function is quadratic and the constraints are linear. The optimization problem is a quadratic problem, which is convex. The optimal solution for the relaxed problem can be found by solving the Karush–Kuhn–Tucker (KKT) conditions [114],

$$
\mathbf{a}^{(0)} = \mathbf{K}^{-1} \begin{bmatrix} \frac{1}{M}\mathbf{p} & 1 \end{bmatrix} \begin{bmatrix} \frac{1}{M^2}\mathbf{p}^T\mathbf{K}^{-1}\mathbf{p} & \frac{1}{M}\mathbf{p}^T\mathbf{K}^{-1}\mathbf{1} \\ \frac{1}{M}\mathbf{1}^T\mathbf{K}^{-1}\mathbf{p} & \mathbf{1}^T\mathbf{K}^{-1}\mathbf{1} \end{bmatrix}^{-1} \begin{bmatrix} \eta \\ 1 \end{bmatrix}. \tag{7.18}
$$

From (7.18), $\mathbf{a}^{(0)}$ is determined by $\eta = [Q^{-1}(\alpha) - Q^{-1}(R_0)]\beta$, which means that the problem to find a good initialization of $\mathbf{a}^{(0)}$ is reduced to finding a good value of β. Furthermore, because (7.18) is derived based on the assumption that $||\Lambda\mathbf{a}||_2$ can be approximated by a constant β, $\mathbf{a}^{(0)}$ is a good initial point if β is approximately $||\Lambda\mathbf{a}^\star||_2$, where $\mathbf{a}^\star$ is the optimal $\mathbf{a}$.

7.3   Strategies with side information

From the analysis in the previous section, we can see that attackers can obtain the risk–distortion curve based on the assumption that the detector uses the fingerprint extracted from the current frame to compute the detection statistics. However, if the detector knows that attackers use a linear filter to further reduce the energy of the embedded fingerprints, the detector will modify the detection statistics to improve the detection performance. Similarly, if attackers are aware that the detector changes the detection statistics, they will change their strategy accordingly. Therefore, with side information, there exist complex dynamics between the attackers and the detector, and the problem can be formulated as a **cat-and-mouse** game, in which the optimal strategy for each player depends on the opponent's strategy. Because the risk–distortion curve is determined by the optimal coefficient $\mathbf{a}^\star$, we consider the optimal coefficient $\mathbf{a}^\star$ as side information to be estimated by both parties.

In this section, we first discuss how the detector and attackers choose their optimal strategies based on the available side information. However, given the fact that the cat-and-mouse game is a game with imperfect information, the detector and attackers are not always able to apply the optimal strategies. Therefore, later in this section, we discuss how colluders choose their collusion parameters under the worst-case scenario.

7.3.1   Optimal strategy for the detector

If the detector knows (estimates) that the attacker uses the linear filter with optimal coefficients $\mathbf{a}_a^\star$ in the intracopy collusion, the detector will modify the detection statistics to improve the detection performance. Assume that the detector uses a linear combination of fingerprints embedded in neighboring frames, $\mathbf{W}^{(k)}\mathbf{a}$, to compute user k's detection statistics; then the detection statistics become

$$
TN_t^{\prime(k)} = \frac{\mathbf{a}^T \mathbf{W}^{(k)T} \left(\mathbf{e}_S + \frac{1}{M}\sum_{i=1}^{M} \mathbf{W}^{(i)} \right) \mathbf{a}_a^\star}{||\mathbf{W}^{(k)}\mathbf{a}||_2}. \tag{7.19}
$$

$TN_t'^{(k)}$ follows Gaussian distribution $\mathcal{N}\left(\frac{1}{M}E\left[\frac{\mathbf{a}^T \mathbf{W}^{(k)T}\mathbf{W}^{(k)}\mathbf{a}_a^\star}{||\mathbf{W}^{(k)}\mathbf{a}||_2}\right], ||\Lambda\mathbf{a}_a^\star||_2\right)$. Therefore, given the false alarm probability $P_{fa} = \alpha$, colluders' risk is

$$
R = Q\left(\frac{Q^{-1}(\alpha)||\Lambda\mathbf{a}_a^\star||_2 - \frac{1}{M}E\left[\frac{\mathbf{a}^T \mathbf{W}^{(k)T}\mathbf{W}^{(k)}\mathbf{a}_a^\star}{||\mathbf{W}^{(k)}\mathbf{a}||_2}\right]}{||\Lambda\mathbf{a}_a^\star||_2}\right). \tag{7.20}
$$

Obviously, a rational detector will choose the optimal coefficient $\mathbf{a}$ to maximize the probability to capture colluders, which is the same as colluders' risk R. According to (7.20), maximizing R is equivalent to maximizing $E\left[\frac{\mathbf{a}^T \mathbf{W}^{(k)T}\mathbf{W}^{(k)}\mathbf{a}_a^\star}{||\mathbf{W}^{(k)}\mathbf{a}||_2}\right]$, which results in new optimal coefficient $\mathbf{a}_d^\star$ as

$$
\mathbf{a}_d^\star = \mathbf{a}_a^\star. \tag{7.21}
$$

Therefore, if the detector knows that colluders use the linear filter $\mathbf{a}_a^\star$ during intracopy collusion, the detector will also use $\mathbf{W}^{(k)}\mathbf{a}_a^\star$ to compute the detection statistics, as this can give the best detection performance.

7.3.2   Optimal strategy for attackers with side information

If colluders know that the detector will use the linear filter $\mathbf{a}_d^\star$ to compute the detection statistics, obviously attackers will try to reduce their risk and use a different linear filter. Let $\mathbf{a}$ be the new coefficient that attackers use, then the detection statistics become

$$
TN_t''^{(k)} = \frac{\mathbf{a}_d^{\star T}\mathbf{W}^{(k)T}\left(\mathbf{e}_S + \frac{1}{M}\sum_{i=1}^M \mathbf{W}^{(i)}\right)\mathbf{a}}{||\mathbf{W}^{(k)}\mathbf{a}_d^\star||_2}. \tag{7.22}
$$

$TN_t''^{(k)}$ follows Gaussian distribution $\mathcal{N}\left(\frac{1}{M}E\left[\frac{\mathbf{a}_d^{\star T}\mathbf{W}^{(k)T}\mathbf{W}^{(k)}\mathbf{a}}{||\mathbf{W}^{(k)}\mathbf{a}_d^\star||_2}\right], ||\Lambda\mathbf{a}||_2\right)$. Therefore, given the false alarm probability $P_{fa} = \alpha$, attackers' risk becomes

$$
R = Q\left(\frac{Q^{-1}(\alpha)||\Lambda\mathbf{a}||_2 - \frac{1}{M}E\left[\frac{\mathbf{a}_d^{\star T}\mathbf{W}^{(k)T}\mathbf{W}^{(k)}\mathbf{a}}{||\mathbf{W}^{(k)}\mathbf{a}_d^\star||_2}\right]}{||\Lambda\mathbf{a}||_2}\right). \tag{7.23}
$$

Surely, a rational attacker will choose the optimal coefficient $\mathbf{a}$ to minimize the risk defined in (7.23). The problem can be formulated as

$$
\min_{\mathbf{a}} \; R = Q\left(\frac{Q^{-1}(\alpha)||\Lambda\mathbf{a}||_2 - \frac{1}{M}E\left[\frac{\mathbf{a}_d^{\star T}\mathbf{W}^{(k)T}\mathbf{W}^{(k)}\mathbf{a}}{||\mathbf{W}^{(k)}\mathbf{a}_d^\star||_2}\right]}{||\Lambda\mathbf{a}||_2}\right),
$$

$$
s.t. \;\; D = \mathbf{a}^T\mathbf{K}\mathbf{a} \le D_0,
$$

$$
\mathbf{1}^T\mathbf{a} = 1, \tag{7.24}
$$

where D_0 is the distortion when $\mathbf{a}_d^\star$ is used for collusion.

Because the Gaussian tail function $Q(x)$ is a monotonically decreasing function, we can rewrite the optimization problem as

$$\min_{\mathbf{a}} \frac{\mathbf{a}_d^{\star T} \mathbf{W}^{(k)T} \mathbf{W}^{(k)} \mathbf{a}}{||\mathbf{W}^{(k)} \mathbf{a}_d^{\star}||_2 \cdot ||\Lambda \mathbf{a}||_2},$$

$$s.t. \quad D = \mathbf{a}^T \mathbf{K} \mathbf{a} \leq D_0,$$

$$\mathbf{1}^T \mathbf{a} = 1, \tag{7.25}$$

which is equivalent to

$$\min_{\mathbf{a}, \xi} \xi,$$

$$s.t. \quad \frac{\mathbf{a}_d^{\star T} \mathbf{W}^{(k)T} \mathbf{W}^{(k)} \mathbf{a}}{||\mathbf{W}^{(k)} \mathbf{a}_d^{\star}||_2} - \xi ||\Lambda \mathbf{a}||_2 \leq 0,$$

$$D = \mathbf{a}^T \mathbf{K} \mathbf{a} \leq D_0,$$

$$\mathbf{1}^T \mathbf{a} = 1. \tag{7.26}$$

Thus, we can find the solution iteratively by solving the following optimization problem:

$$\min_{\mathbf{a}^l} \frac{\mathbf{a}_d^{\star T} \mathbf{W}^{(k)T} \mathbf{W}^{(k)} \mathbf{a}^l}{||\mathbf{W}^{(k)} \mathbf{a}_d^{\star}||_2} - \xi^l ||\Lambda \mathbf{a}^l||_2,$$

$$s.t. \quad D = \mathbf{a}^{lT} \mathbf{K} \mathbf{a}^l \leq D_0,$$

$$\mathbf{1}^T \mathbf{a}^l = 1, \tag{7.27}$$

with $\xi^{(l+1)} = \frac{\mathbf{a}_d^{\star T} \mathbf{W}^{(k)T} \mathbf{W}^{(k)} \mathbf{a}^l}{||\mathbf{W}^{(k)} \mathbf{a}_d^{\star}||_2 \cdot ||\Lambda \mathbf{a}||_2}$, and l being the iteration index. At each iteration, if $\xi^l \leq 0$, the optimization problem in (7.27) is convex, the global optimal solution can be found using a numerical method. However, if $\xi^l > 0$, the optimization problem in (7.27) is nonconvex. Then, we need to use CCCP to find the locally optimal solution.

7.3.3   Min–max strategy for the attacker: worst-case scenario

Based on the previous discussion, we know the attackers' and the detector's optimal strategies with side information – that is, how the detector and attackers should react based on the knowledge of their opponent's strategy. However, in reality, attackers need to choose their strategy first. Then, the detector will choose a strategy to detect the attacker. In this game, the best-case scenario for attackers occurs when the detector uses a fixed strategy that is known to the attacker. In such a best-case scenario, the attackers' optimal strategy can be found by solving (7.27). Meanwhile, the worst-case scenario for attackers occurs when the detector has full knowledge of the attackers' strategy and choose an optimal strategy based on the attackers' strategy. In such a worst-case scenario, the attackers' optimal strategy is the min–max strategy, to minimize the worst-case risk.

7.3.4   Without additive noise

If attackers use a linear filter with coefficients $\mathbf{a}_a$ during intracopy collusion and the detector uses $\mathbf{W}^{(k)}\mathbf{a}_d$ to compute the detection statistics, then the detection statistics become

$$TN_t'''^{(k)} = \frac{\mathbf{a}_d{}^T \mathbf{W}^{(k)T}\left(\mathbf{e}_S + \frac{1}{M}\sum_{i=1}^M \mathbf{W}^{(i)}\right)\mathbf{a}_a}{||\mathbf{W}^{(k)}\mathbf{a}_d||_2}. \tag{7.28}$$

Similar to (7.20) and (7.23), the risk of the attackers being detected becomes

$$R = Q\left(\frac{Q^{-1}(\alpha)||\Lambda\mathbf{a}_a||_2 - \frac{1}{M}E\left[\frac{\mathbf{a}_d{}^T\mathbf{W}^{(k)T}\mathbf{W}^{(k)}\mathbf{a}_a}{||\mathbf{W}^{(k)}\mathbf{a}_d||_2}\right]}{||\Lambda\mathbf{a}_a||_2}\right). \tag{7.29}$$

Obviously, a rational detector will always choose the optimal coefficients $\mathbf{a}_d$ to maximize the chance of capturing colluders. In the worst-case scenario for attackers, the detector has the full knowledge of the attackers' strategy $\mathbf{a}_a$. Therefore, when the detector uses the optimal strategy, the attackers' risk becomes

$$R(\mathbf{a}_d^\star) = \max_{\mathbf{a}_d} Q\left(\frac{Q^{-1}(\alpha)||\Lambda\mathbf{a}_a||_2 - \frac{1}{M}E\left[\frac{\mathbf{a}_d{}^T\mathbf{W}^{(k)T}\mathbf{W}^{(k)}\mathbf{a}_a}{||\mathbf{W}^{(k)}\mathbf{a}_d||_2}\right]}{||\Lambda\mathbf{a}_a||_2}\right). \tag{7.30}$$

Because the attackers know that the detector chooses the optimal strategy based on the strategy chosen by the attackers, rational attackers will choose the optimal coefficients $\mathbf{a}_a$ to minimize their risk in (7.30). Therefore, the problem of finding the optimal $\mathbf{a}_a$ can be formulated as

$$\min_{\mathbf{a}_a} R(\mathbf{a}_d^\star) = \min_{\mathbf{a}_a}\max_{\mathbf{a}_d} Q\left(\frac{Q^{-1}(\alpha)||\Lambda\mathbf{a}_a||_2 - \frac{1}{M}E\left[\frac{\mathbf{a}_d{}^T\mathbf{W}^{(k)T}\mathbf{W}^{(k)}\mathbf{a}_a}{||\mathbf{W}^{(k)}\mathbf{a}_d||_2}\right]}{||\Lambda\mathbf{a}_a||_2}\right),$$

$$s.t. \quad D = \mathbf{a}_a{}^T\mathbf{K}\mathbf{a}_a \le D_0,$$

$$\mathbf{1}^T\mathbf{a}_a = 1. \tag{7.31}$$

Therefore, the attackers' optimal strategy in the worst-case scenario is the min–max strategy.

According to (7.21), we know that the optimal $\mathbf{a}_d$ equals $\mathbf{a}_a$. Let $\mathbf{a}_d = \mathbf{a}_a = \mathbf{a}$; then the optimization problem in (7.31) becomes

$$\min_{\mathbf{a}} \ Q\left(\frac{Q^{-1}(\alpha)||\Lambda\mathbf{a}||_2 - \frac{1}{M}E\left[||\mathbf{W}^{(k)}\mathbf{a}||_2\right]}{||\Lambda\mathbf{a}||_2}\right),$$

$$s.t. \quad D = \mathbf{a}^T\mathbf{K}\mathbf{a} \le D_0,$$

$$\mathbf{1}^T\mathbf{a} = 1, \tag{7.32}$$

As with (7.27), we can find the solution iteratively by solving the following optimization problem:

$$\min_{\mathbf{a}^l} \ ||\mathbf{W}^{(k)}\mathbf{a}^l||_2 - \xi^l||\mathbf{\Lambda}\mathbf{a}^l||_2,$$

$$s.t. \ \ D = \mathbf{a}^{l^T}\mathbf{K}\mathbf{a}^l \le D_0,$$

$$\mathbf{1}^T\mathbf{a}^l = 1, \tag{7.33}$$

with $\xi^{l+1} = \frac{||\mathbf{W}^{(k)}\mathbf{a}^l||_2}{||\mathbf{\Lambda}\mathbf{a}^l||_2}$. Because $\xi^l > 0$, the optimization problem in (7.33) is nonconvex. Therefore, we need to use CCCP to find the locally optimal solution.

7.3.5   Risk reduction using additive white gaussian noise

Previously, we discussed the attackers' min–max strategy when colluders do not introduce AWGN into the colluded copy. In this subsection, we consider the scenario in which the colluded copy is further distorted by AWGN to deter the fingerprint detection performance.

Assume that after the intracopy and intercopy collusion attacks, colluders introduce AWGN to the colluded copy to further reduce their risk of being detected. Then, the colluded frame t becomes

$$V_t = \sum_{k=1}^{M} \frac{1}{M}\left[\sum_{i=-n}^{n} a_i X_{t+i}^{(k)}\right] + \mathbf{n}, \tag{7.34}$$

where $\mathbf{n}$ is AWGN with zero mean and σ_n^2 variance – that is, $\mathbf{n} \sim \mathcal{N}(0, \sigma^2)$.

In the worst case scenario, similar to (7.28), the detection statistics can be computed by

$$TN_t''''^{(k)} = \frac{\mathbf{a}^T\mathbf{W}^{(k)T}\left(\mathbf{e}_S + \frac{1}{M}\sum_{i=1}^{M}\mathbf{W}^{(i)} + \mathbf{n}\right)\mathbf{a}}{||\mathbf{W}^{(k)}\mathbf{a}||_2}. \tag{7.35}$$

Similar to the analysis in the previous subsection, we can find the min–max strategy iteratively by solving the following optimization problem:

$$\min_{\mathbf{a}^l} \ ||\mathbf{W}^{(k)}\mathbf{a}^l||_2^2 - \xi^l(||\mathbf{\Lambda}\mathbf{a}^l||_2^2 + \sigma_n^2),$$

$$s.t. \ \ D = \mathbf{a}^{l^T}\mathbf{K}\mathbf{a}^l \le D_0,$$

$$\mathbf{1}^T\mathbf{a}^l = 1, \tag{7.36}$$

with $\xi^{l+1} = \frac{||\mathbf{W}^{(k)}\mathbf{a}^l||_2^2}{||\mathbf{\Lambda}\mathbf{a}^l||_2^2 + \sigma_n^2}$. Because $\xi^l > 0$, we need to use CCCP to find the locally optimal solution.

7.4   Parameter estimation

From the preceding analysis, the risk–distortion relationship is determined by three parameters $\mathbf{p}$, $\mathbf{W}^{(k)T}\mathbf{W}^{(k)}$, and $\mathbf{K}$. Now, we study how to estimate these three parameters. Attackers have no knowledge of either the original source signal S_t or the fingerprint signal $W_t^{(k)}$. Instead, they have only the fingerprinted signal $X_{t-n}^{(k)}, \ldots, X_{t+n}^{(k)}$. To obtain the risk–distortion relationship, we need to first estimate the parameters $\mathbf{p}$, $\mathbf{W}^{(k)T}\mathbf{W}^{(k)}$, and $\mathbf{K}$ from $X_{t-n}^{(k)}, \ldots, X_{t+n}^{(k)}$.

Let $\mathbf{X}^{(k)} = [X_{t-n}^{(k)}, \ldots, X_{t+n}^{(k)}]$. From (7.1), we can see that the fingerprinted signal is the sum of the original signal and the fingerprint, based on which the difference between $\mathbf{X}^{(k)}$ and its smooth version among all the colluders can be expressed as

$$\mathbf{X}^{(k)} - \frac{1}{M}\sum_{i=1}^{M}\mathbf{X}^{(i)} = \frac{M-1}{M}\mathbf{W}^{(k)} - \frac{1}{M}\sum_{i=1,i\neq k}^{M}\mathbf{W}^{(i)}. \tag{7.37}$$

This means that we can use the fingerprinted signal $\mathbf{X}^{(k)}$ to compute the correlation matrix of the fingerprint signal $\mathbf{W}^{(k)}$ by

$$E[\mathbf{W}^{(k)T}\mathbf{W}^{(k)}] = \frac{1}{M-1}\sum_{j=1}^{M} E\left[\mathbf{X}^{(j)} - \frac{1}{M}\sum_{i=1}^{M}\mathbf{X}^{(i)}\right]^T \left[\mathbf{X}^{(j)} - \frac{1}{M}\sum_{i=1}^{M}\mathbf{X}^{(i)}\right]. \tag{7.38}$$

The parameter $\mathbf{p}$ can be estimated using

$$\mathbf{p}(i) = \frac{E[\mathbf{W}^{(k)T}\mathbf{W}^{(k)}]_{i,n}}{\sqrt{\frac{1}{2n+1}\sum_{j=1}^{2n+1} E[\mathbf{W}^{(k)T}\mathbf{W}^{(k)}]_{j,j}}}, \tag{7.39}$$

where $E[\mathbf{W}^{(k)T}\mathbf{W}^{(k)}]_{i,j}$ is the ith row and jth column element of $E[\mathbf{W}^{(k)T}\mathbf{W}^{(k)}]$.

To estimate the parameter $\mathbf{K}$, we need to first estimate $E[||\mathbf{e}_S||^2]$ using

$$\begin{aligned} E[||\mathbf{e}_S||^2]_{i,j} &= E[S_{t-n+i} - S_t]^T[S_{t-n+j} - S_t] \\ &= E[X_{t-n+i}^{(k)T}X_{t-n+j}^{(k)}] - E[X_{t-n+j}^{(k)T}X_t^{(k)}] \\ &\quad - E[X_{t-n+i}^{(k)T}X_{t-n+j}^{(k)}] + E[X_t^{(k)T}X_t^{(k)}] - E[\mathbf{W}^{(k)T}\mathbf{W}^{(k)}]_{i,j} \\ &\quad + E[\mathbf{W}^{(k)T}\mathbf{W}^{(k)}]_{i,n} + E[\mathbf{W}^{(k)T}\mathbf{W}^{(k)}]_{j,n} - E[\mathbf{W}^{(k)T}\mathbf{W}^{(k)}]_{n,n}. \end{aligned} \tag{7.40}$$

7.5   Simulation results

We simulate and evaluate the risk–distortion model discussed in the previous sections on real videos. Two video sequences (Akiyo and Foreman) in quarter common intermediate format (QCIF) are tested. We use the human visual model-based spread spectrum embedding [75], and embed the fingerprint in the DCT domain. We generate independent vectors of length $N = 176 \times 144$ from Gaussian distribution $N(0, 1)$, and then apply Gram–Schmidt orthogonalization to produce strictly orthogonal fingerprints. Then, we scale the fingerprint to let the variance be σ_w^2, followed by the inverse Gram–Schmidt orthogonalization to ensure that the fingerprint of each user satisfies (7.2) strictly with $\gamma = 0.5$. We assume that the collusion attacks are also in the DCT domain. On the detector's side, a nonblind detection is performed in which the host signal is first removed from the colluded copy, and the detector uses correlation-based detection statistics to identify attackers. In all the following simulations, the parameter n is set to be 5, which means that the ten temporally adjacent frames are involved in the intracopy attack process for each attacker. The false-alarm probability is set to be $\alpha = 10^{-4}$.

We first evaluate the accuracy of the risk–distortion model by comparing it with the baseline curve, which is the experimental risk–distortion curve. Here, the experimental risk is defined as the average positive detection probability by averaging over 400 runs of simulation. For each video sequence, the number of attackers $M = 2$ and $M = 4$ are tested. As shown in Figure 7.2, the risk–distortion curve derived by the model coincides with the baseline curve with a small mismatch for both sequences, which demonstrates the effectiveness of the risk–distortion model. The mismatch comes mainly from the Gaussian model error for the residue and the parameter estimation error. In the rest of this chapter, we denote the risk-distortion curve obtained by our model as the *absolute risk–distortion curve*.

The convergence performances of the CCCP process are shown in Figure 7.3. From Figure 7.3, we can see that with CCCP, for any fixed risk constraint, the distortion converges in a few iterations (fewer than 8 in the examples). Because of page limitations, we show only the cases in which risk is fixed at 0.05 and 0.2. Similar behaviors are observed for different risk constraints.

We then study the risk–distortion curve that results when the side information is available, which we denote as the *relative risk–distortion curve*. In such a case, the optimal strategy for the attacker or detector lies on the opponent's strategy. Based on the action that the opponent took, the attackers or detector can choose the best response using (7.21) or (7.27). In Figure 7.4, we show the result of the absolute risk–distortion curve and relative risk–distortion curve. We start with the absolute risk–distortion curve, which is obtained using (7.14). Then, if the detector has the perfect knowledge of the attackers' strategy, the detector chooses the optimal strategy based on the side information. The resulting risk–distortion curve is denoted as *relative risk–distortion curve stage 1*. On the other hand, if the attackers know that the detector uses the side information of the attacker in previous stage, they will change their optimal strategy accordingly. The resulting risk distortion curve is denoted as *relative risk–distortion curve stage 2*. We repeat these detection and attack processes until stage 5. As shown in Figure 7.4, when the detector has the perfect side information of the attackers' strategy, the risk of the attackers to be detected increases and the risk–distortion curve moves up in the left arrow's direction. On the other hand, if the attackers have the perfect side information of the detector's strategy, the risk of the attackers to be detected decreases and the risk–distortion curve moves down in the right arrow's direction. This phenomenon shows the importance of the side information. The one who has the perfect side information about the opponent will lead the game and pull the risk–distortion curve in a beneficial direction.

Moreover, from Figure 7.4, we can see that when the distortion is larger than 1.5, the *relative risk–distortion curve stage 5* curve increases as the distortion increases. This

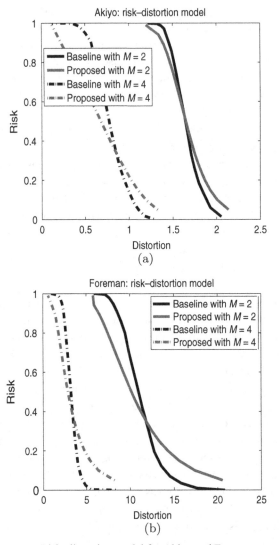

Fig. 7.2 Risk–distortion model for Akiyo and Foreman sequences: (a) Akiyo; (b) Foreman

phenomenon is partly because only the locally optimal solution is found using CCCP when the optimization problem in (7.27) is nonconvex.

In reality, the attackers need to choose their strategy first. In case of a "naive" detector with a fixed strategy, if the attackers know the perfect side information of the detector, they can choose their optimal strategy based on the side information. On the other hand, if the detector is a powerful detector that can always estimate the attackers' strategy, the best strategy for the attackers is to minimize the risk of the worst-case scenario – that is, the min–max strategy. In Figure 7.5, we show the risk–distortion curve with min–max strategy. We can see that although the risk–distortion curve with min–max strategy achieves the lowest risk among all the cases in which the detector has the perfect

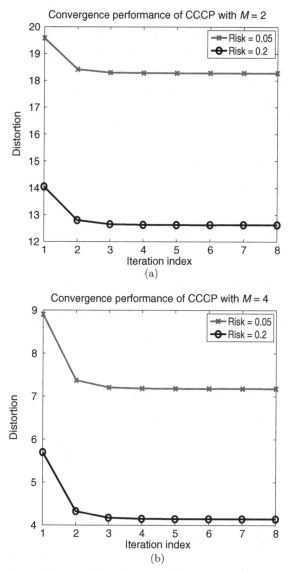

Fig. 7.3 Convergence performance of CCCP: (a) $M = 2$; (b) $M = 4$

side information, there is a big risk gap compared with the *absolute risk–distortion curve*.

In Figure 7.6, we show the risk–distortion curves with min–max strategy and AWGN. We can see that as the noise variance increases, the risk–distortion curve moves along the red arrow direction. This is because when the noise variance increases, the distortion increases but the risk decreases. We can also see that when the noise variance is equal to 1.5, the risk–distortion curve with min–max strategy meets the *absolute risk–distortion curve* for all distortions larger than 2.2. Therefore, with a proper noise variance, we can reach the *absolute risk–distortion curve* even with the min–max strategy.

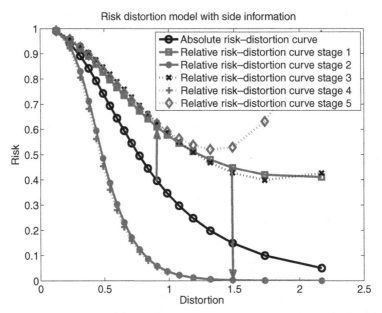

Fig. 7.4 The *absolute risk–distortion curve* and the *relative risk–distortion curve*

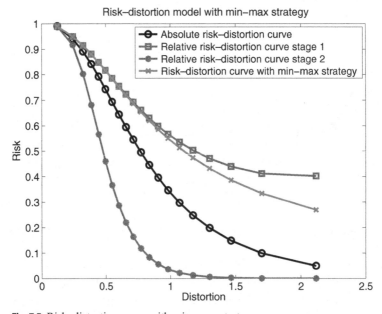

Fig. 7.5 Risk–distortion curve with min–max strategy

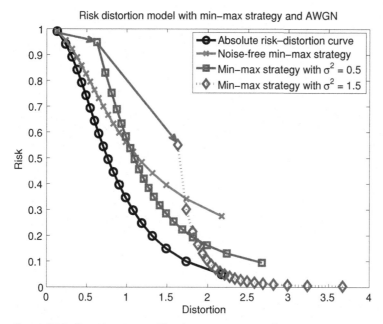

Fig. 7.6 Risk-distortion curve with min–max strategy and AWGN

7.6   Chapter summary and bibliographical notes

Different types of social networks may have different types of side information that can be employed by the users, which can change the user dynamics. In this chapter, we illustrated how video fingerprinting in social networks is influenced by the knowledge of collusion parameters. We first provided a theoretical analysis on the risk–distortion relationship for the linear video collusion attack with Gaussian fingerprint, and then analyzed how the colluders and fingerprint detector change their optimal strategies when side information is available. We conducted several experiments on real video sequences to verify our analysis. From the experimental results, we could see that if the attackers have the side information on the detector's strategy, they can choose the corresponding optimal strategy to destroy the fingerprint with a small distortion. However, if the detector is so powerful that it can always estimate the collusion parameters as side information, the best strategy for the attacker is the min–max strategy. Moreover, we showed that the attackers can further reduce the risk of being detected by introducing AWGN, with a cost of larger distortion.

In the literature, little effort has been made to explicitly study the relationship between risk, such as the probability of the colluders being detected, and the distortion of the colluded signal. Interested readers can refer to references [65–67,117,118] for the analytical study on the performance of Gaussian fingerprints under various collusion attacks and references [110,111] for robust video fingerprinting schemes.

Part III

Fairness and cooperation stimulation

8 Game-theoretic modeling of colluder social networks

As shown in Chapters 5 and 6, cooperation enables users in a social network to access extra resources from others and thus to receive higher payoffs. Meanwhile, each user also contributes his or her own resources to help others. However, because the nature of participation nature in many media-sharing social networks is often voluntary and unregulated, users' full cooperation cannot be guaranteed, and a critical issue to be resolved first is to analyze when users will cooperate with each other and design cooperation strategies. In this chapter, we use colluder social networks in multimedia fingerprinting as an example, and analyze when users will collaborate with each other and how they reach agreements.

For a colluder in multimedia fingerprinting systems, the first issue to address is to decide whether he or she would like to participate in collusion and with whom he or she would like to collude. When colluders' goal is to minimize their probability of being detected, a collusion attack with more attackers reduces the energy of each contributing fingerprint by a larger ratio and, therefore, each colluder has a smaller chance of being caught. Thus, to minimize the risk, colluders are always willing to cooperate with one another because it reduces all colluders' risk, and a colluder should find as many fellow attackers as possible.

Nevertheless, colluding with more attackers also means sharing with more people the reward from illegal usage of multimedia and, therefore, colluders may not always want to cooperate. Furthermore, to accommodate heterogenous networks and diverse devices with different capability, scalable video coding is widely adopted, which encodes the video stream into different layers. In such a scenario, some users may subscribe to higher-resolution copies of the video, and some users may subscribe only to lower-resolution copies. Hence, when colluders receive copies of different resolutions, an attacker also needs to decide with whom to collude. Should the attacker collude with people who have the high resolution copy, or is it better to cooperate with those who have the base layer only? It is of ample importance to investigate the conditions under which attackers cooperate and to study how a colluder selects his or her fellow attackers to form a coalition. Furthermore, before a collusion relationship can be established, an agreement must be reached regarding how to distribute risk and reward. Nevertheless, each colluder prefers the collusion that favors his or her own payoff the most with the lowest risk and the highest reward, and different colluders have different preferences of strategies. To resolve this conflict, a critical issue for colluders is to decide how to fairly distribute the risk and the reward. Therefore, understanding how colluders

Table 8.1 List of symbols used in this chapter

Set of frame indices encoded in base layer	F_b
Set of frame indices encoded in enhancement layer	F_e
Set of frame indices of user $u^{(i)}$'s copy	$F^{(i)}$
Normalized temporal resolution of $u^{(i)}$ ($|F^{(i)}|/(|F_b|+|F_e|)$)	$f^{(i)}$
Normalized temporal resolution of a low-resolution copy ($|F_b|/(|F_b|+|F_e|)$)	f_b
Lengths of the fingerprints embedded in the base layer	N_b
Lengths of the fingerprints embedded in the enhancement layer	N_e
Set of colluders who receive a low-resolution copy	SC^b
Set of colluders who receive a high-resolution copy	SC^{be}
Total number of colluders ($K^b + K^{be}$)	K
Set of frame indices of the colluded copy	F_c
Normalized resolution of the colluded copy	f_c

negotiate with one another to achieve fairness of the attack is a must when analyzing colluder behaviors. Such analysis helps us have a better understanding of multiuser collusion, and offers important guidelines for better design of collusion-resistant multimedia fingerprinting.

In this chapter, we first investigate the conditions under which colluders will cooperate with one another, study how they choose the collusion parameters to ensure that all colluders increase their payoffs, and analyze how colluders form a coalition to maximize their payoffs. We then consider different definitions of fairness and investigate how colluders share risk and reward "fairly." One specific property of multimedia data is that their market value is very time-sensitive. For instance, if the pirated version of a movie is available right after it is first released, it would have a much higher market price and, therefore, all colluders have the incentive to mount collusion as soon as possible. In this chapter, we also address the time-sensitivity of the received reward in multiuser collusion, and investigate the time-sensitive bargaining among colluders.

8.1   Multiuser collusion game

In this section, we consider the scalable fingerprinting system in Chapter 2, in which the video stream is divided into two layers: the base layer and the enhancement layer. Colluders follow the two-stage collusion model in Section 2.3.2, and select the collusion parameter β to achieve fairness of collusion. Let SC^b be the set of colluders who subscribed to the low-resolution copies and SC^{be} be the set of colluders who possess the high-resolution copies with both layers. Here, we consider the scenario in which colluders who receive fingerprinted copies of the same resolution agree to have the same probability of being detected and receive the same reward. In addition, we assume that colluders generate a high-resolution copy whenever possible – that is, $f_c = 1$ when $K^{be} \geq 1$. Table 8.1 lists the symbols that we use in this chapter.

8.1.1   Game model

We model the colluder dynamics as a bargaining process, in which colluders negotiate with one another on the fair distribution of the risk and the reward.

- **Players:** Because colluders who receive fingerprinted copies of the same resolution agree to have the same risk and same reward, the game can be simplified to a two-user game. Colluders who receive the low-resolution copies act as a single player, and colluders who have the high-resolution copies act as a single player during the bargaining process.
- **Strategies:** Based on the two-stage collusion model shown in Section 2.3.2, the collusion parameter β controls the risk for both SC^b and SC^{be} and, thus, their utilities.
- **Utility function:** For colluder $\mathbf{u}^{(i)}$, the payoff function $\pi^{(i)}$ should contain two terms: colluder i's loss of being detected and the colluder's reward received from illegal usage of multimedia content. In our game-theoretic framework, all colluders in SC^b have the same payoff π^b, and all colluders in SC^{be} have the same payoff π^{be}.

8.1.2   Utility function definition

In our game, for colluder i, we use the expected payoff as the colluder's utility function; the utility function contains two terms, colluder i's loss from being detected and the reward received from illegal usage of multimedia content.

As discussed before, the reward depends not only on the colluded copy's quality, but also on the time at which the colluded copy is released. In addition, in some scenarios, the market value of a low-quality copy decreases much faster than that of the higher-resolution copy. For instance, when a movie is available only in theaters, people might be interested in the low-resolution colluded copy. Once its DVD copies are for sale on the market, people might still be interested in the high-resolution pirated copy because of its low price, but they may have little incentive to purchase a low-resolution pirated copy. Therefore, to receive higher rewards, colluders should reach an agreement on the fair distribution of the risk and the reward as soon as possible, and have the colluded copy ready at the earliest possible time.

Here we use the exponential-decay model for the market value of a colluded copy, and the reward that player i receives is discounted by a constant factor $\delta_g \in (0, 1]$ for each round that colluders fail to reach an agreement. Here, the subscript g is the subgroup index, and is either b or be. That is, the discount factor is δ_b when colluder $i \in SC^b$ and is δ_{be} when $i \in SC^{be}$. In this chapter, we consider the scenarios in which the market value of the high-resolution copy decays more slowly than that of the low-resolution one, and assume that $\delta_{be} \geq \delta_b$. When $\delta_b = \delta_{be} = 1$, it corresponds to the scenario in which the market value of the colluded copy is not time-sensitive and colluders can take an infinite number of rounds to negotiate.

Therefore, if colluders reach an agreement on the collusion parameter β at the kth round, colluder i's utility function is

$$\pi_k^{(i)} = -P_d^{(i)} L^{(i)} + \left(1 - P_d^{(i)}\right) (\delta_g)^{k-1} R^{(i)}. \tag{8.1}$$

Here, the subscript k is the round index and the superscript i is the colluder index. In (8.1), $P_d^{(i)}$ is colluder i's probability of being detected, $L^{(i)}$ is colluder i's loss if he or she is detected, and $R^{(i)}$ is the reward that i receives if he or she is not caught by the digital rights enforcer and if colluders reach an agreement at the first round. Here, we consider a simple scenario in which all colluders in the same subgroup has the same loss term – that is, $L^{(i)} = L^b$ for all $i \in SC^b$ and $L^{(j)} = L^{be}$ for all $j \in SC^{be}$. In the following sections, we analyze $P_d^{(i)}$ and $R^{(i)}$ in details.

8.1.2.1   Probability of being detected $P_d^{(i)}$

Let N_b and N_e be the lengths of the fingerprints embedded in the base layer and the enhancement layer, respectively. If all colluders receive fingerprinted copies of the same resolution and they agree to equally share the risk, with orthogonal fingerprint modulation, if the detection noise is i.i.d. Gaussian with zero mean and variance σ_n^2, then a guilty colluder i's probability of being detected is

$$P_d^{sr} = Q\left(\frac{h - \sqrt{N_c}\sigma_w/K}{\sigma_n}\right), \tag{8.2}$$

and the probability of accusing an innocent user is $P_{fa}^{(i)} = Q(h/\sigma_n)$. Here, $Q(\cdot)$ is the Gaussian tail function, h is a predetermined threshold, N_c is the length of the fingerprint $\mathbf{Y}$ extracted from the colluded copy, and σ_w^2 is the variance of i's fingerprint $\mathbf{W}^{(i)}$. $N_c = N_b$ if the colluded copy contains the base layer only, and $N_c = N_b + N_e$ when the colluded copy is of high resolution with both layers. The superscript sr denotes that all fingerprinted copies in collusion have the same resolution.

Because the scalable video coding is applied, colluders may receive fingerprinted copies of different resolutions – some colluders receive copies with the base layer only, and some colluders receive both the base layer and the enhancement layer. In such a scenario, from the analysis in Chapter 5, with orthogonal fingerprint modulation, if the colluders colluders in the same subgroup (receive the same-quality copies) agree to share the same risk and the detection noise is i.i.d. Gaussian $\mathcal{N}\left(0, \sigma_n^2\right)$, for colluder i who receives a low-resolution fingerprinted copy, the chance of being captured is

$$P_{d,c}^b = Q\left(\frac{h - \beta\sqrt{N_b}\sigma_w/K}{\sigma_n}\right), \tag{8.3}$$

where β is the collusion parameter and the superscript b indicates that colluder i is the subgroup SC^b. If colluder i is innocent, his or her chance of being falsely accused is $P_{fa}^{(i)} = Q\left(h/\sigma_n\right)$.

For user i who receives both layers from the content owner, the self-probing detector in Chapter 6 has approximately the same performance as the optimum fingerprint detector with perfect information about the statistical means of the detection statistics. Therefore, during collusion, attackers should consider the worst-case scenario and assume that the fingerprint detector can always select the optimum detection statistic with the largest mean.

For colluder i, following the analysis in Chapter 6, the probability of being detected is

$$P_d^{(i)} = Q\left(\frac{h - \mu_{be}}{\sigma_n}\right),$$

where $\mu_{be} \triangleq \max\{\mu_{be}^b, \mu_{be}^e, \mu_{be}^c\}$ for $i \in SC^{be}$,

where $\mu_{be}^b = \dfrac{(1-\beta)\sqrt{N_b}}{K^{be}}\sigma_w$, $\mu_{be}^e = \dfrac{\sqrt{N_e}}{K^{be}}\sigma_w$,

and $\mu_{be}^c = \dfrac{(1-\beta)N_b + N_e}{K^{be}\sqrt{N_b + N_e}}\sigma_w$. (8.4)

To determine colluder i's probability of being detected, we need to first analyze the relationship between β and μ_{be}. Following the same analysis as in Section 6.3.2, there are three possibilities.

- **Case 1, $\mu_{be} = \mu_{be}^b$:** This happens if and only if $\mu_{be}^b \geq \mu_{be}^e$, and $\mu_{be}^b \geq \mu_{be}^c$, that is,

$$(1-\beta) \geq \max\left\{\frac{\sqrt{N_e}}{\sqrt{N_b}}, \frac{N_e}{\sqrt{N_b}(\sqrt{N_b + N_e} - \sqrt{N_b})}\right\} \quad (8.5)$$

Note that $\sqrt{N_b} + \sqrt{N_e} \geq \sqrt{N_b + N_e}$; that is, $\frac{\sqrt{N_e}}{\sqrt{N_b}} \leq \frac{N_e}{\sqrt{N_b}(\sqrt{N_b + N_e} - \sqrt{N_b})}$. Thus, $\mu_{be} = \mu_{be}^b$ if and only if

$$0 \leq \beta \leq 1 - \frac{N_e}{\sqrt{N_b}(\sqrt{N_b + N_e} - \sqrt{N_b})}. \quad (8.6)$$

However, note that

$$1 - \frac{N_e}{\sqrt{N_b}(\sqrt{N_b + N_e} - \sqrt{N_b})} = \frac{\sqrt{N_b}\sqrt{N_b + N_e} - N_b - N_e}{\sqrt{N_b}(\sqrt{N_b + N_e} - \sqrt{N_b})}$$

$$= \frac{\sqrt{N_b + N_e}(\sqrt{N_e} - \sqrt{N_b + N_e})}{\sqrt{N_b}(\sqrt{N_b + N_e} - \sqrt{N_b})} < 0. \quad (8.7)$$

Hence, for all $N_b > 0$, μ_{be}^b cannot be the largest among the three, μ_{be}^b, μ_{be}^e, and μ_{be}^c, and scenario 1 can never happen.

- **Case 2, $\mu_{be} = \mu_{be}^e$:** This scenario happens if and only if $\mu_{be}^e \geq \mu_{be}^b$ and $\mu_{be}^e \geq \mu_{be}^c$. That is,

$$(1-\beta) \leq \min\left\{\frac{\sqrt{N_e}}{\sqrt{N_b}}, \frac{\sqrt{N_e}(\sqrt{N_b + N_e} - \sqrt{N_e})}{N_b}\right\}. \quad (8.8)$$

Using the same analysis as in (8.6), the necessary and sufficient condition for case 2 is:

$$\mu_{be} = \mu_{be}^e \Leftrightarrow \beta^+ \triangleq 1 - \frac{\sqrt{N_e}(\sqrt{N_b + N_e} - \sqrt{N_e})}{N_b} \leq \beta \leq 1. \quad (8.9)$$

- **Case 3,** $\mu_{be} = \mu_{be}^c$: Using the same analysis as above, the necessary and sufficient condition for case 3 is

$$\mu_{be} = \mu_{be}^c \Leftrightarrow 0 \leq \beta \leq \beta^+ = 1 - \frac{\sqrt{N_e}(\sqrt{N_b + N_e} - \sqrt{N_e})}{N_b}. \tag{8.10}$$

To summarize, if colluder i receives a high-resolution copy and participates in collusion, his or her chance of being detected is

$$P_{d,c}^{be} = \begin{cases} Q\left(\frac{h}{\sigma_n} - \frac{(1-\beta)N_b + N_e}{K^{be}\sqrt{N_b + N_e}} \cdot \frac{\sigma_w}{\sigma_n}\right), & \text{if } \beta \leq \beta^+, \\ Q\left(\frac{h}{\sigma_n} - \frac{\sqrt{N_e}}{K^{be}} \cdot \frac{\sigma_w}{\sigma_n}\right), & \text{if } \beta > \beta^+. \end{cases} \tag{8.11}$$

Here, the superscript be indicates that colluder i is in subgroup SC^{be}. If user i is innocent, then the probability of falsely accusing him or her is $P_{fa}^{(i)} = Q(h/\sigma_n)$.

8.1.2.2   Reward $R^{(i)}$

In this chapter, we consider the scenario in which colluders receive more reward from the illegal usage of multimedia when the colluded copy has higher resolution. For instance, the pirated video with DVD quality would have higher value than the video with VCR quality. With temporal scalable video coding, we use the frame rate to quantify the video quality. Therefore, the reward function $R^{(i)}$ is an increasing function of f_c, the normalized temporal resolution of the colluded copy. In addition, when colluders receive fingerprinted copies of different resolutions, we consider the scenario in which colluders distribute the reward based on the resolution of each contributing copy, and an attacker receives more reward if he or she contributes a copy of higher resolution. This is because, if colluders in SC^{be} do not participate in collusion, the quality of the colluded copy and the total received reward will be lower. Thus, we let $R^{(i)}$ be an increasing function of $f^{(i)}$, the normalized resolution of the fingerprinted copy from colluder i. Furthermore, some colluders may wish to receive a higher reward at a cost of higher risk, and we also let $R^{(i)}$ be a nondecreasing function of $P_d^{(i)}$.

Based on this discussion, the reward function we use is

$$R^{(i)} = \frac{\left(f^{(i)}\right)^\gamma D\left(P_d^{(i)}\right)}{\sum_{j=1}^K \left(f^{(j)}\right)^\gamma D\left(P_d^{(j)}\right)} f_c \theta. \tag{8.12}$$

Here, θ is a parameter to address the tradeoff between the risk that a colluder takes and the reward that he or she receives, and it has a smaller value when colluders place more emphasis on risk minimization. $\theta = 0$ corresponds to the scenario in which colluders' only goal is to minimize their risk, which has been well studied in the literature. In (8.12), $\gamma \geq 0$ is a parameter that colluders use to adjust how they distribute the reward based on the resolution of each contributing copy. For example, if $\gamma = 0$, then the reward is distributed equally among colluders with the same quality copies. For colluder $i \in SC^{be}$ who contributes a high-resolution copy, $R^{(i)}$ is an increasing function of γ and colluder i receives more reward when γ takes a larger value. $D\left(P_d^{(i)}\right)$ is a nondecreasing function of $P_d^{(i)}$ and allows colluders who take a higher risk to receiver more reward from collusion. The denominator in (8.12) is the normalization term.

In (8.12), when all colluders receive fingerprinted copies of the same resolution, the reward function is $R^{(i)} = f_c\theta/K$ – that is, colluders equally distribute the risk as well as the reward among themselves. When $D(P_d^{(i)}) = 1$ for all colluders, the total reward is distributed among colluders based only on the resolution of the fingerprinted copies they contribute, but not on their risk. In such a scenario, if the colluded copy has high resolution and $f_c = 1$, colluder i's reward can be simplified to

$$R^{(i)} = \frac{(f^{(i)})^\gamma}{K^b(f_b)^\gamma + K^{be}}\theta. \tag{8.13}$$

To summarize, we model the colluder dynamics as a bargaining process, wherein colluders in SC^b and SC^{be} negotiate on the selection of the collusion parameter β. For colluder i, the utility function is defined in (8.1), and the reward function that we use is in (8.12).

8.2 Feasible and Pareto optimal collusion

In the following, we first study when colluders will cooperate with each other; this section analyzes the feasible region and the Pareto optimal set. In this section, we consider the scenario in which the market value of the colluded copy is not time-sensitive and $\delta_b = \delta_{be} = 1$, and we let $D(P_d^{(i)})$ in (8.1) equal to 1 for all colluders. In this simplified colluder game, when colluders receive fingerprinted copies of the same resolution, all colluders have the same payoff,

$$\pi^{sr} = -P_d^{sr}L + \left(1 - P_d^{sr}\right)\frac{f_c\theta}{K}, \tag{8.14}$$

where P_d^{sr} is in (8.2). In (8.14), $L = L^b$ when $f^c = f_b$ and the colluded copy has low resolution, and $L = L^{be}$ when $f_c = 1$. If colluders receive copies of different resolutions, colluder i's utility function is

$$\pi^g = -P_{d,c}^g L^g + \left(1 - P_{d,c}^g\right)\frac{(f^g)^\gamma}{K^b(f_b)^\gamma + K^{be}}\theta, \tag{8.15}$$

where $P_{d,c}^g$ is in (8.3) if $i \in SC^b$ and $P_{d,c}^g$ is in (8.11) if $i \in SC^{be}$. Here, the superscript g is the colluder subgroup index, and it is either b if $i \in SC^b$ or be if $i \in SC^{be}$.

8.2.1 Feasible set

Given an N-person general-sum game, there is a certain subset $\mathbb{S}$ of $\mathbb{R}^N$, called the *feasible set*. It is feasible in the sense that, given any $(\pi_1, \pi_2, ..., \pi_N) \in S$, it is possible for the N players to act together and obtain the utilities $\pi_1, \pi_2, ..., \pi_N$, respectively. From the preceding analysis, for a given β, we can calculate the payoffs $\pi^{(i)}$ as in (8.15) for all colluders. From the definition of the payoff function, colluders who receive fingerprinted copies of the same quality have the same payoff, and π^b and π^{be} are the

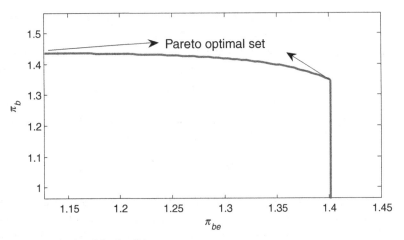

Fig. 8.1 Example of the feasible set

payoffs for colluders in SC^b and SC^{be}, respectively. Figure 8.1 plots π^b versus π^{be}, and the feasible set is the solid line. In Figure 8.1, the straight line segment corresponds to scenario 2, in which $\mu_{be} = \mu_{be}^e = \frac{\sqrt{N_e}}{K^{be}}$ and is independent of β. Therefore, when $\beta > \beta^+$, π^{be} remains the same while π^b keeps decreasing as β increases. Similarly, the curve segment in Figure 8.1 corresponds to scenario 3, with $0 \le \beta \le \beta^+$ and $\mu_{be} = \mu_{be}^c$. In this scenario, π^{be} in an increasing function of β, and π^b decreases as β increases.

8.2.2   Pareto optimality

Among all possible solutions in the feasible set, attackers are especially interested in those in the Pareto optimal set $\mathbb{S}^+ \subseteq \mathbb{S}$. A solution is Pareto optimal if no one can further increase his or her utility without decreasing others'. Players would always like to settle at a Pareto optimal outcome, because if colluders select a point that is not Pareto optimal, there exists another solution for which at least one player can have a larger payoff without hurting other players. Therefore, the player who can achieve a higher payoff without hurting others' has the incentive to push other players to deviate from the non–Pareto optimal solution. In addition, the other rational players will agree with this player, because their own utilities are not influenced. Pareto optimal solutions are not unique in most cases. In this subsection, we investigate the Pareto optimal solutions in the colluder social network and analyzes the necessary and sufficient conditions for a point to be Pareto optimal.

- *Necessary condition:* For a point in the feasible set to be Pareto optimal, decreasing μ_b and increasing π^b must result in a larger μ_{be} and a smaller π^{be}. Note that from (8.4), μ_b is an increasing function of β. Thus, if a point is a Pareto optimal point, μ_{be} must be a decreasing function of β, which happens only when $\mu_{be} = \mu_{be}^c$. Consequently, if a point is Pareto optimal, β must satisfy (8.10), and (8.10) is the necessary condition for Pareto optimality.

- **Sufficient condition:** If $\mu_{be} = \mu_{be}^c$, to increase the payoff of the colluders in SC^{be}, colluders must select a larger β and decrease μ_{be}. However, a larger β implies a larger μ_b, and thus a smaller π^b. Consequently, those points that satisfy (8.4) are Pareto optimal points, and (8.4) is the sufficient condition for Pareto optimality.

To conclude, the multiuser collusion is Pareto optimal if and only if $\mu_{be} = \mu_{be}^c$ and (8.4) is satisfied, which corresponds to the curve segment in Figure 8.1. Mathematically speaking, the Pareto optimal set corresponds to the solutions in which attackers select $0 \leq \beta \leq \beta^+$; that is, $\mathbb{S}^+ = \{(\pi^b, \pi^{be}) \in \mathbb{S} : 0 \leq \beta \leq \beta^+\}$.

8.2.3   Feasible collusion

Attackers will collude with one another if and only if collusion helps increase their utilities, and they are interested in solutions in the Pareto optimal set $\mathbb{S}^+$ that give them higher payoffs when compared with the scenario in which they do not cooperate with each other.

- First, if attacker i does not participate in collusion and does not use multimedia content illegally, his or her payoff is zero. Thus, attacker i colludes with other attackers only if he or she receives positive payoff from collusion, and colluders are interested only in solutions in $\mathbb{S}^+$ where $\pi^b \geq 0$ and $\pi^{be} \geq 0$.
- Furthermore, one possible outcome of the bargaining between SC^b and SC^{be} is that they do not reach an agreement. In such a scenario, attackers will collude only with their fellow attackers in the same subgroup, and SC^b and SC^{be} do not cooperate with each other. Let π_{nc}^b denote the utility of a colluder in SC^b if he or she colludes with attackers in SC^b only, but not those in SC^{be}; and similarly, π_{nc}^{be} is the utility of an attacker in SC^{be} if he or she colludes with attackers in SC^{be} only, but not those on SC^b. Here, the subscript nc means no cooperation. Therefore, SC^b and SC^{be} will collude with each other only if the two-stage collusion increases both players' payoffs, and they look for solutions in $\mathbb{S}^+$ where $\pi^b \geq \pi_{nc}^b$ and $\pi^{be} \geq \pi_{nc}^{be}$.

This analysis helps colluders further narrow down the feasible set to

$$\mathbb{S}_p = \{(\pi^b, \pi^{be}) \in \mathbb{S} : 0 \leq \beta \leq \beta^+, \pi^b \geq \underline{\pi}^b \triangleq \max(\pi_{nc}^b, 0),$$

$$\pi^{be} \geq \underline{\pi}^{be} \triangleq \max(\pi_{nc}^{be}, 0)\}. \tag{8.16}$$

8.3   When to collude

In this section, we analyze *when* attackers will collude with other attackers and discuss the minimum and the maximum numbers of colluders that make the collusion attack feasible and Pareto optimal. We also provide the optimal size of the social network that maximizes every user's utility. In this section, we consider the scenario in which $L^{(i)} = 1$

for all $i \in SC$ and $D(\cdot) = 1$. The analysis for other scenarios is similar, and the same methodology can be used.

8.3.1   Single-resolution collusion

We first study how colluders form a coalition when all of them receive copies of the same quality.

8.3.1.1   Colluders' utility functions

As an example and without loss of generality, we assume that all attackers receive high-resolution copies with both layers, and they generate a colluded copy of high resolution; that is, $K = K^{be}$ and $f_c = 1$. The analysis is similar for the scenario in which all fingerprinted copies have the base layer only and thus, that scenario omitted. In such a scenario, because all copies have the same resolution, there is no bargaining in collusion, and attackers simply average all copies that they have with equal weights. From Section 8.1, colluder i's utility function is

$$\pi^{(i)} = -P_d^{(i)} + \left(1 - P_d^{(i)}\right) \frac{\theta}{K},$$

$$\text{where} \quad P_d^{(i)} = Q\left(\frac{h - \sqrt{N_b + N_e}\sigma_w/K}{\sigma_n}\right). \tag{8.17}$$

Figure 8.2 shows an example of $\pi^{(i)}$ versus the total number of colluders K. In Figure 8.2, the lengths of the fingerprints embedded in the base layer and the enhancement layer are $N_b = 50000$ and $N_e = 50000$, respectively. In Figure 8.2, we use $\theta = 50$ as an example to illustrate colluders' payoffs, and we observe similar trends with other values of θ. $\sigma_w^2 = \sigma_n^2 = 1$ and h is selected so that the probability of falsely accusing an innocent is $P_{fa} = 10^{-3}$ in Figure 8.2(a) and $P_{fa} = 10^{-8}$ in Figure 8.2(b) and (c). From Figure 8.2, when there are only a few colluders, $\pi^{(i)} < 0$ because of the high probability of being detected. For example, in Figure 8.2(a), colluders receive a negative payoff when $K < 126$. In this scenario, colluders may not want to use multimedia illegally, as it is too risky. Furthermore, from Figure 8.2, colluding with more attackers does not always increase a colluder's payoff, and in Figure 8.2(a), $\pi^{(i)}$ becomes a decreasing function of K when there are more than 206 attackers.

Let $K_0 \triangleq \{K : \pi^{(i)}(K-1) < 0, \pi^{(i)}(K) \geq 0\}$ be the smallest K that gives colluders a nonnegative payoff. Attackers will collude with one another if and only if there are more than K_0 colluders and when they receive positive payoffs from collusion. Also, we define $K_{max} \triangleq \arg_{K \geq K_0} \max \pi^{(i)}$ as the optimum K that maximizes colluder's utility when all attackers receive copies of the same resolution. A colluder should find a total of K_{max} attackers, if possible, to maximize his or her payoff. In the example in Figure 8.2(a), $K_0 = 126$ and $K_{max} = 206$. In this section, we analyze K_0 and study when attackers will collude with each other. We will analyze K_{max} in Section 8.6 and study the optimal number of colluders that maximizes their payoffs.

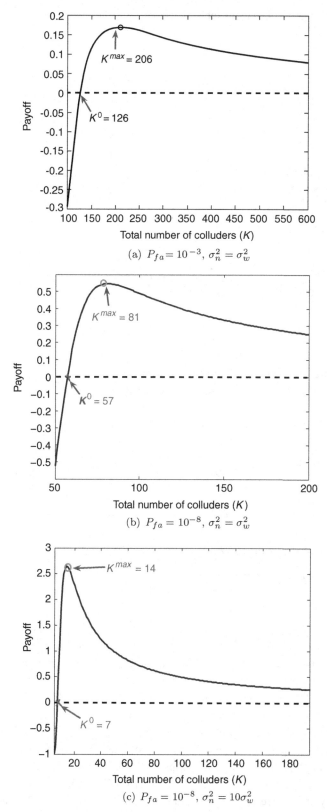

Fig. 8.2 $\pi^{(i)}$ when all colluders receive fingerprinted copies of high resolution with different P_{fa} and σ_n^2. $N_b = N_e = 50000$ and $\theta = 50$

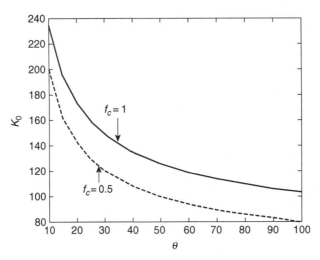

Fig. 8.3 K_0 versus θ. $N_b = N_e = 50000$, $\sigma_w^2 = \sigma_n^2 = 1$, and $P_{fa} = 10^{-3}$

8.3.1.2   Analysis of the minimal number of colluders

Given N_b, N_e, and θ, to find K_0, we solve the equation

$$\pi^{(i)}(K) = -Q\left(\frac{h - \sqrt{N_b + N_e}\sigma_w/K}{\sigma_n}\right)$$

$$+ \left[1 - Q\left(\frac{h - \sqrt{N_b + N_e}\sigma_w/K}{\sigma_n}\right)\right]\frac{\theta}{K} = 0,$$

$$\text{or equivalently,}\quad Q\left(\frac{h - \sqrt{N_b + N_e}\sigma_w/K}{\sigma_n}\right) = \frac{\theta}{K + \theta}. \tag{8.18}$$

Define $x = 1/K$ as the inverse of K. Then (8.18) can be rewritten as

$$Q\left(\frac{h - \sqrt{N_b + N_e}\sigma_w x}{\sigma_n}\right) = Q(a - b_s x) = 1 - \frac{1}{1 + \theta x}, \tag{8.19}$$

where $a = h/\sigma_n$ and $b_s = \sqrt{N_b + N_e}\sigma_w/\sigma_n$. Note that $K \geq 1$ is a positive integer. To gain insights into K_0, we first extend the support region of K from positive integers to positive real numbers, find the solution x^* to (8.19), and then let $K_0 = \lceil 1/x^* \rceil$. Consider the example in Figure 8.2(a) where $N_b = N_e = 50000$, $\sigma_w^2 = \sigma_n^2 = 1$, $\theta = 50$, and $P_{fa} = 10^{-3}$. The numerical solution to (8.19) is $x^* = 0.00798$ and $K_0 = \lceil 1/x^* \rceil = 126$, which agrees with our exhaustive search results.

Figure 8.3 shows K_0 for different values of θ and f_c. The system setup in Figure 8.3 is similar to that in Figure 8.2(a). Figure 8.3 suggests that K_0 is a decreasing function of θ. As an example, when $f_c = 1$, K_0 drops from 235 to 103 when θ increases from 10 to 100. In addition, from Figure 8.3, K_0 takes a smaller value if colluders generate a colluded copy of lower resolution. For example, with $\theta = 50$, $K_0 = 126$ when $f_c = 1$ and $K_0 = 100$ when $f_c = 0.5$.

8.3.2 Multiresolution collusion

Next, we consider the scenario in which colluders receive fingerprinted copies of different resolutions, analyze when attackers will collude with other attackers with different-quality copies, and investigate how an attacker selects fellow attackers to maximize his or her own payoff.

In Section 8.2.3, multiuser collusion is modeled as a two-player game, in which the two subgroups of colluders, SC^b and SC^{be}, negotiate with each other to reach an agreement on fair distribution of the risk and the reward. To understand the complicated dynamics among colluders, the first step is to analyze $\mathbb{S}_p$ and to investigate under what conditions attackers will collude with each other.

8.3.2.1 Colluders' payoff functions

From Section 8.2.3, one possible outcome of the bargaining between SC^b and SC^{be} is that they do not reach an agreement. In such a scenario, attackers collude only with their fellow attackers in the same subgroup, and SC^b and SC^{be} do not cooperate with each other. Given N_b, N_e, K^b, and K^{be}, if an attacker in SC^b colludes only with those in SC^b, following the same analysis as in Section 8.2, his or her utility is

$$\pi_{nc}^b = -P_{d,nc}^b + \left(1 - P_{d,nc}^b\right) R_{nc}^b$$

$$\text{where} \quad P_{d,nc}^b = Q\left(\frac{h}{\sigma_n} - \frac{\sqrt{N_b}\sigma_w}{K^b\sigma_n}\right) = Q\left(a - b_b/K^b\right)$$

$$\text{and} \quad R_{nc}^b = \frac{\theta f_b}{K^b}. \tag{8.20}$$

In (8.20), $a = h/\sigma_n$ and $b_b = \sqrt{N_b}\sigma_w/\sigma_n$. Similarly, if an attacker in SC^{be} colludes only with those in SC^{be}, his or her payoff is

$$\pi_{nc}^{be} = -P_{d,nc}^{be} + \left(1 - P_{d,nc}^{be}\right) R_{nc}^{be}$$

$$\text{where} \quad P_{d,nc}^{be} = Q\left(\frac{h}{\sigma_n} - \frac{\sqrt{N_b + N_e}\sigma_w}{K^{be}\sigma_n}\right) = Q\left(a - b_s/K^{be}\right)$$

$$\text{and} \quad R_{nc}^{be} = \frac{\theta}{K^{be}}. \tag{8.21}$$

In (8.21), $b_s = \sqrt{N_b + N_e}\sigma_w/\sigma_n$.

If SC^b and SC^{be} collaborate with each other and select the collusion parameter β, following the analysis in Section 8.1, for an attacker $i \in SC^b$, his or her utility is

$$\pi^b = -P_{d,c}^b + \left(1 - P_{d,c}^b\right) R_c^b,$$

$$\text{where} \quad P_{d,c}^b = Q\left(\frac{h}{\sigma_n} - \frac{\beta\sqrt{N_b}}{K^b} \cdot \frac{\sigma_w}{\sigma_n}\right) = Q\left(a - \beta\frac{b_b}{K^b}\right)$$

$$\text{and} \quad R_c^b = \frac{(f^c)^\gamma \theta}{K^b(f^c)^\gamma + K^{be}}. \tag{8.22}$$

Similarly, for $0 \leq \beta \leq \beta^+$, an attacker $i \in SC^{be}$'s payoff is

$$\pi^{be} = -P_{d,c}^{be} + \left(1 - P_{d,c}^{be}\right) R_c^{be}, \text{ where}$$

$$P_{d,c}^{be} = Q\left(\frac{h}{\sigma_n} - \frac{(1-\beta)N_b + N_e}{K^{be}\sqrt{N_b + N_e}} \cdot \frac{\sigma_w}{\sigma_n}\right) = Q\left(a - \frac{b_s}{K^{be}} + \beta\frac{b_{be}}{K^{be}}\right)$$

and $R_c^{be} = \dfrac{\theta}{K^b(f^c)^\gamma + K^{be}}.$ \hfill (8.23)

In (8.23), $b_{be} = \frac{N_b \sigma_w}{\sqrt{N_b + N_e}\sigma_n}$.

From Section 8.2.3, among all the possible solutions $\{(\pi^b, \pi^{be})\}$ in the feasible set $\mathbb{S}$, colluders are interested only in those in

$$\mathbb{S}_p = \{(\pi^b, \pi^{be}) \in \mathbb{S} : \pi^b \geq \underline{\pi}^b = \max(0, \pi_{nc}^b),$$

$$\pi^{be} \geq \underline{\pi}^{be} = \max(0, \pi_{nc}^b), 0 \leq \beta \leq \beta^+\}, \hfill (8.24)$$

where cooperation helps both SC^b and SC^{be} increase their payoffs.

- From (8.22), $P_{d,c}^b$ is an increasing function of β and, therefore, π^b is a decreasing function of β. Let $\bar{\beta}$ be the β that makes π^b equal to $\underline{\pi}^b$; that is, $\pi^b(\bar{\beta}) = \underline{\pi}^b$. Then, the constraint $\pi^b \geq \underline{\pi}^b$ is equivalent to letting $\beta \leq \bar{\beta}$.
- Similarly, from (8.23), $P_{d,c}^{(be)}$ is a decreasing function of β, and thus π^{be} is an increasing function of β. Let $\underline{\beta}$ be the β that makes π^{be} equal to $\underline{\pi}^{be}$; that is, $\pi^{be}(\underline{\beta}) = \underline{\pi}^{be}$. Therefore, the constraint $\pi^{be} \geq \underline{\pi}^{be}$ is equivalent to selecting $\beta \geq \underline{\beta}$.
- Furthermore, if we compare (8.21) and (8.23), $P_{d,c}^{be} = P_{d,nc}^{be}$ when $\beta = 0$. That is, if colluders select $\beta = 0$, then collaborating with SC^b does not help SC^{be} further reduce their risk of being detected. Meanwhile, $R_c^{be} < R_{nc}^{be}$ and colluders in SC^{be} receive a smaller reward if they cooperate with SC^b. Consequently, $\pi^{be}(0) < \pi_{nc}^{be} \leq \underline{\pi}^{be} = \max\left(\pi_{nc}^{be}, 0\right) = \pi^{be}(\underline{\beta})$. Thus, $\underline{\beta} > 0$, because π^{be} is an increasing function of β.

From this analysis, we can rewrite $\mathbb{S}_p$ as

$$\mathbb{S}_p = \{(\pi^b, \pi^{be}) \in \mathbb{S} : \underline{\beta} \leq \beta \leq \min(\bar{\beta}, \beta^+)\}. \hfill (8.25)$$

When attackers receive fingerprinted copies of different resolutions, the two subgroups of colluders SC^b and SC^{be} will collude with each other if and only if there exists at least one β such that $\underline{\beta} \leq \beta \leq \min(\bar{\beta}, \beta^+)$, or equivalently, when $\mathbb{S}_p$ is not empty.

8.3.2.2   Lower and upper bounds of the collusion parameter β

To further understand under what conditions SC^b and SC^{be} will cooperate with each other, we will first analyze $\underline{\beta}$ and $\bar{\beta}$.

From the previous discussion, given N_b, N_e, K^b, and K^{be}, colluders should select β such that

$$\pi^b(\beta) = -P_{d,c}^b(\beta) + \left[1 - P_{d,c}^b(\beta)\right] R_c^b \geq \underline{\pi}^b = \max\left(0, \pi_{nc}^b\right), \hfill (8.26)$$

where $P_{d,c}^b(\beta)$ and R_c^b are in (8.22). Consequently, we have

$$P_{d,c}^b(\beta) = Q\left(a - \frac{\beta b_b}{K^b}\right) \le \frac{R_c^b - \pi^b}{R_c^b + 1}. \tag{8.27}$$

Because $Q(x)$ is a decreasing function of x, therefore, we have

$$a - \beta\frac{b_b}{K^b} \ge Q^{-1}\left(\frac{R_c^b - \pi^b}{R_c^b + 1}\right),$$

or equivalently, $\quad \beta \le \bar\beta = \left[a - Q^{-1}\left(\frac{R_c^b - \pi^b}{R_c^b + 1}\right)\right]\frac{K^b}{b_b}. \tag{8.28}$

Similarly, given N_b, N_e, K^b, and K^{be}, colluders should select β such that

$$\pi^{be}(\beta) = -P_{d,c}^{be}(\beta) + \left[1 - P_{d,c}^{be}(\beta)\right] R_c^{be} \ge \underline\pi^{be} = \max\left(0, \pi_{nc}^{be}\right), \tag{8.29}$$

where $P_{d,c}^{be}(\beta)$ and R_c^{be} are in (8.23). Therefore, we have

$$P_{d,c}^{be}(\beta) = Q\left(a - \frac{\sqrt{N_b + N_e}\sigma_w}{K^{be}\sigma_n} + \frac{\beta b_{be}}{K^{be}}\right) \ge \frac{R_c^{be} - \pi^{be}}{R_c^{be} + 1},$$

or equivalently, $\beta \ge \underline\beta = \left[Q^{-1}\left(\frac{R_c^{be} - \pi^{be}}{R_c^{be} + 1}\right) - a + \frac{\sqrt{N_b + N_e}\sigma_w}{K^{be}\sigma_n}\right]\frac{K^{be}}{b_{be}}$

$$= \frac{N_b + N_e}{N_b} + \left[Q^{-1}\left(\frac{R_c^{be} - \pi^{be}}{R_c^{be} + 1}\right) - a\right]\frac{K^{be}}{b_{be}}. \tag{8.30}$$

Figure 8.4 shows examples of $\underline\beta$ and $\bar\beta$. The simulation setup is similar to that in Figure 8.2(a). $K^{be} = 120$ in Figure 8.4(a), and $K^b = 50$ in Figure 8.4(b). From Figure 8.4(a), $\bar\beta < \underline\beta$ when $K^b < 9$, and $\underline\beta > \beta^+$ when $K^b > 358$. Therefore, in this example, where K^{be} is fixed as 120, $\mathbb{S}_p \ne \emptyset$ if and only if $9 \le K^b \le 358$. Similarly, from Figure 8.4(b), $\underline\beta > \bar\beta$ if $K^{be} < 94$ or $K^{be} > 207$. Thus, when $K^b = 50$ is fixed, SC^{be} and SC^b will collude with each other if and only if $94 \le K^{be} \le 207$.

8.3.2.3 Analysis of the number of colluders

From Figure 8.4, given N_b, N_e, and θ, for some pairs of (K^b, K^{be}), $\mathbb{S}_p$ may be empty and thus, SC^b and SC^{be} will not cooperate. Define $\mathbb{K}_p \triangleq \{(K^b, K^{be}) : \mathbb{S}_p \ne \emptyset\}$ as the set including all pairs of (K^b, K^{be}) where $\mathbb{S}_p$ is not empty and where SC^b and SC^{be} will collude with each other.

Given N_b, N_e and θ, SC^b and SC^{be} will collude with each other if and only if $\mathbb{S}_p \ne \emptyset$ – that is, when $\underline\beta \le \beta^+$ and $\underline\beta \le \bar\beta$. Because π^{be} in (8.23) is an increasing function of β, if $\underline\beta \le \beta^+$, then we have

$$\pi^{be}(\beta^+) = -P_{d,c}^{be}(\beta^+) + \left[1 - P_{d,c}^{be}(\beta^+)\right] R_c^{be} \ge \pi^{be}(\underline\beta) = \underline\pi^{be},$$

or equivalently, $\quad R_c^{be} = \frac{\theta}{K^b(f_c)^\gamma + K^{be}} \ge \frac{\underline\pi^{be} + P_{d,c}^{be}(\beta^+)}{1 - P_{d,c}^{be}(\beta^+)}. \tag{8.31}$

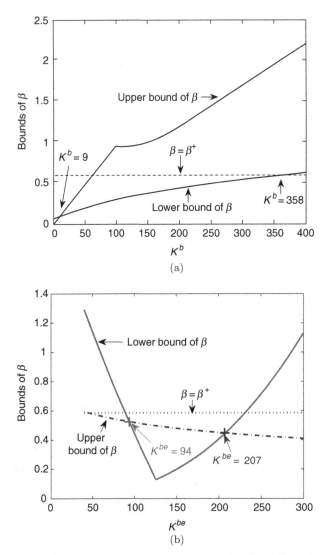

Fig. 8.4 $\bar{\beta}$ in (8.28) and $\underline{\beta}$ in (8.30). $N_b = N_e = 50000$, $\sigma_w^2 = \sigma_n^2 = 1$, $P_{fa} = 10^{-3}$, and $\theta = 50$. (a) $K^{be} = 120$. (b) $K^b = 50$

Consequently, to ensure $\underline{\beta} \leq \beta^+$, (K^b, K^{be}) must satisfy

$$K^b \leq K^{b'}(K^{be}) \triangleq \frac{\theta\left(1 - P_{d,c}^{be}(\beta^+)\right)}{\left(\underline{\pi}^{be} + P_{d,c}^{be}(\beta^+)\right)(f_b)^\gamma} - \frac{K^{be}}{(f_b)^\gamma}. \qquad (8.32)$$

From (8.28) and (8.30), to ensure $\underline{\beta} \leq \bar{\beta}$, (K^b, K^{be}) must satisfy

$$\underline{\beta} = \frac{N_b + N_e}{N_b} + \left[Q^{-1}\left(\frac{R_c^{be} - \underline{\pi}^{be}}{R_c^{be} + 1}\right) - a\right]\frac{K^{be}}{b_{be}}$$

$$\leq \bar{\beta} = \left[a - Q^{-1}\left(\frac{R_c^b - \underline{\pi}^b}{R_c^b + 1}\right)\right]\frac{K^b}{b_b}. \qquad (8.33)$$

Combining (8.32) and (8.33), we have

$$\mathbb{K}_p = \left\{ (K^b, K^{be}) : K^b \leq \frac{\theta \left(1 - P_{d,c}^{be}(\beta^+)\right)}{\left(\pi^{be} + P_{d,c}^{be}(\beta^+)\right)(f_b)^{\gamma}} - \frac{K^{be}}{(f_b)^{\gamma}}, \right.$$

$$\left. \frac{N_b + N_e}{N_b} + \left[Q^{-1} \left(\frac{R_c^{be} - \pi^{be}}{R_c^{be} + 1} \right) - a \right] \frac{K^{be}}{b_{be}} \leq \left[a - Q^{-1} \left(\frac{R_c^b - \pi^b}{R_c^b + 1} \right) \right] \frac{K^b}{b_b} \right\}.$$

$$(8.34)$$

The shaded area in Figure 8.5(a) shows an example of $\mathbb{K}_p$. At point A in Figure 8.5(a), when $K^{be} < 91$, no matter which value K^b takes, $\mathbb{S}_p$ is always empty and attackers will not collude with one another. Similarly, when $K^{be} > 226$ (point B in Figure 8.5(a)), no matter how many attackers are in SC^b and how they select β, cooperation between SC^b and SC^{be} cannot improve all colluders' payoffs. Furthermore, when $K^b > 431$ (point C in Figure 8.5(a)), SC^b and SC^{be} will not collude with each other. In addition, when $125 \leq K^{be} \leq 226$, $(K^b = 1, K^{be})$ is in the feasible region $\mathbb{K}_p$ and the lower bound of K^b is 1. To quantify these boundary points of $\mathbb{K}_p$, we define

$$\underline{K}^{be} \overset{\Delta}{=} \min \left\{ K^{be} : \exists K^b \text{ s.t. } (K^b, K^{be}) \in \mathbb{K}_p \right\},$$

$$\bar{K}^{be} \overset{\Delta}{=} \max \left\{ K^{be} : \exists K^b \text{ s.t. } (K^b, K^{be}) \in \mathbb{K}_p \right\},$$

$$\text{and} \quad \bar{K}^b \overset{\Delta}{=} \max \left\{ K^b : \exists K^{be} \text{ s.t. } (K^b, K^{be}) \in \mathbb{K}_p \right\}. \quad (8.35)$$

In the example in Figure 8.5(a), $\underline{K}^{be} = 91$, $\bar{K}^{be} = 226$, and $\bar{K}^b = 431$.

Figure 8.5(b) and 8.5(c) show the feasible region $\mathbb{K}_p$ when $\theta = 150$ and $\theta = 10$, respectively, where we observe the same trend as in Figure 8.5(a). In addition, when θ takes a smaller value and colluders place more emphasis on risk minimization, they prefer to collude with more people to reduce their risk of being detected, and more colluders join the coalition. This is similar to the single-resolution case.

From Figure 8.5 and (8.35), if $K^b > \bar{K}^b$, $K^{be} < \underline{K}^{be}$, or $K^{be} > \bar{K}^{be}$, then it is impossible to find a β that increases all colluders' payoffs, and SC^b and SC^{be} will not cooperate with each other. Therefore, during collusion, as a preliminary step, colluders should first check that $K^b \leq \bar{K}^b$ and $\underline{K}^{be} \leq K^{be} \leq \bar{K}^{be}$. Then, they should ensure that (K^b, K^{be}) is in the set $\mathbb{K}_p$ defined in (8.35), and guarantee that there exists at least one β that increases both SC^b and SC^{be}'s payoffs. In the following, we analyze the boundary points of $\mathbb{K}_p$ $(\underline{K}^{be}, \bar{K}^{be} \text{ and } \bar{K}^b)$ in detail.

- $\underline{K}^{be}$: Using an exhaustive search, we find that at point A in Figure 8.5(a), $K^b = 56$ and $K^{be} = 91$. Because $K^b < K_0(f_c = f_b) = 100$ and $K^{be} < K_0(f_c = 1) = 126$, we have $\pi_{nc}^b < 0$, $\pi_{nc}^{be} < 0$, and $\underline{\pi}^{be} = \underline{\pi}^b = 0$. To have a better understanding of $\underline{K}^{be}$, Figure 8.6 plots $\underline{\beta}$ and $\bar{\beta}$ around the point $(K^b = 56, K^{be} = 91)$. As can be seen from Figure 8.6, at point A, $\underline{\beta} = \bar{\beta} = \beta^+$, and $\mathbb{S}_p$ has only one item, which is $\mathbb{S}_p = \left\{ (\pi^b, \pi^{be}) : \beta = \beta^+ \right\}$. From the analysis in Section 8.3.2.1, $\pi^{be} = \underline{\pi}^{be}$ when $\beta = \underline{\beta}$, and $\pi^b = \underline{\pi}^b$ when $\beta = \bar{\beta}$. Therefore, at the boundary point A, (K^b, K^{be}) satisfies

$$\begin{cases} \pi^b(K^b, K^{be}, \beta^+) = \underline{\pi}^b = 0, \\ \pi^{be}(K^b, K^{be}, \beta^+) = \underline{\pi}^{be} = 0. \end{cases} \quad (8.36)$$

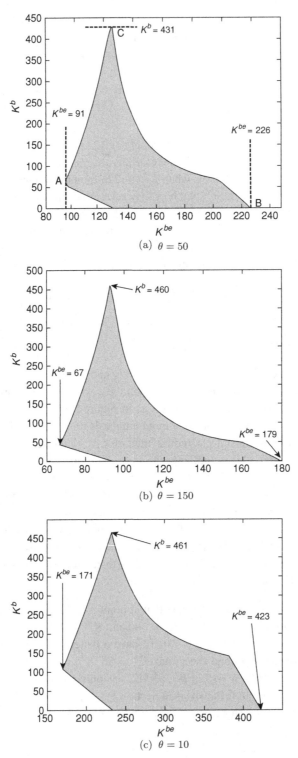

Fig. 8.5 An example of $\mathbb{K}_p$. $N_b = N_e = 50000$ and the probability of falsely accusing an innocent user is $P_{fa} = 10^{-3}$

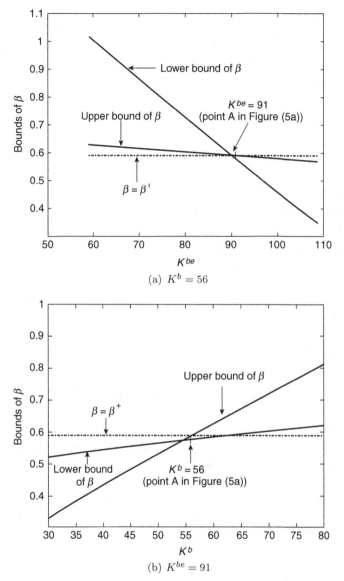

Fig. 8.6 $\underline{\beta}$ and $\bar{\beta}$ at point A in Figure 8.5(a)

To find $\underline{K}^b$, we should first find the solution (K^b, K^{be}) to (8.36) and then select $\underline{K}^{be} = \lceil K^{be} \rceil$. Using Figure 8.5(a) as an example, given the parameters $N_b = N_e = 50000$, $\gamma = 1/3$, $\theta = 50$ and $P_{fa} = 10^{-3}$, we first find the solution to (8.36) numerically, which is $(K^b = 55.88, K^{be} = 90.15)$. We then calculate $\underline{K}^{be} = \lceil K^{be} \rceil = 91$, which is consistent with the result we find using an exhaustive search.

- $\bar{K}^{be}$:   To analyze $\bar{K}^{be}$, using an exhaustive search, we find that at point B in Figure 8.5(a), $K^b = 1 < K_0(f_c = f_b) = 100$ and $K^{be} = 226 > K_0(f_c = 1) = 126$. Therefore, at this point, $\underline{\pi}^b = 0$ and $\underline{\pi}^{be} = \pi_{nc}^{be} > 0$. As shown in Figure 8.7, at this point,

$\beta = \bar{\beta}$, and $\mathbb{S}_p = \{(\pi^b, \pi^{be}) : \beta = \underline{\beta} = \bar{\beta}\}$ has only one entry. Also, from Figure 8.7(b), when $K^{be} = \bar{K}^{be}$, if SC^b has more than one attacker (that is, $K^b \geq 2$), there is no β that can improve both SC^{be} and SC^b's payoffs. Therefore, point B corresponds to the scenario in which $K^b = 1$, $K^{be} = \bar{K}^{be}$ and $\beta = \bar{\beta}$. Thus, from (8.28) and (8.30), to find $\bar{K}^{be}$, we first solve

$$\underline{\beta} = \frac{N_b + N_e}{N_b} + \left[Q^{-1} \left(\frac{R_c^{be} - \pi_{nc}^{be}}{R_c^{be} + 1} \right) - a \right] \frac{K^{be}}{b_{be}}$$

$$= \bar{\beta} = \left[a - Q^{-1} \left(\frac{R_c^b}{R_c^b + 1} \right) \right] \frac{1}{b_b}, \tag{8.37}$$

and then let $\bar{K}^{be} = \lfloor K^{be} \rfloor$. In (8.37), $R_c^{be} = \theta / [(f_b)^\gamma + K^{be}]$, $R^b = \theta(f_b)^\gamma / [(f_b)^\gamma + K^{be}]$, and π_{nc}^{be} is in (8.21). As an example, given the system setup in Figure 8.5(a), the numerical solution to (8.37) is $K^{be} = 226.64$ and thus $\bar{K}^{be} = \lfloor 226.64 \rfloor = 226$. It is consistent with the result we found using an exhaustive search.

- $\bar{K}^b$: At point C in Figure 8.5(a), we find $K^b = 431$ and $K^{be} = 125$ using exhaustive search and $\underline{\beta} = \beta^+$, as shown in Figure 8.8. From the analysis in Section 8.3.2.3, for a given K^{be}, to satisfy the constraint $\underline{\beta} \leq \beta^+$, it is required that $K^b \leq K^{b'}$, where $K^{b'}$ is defined in (8.32). Therefore, we have $\bar{K}^b = \lfloor \max_{K^{be}} K^{b'} \rfloor$. Using the system setup in Figure 8.5(a) as an example, Figure 8.9 plots $K^{b'}$ versus K^{be}, and $K^{b'}$ achieves a maximum of 431.88 when $K^{be} = 125$. Consequently, $\bar{K}^b = \lfloor 431.88 \rfloor = 431$, which agrees with the result we found using an exhaustive search.

To summarize, given N_b, N_e, and other parameters including θ and γ, to ensure that cooperation can help both SC^b and SC^{be} improve their payoffs, colluders should first follow the preceding analysis to find $\underline{K}^{be}$, $\bar{K}^{be}$, and $\bar{K}^b$, and ensure that $\underline{K}^{be} \leq K^{be} \leq \bar{K}^{be}$ and $K^b \leq \bar{K}^b$. Then, attackers should further check whether (K^b, K^{be}) satisfies the constraints in (8.34) and whether $\mathbb{S}_p$ is not empty. If $(K^b, K^{be}) \in \mathbb{K}_p$, colluders should use (8.28) and (8.30) to calculate $\bar{\beta}$ and $\underline{\beta}$, respectively, and find $\mathbb{S}_p = \{(\pi^b, \pi^{be}) : \underline{\beta} \leq \beta \leq \min(\bar{\beta}, \beta^+)\}$. By doing so, no matter which pair (π^b, π^{be}) that colluders select in $\mathbb{S}_p$, it is a Pareto optimal solution and all colluders increase their payoffs by cooperating with one another.

8.4   How to collude: the bargaining model

The previous section analyzes when colluders should form a coalition. In this section, we discuss how colluders bargain with one another and reach an agreement. We model the colluders' bargaining process as follows:

- In the first bargaining stage, SC^{be} offers the collusion parameter β_1 that uniquely maps to the utility pair (π_1^b, π_1^{be}) in the Pareto optimal set.
- On receiving the offer, SC^b has the choice to accept this offer and gets the payoff π_1^b, or to reject it and offer back β_2, which corresponds to another payoff pair (π_2^b, π_2^{be}) in the Pareto optimal set.

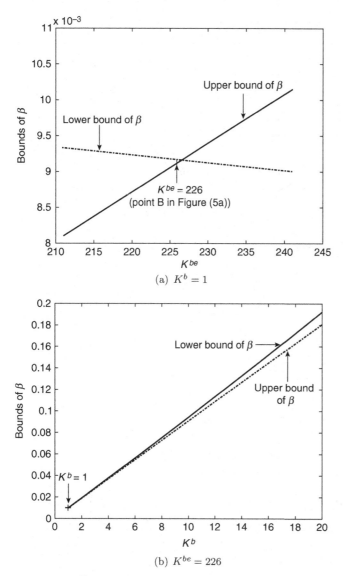

Fig. 8.7 $\underline{\beta}$ and $\bar{\beta}$ at point B in Figure 8.5(a)

- If SC^b decides to offer back, SC^{be} again has the choice to accept the offer (π_2^b, π_2^{be}) or offer back.
- This bargaining process continues until both groups of colluders agree on one offer. Each time a subgroup rejects the offer and the negotiation process goes to the next round, the rewards that SC^b and SC^{be} that receive are decayed by δ_b and δ_{be}, respectively.

In this model, SC^{be} makes an offer first and takes the advantage of the first move. This is because, without SC^{be}, colluders in SC^b can generate only a low-resolution

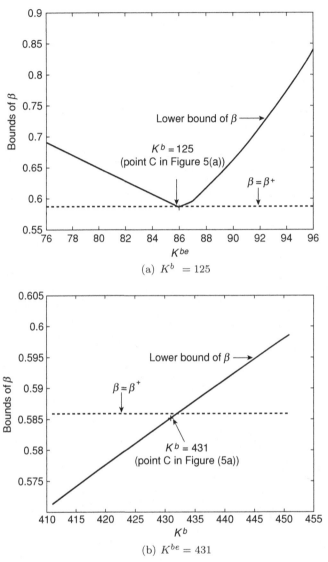

Fig. 8.8 $\underline{\beta}$ and $\bar{\beta}$ at point C in Figure 8.5(a)

colluded copy with much lower reward, and thus, SC^{be} has more bargaining power than SC^{b}.

Given this bargaining model, we first consider the scenario in which the reward received from collusion is not time-sensitive and in which colluders can use an infinite number of rounds to reach an agreement. We then consider time-restricted bargaining, in which the reward decays as the number of bargaining rounds increases. For these two different scenarios, we discuss how colluders reach an agreement and find the Nash equilibrium of the colluder bargaining game.

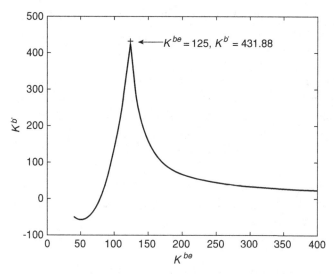

Fig. 8.9 $K^{b'}$ versus K^{be}. $N_b = N_e = 50000$, $\sigma_w^2 = \sigma_n^2 = 1$, $\gamma = 1/3$, $P_{fa} = 10^{-3}$, and $\theta = 50$

8.4.1 Fair collusion with non–time-restricted bargaining

We start with the simple scenario in which the reward received from illegal usage of multimedia is not time-sensitive – that is, both δ_b and δ_{be} are equal to 1. In such a scenario, colluders will bargain until both groups are satisfied and the selected collusion is fair to all colluders. Depending on the definition of fairness and the objectives of collusion, colluders select different collusion strategies. In this section, we consider four commonly used fairness criteria and demonstrate how colluders achieve fair collusion.

Absolute fairness: The most straightforward fairness criterion is absolute fairness, under which all colluders have the same utility. That is, it selects $(\pi^b, \pi^{be}) \in \mathbb{S}_p$ that satisfies

$$\pi_{abs} = \pi^b = \pi^{be}. \tag{8.38}$$

Even if $\mathbb{S}_p$ is not empty, it is possible that the absolute fairness solution does not exist – that is, there is no pair $(\pi^b, \pi^{be}) \in \mathbb{S}_p$ where $\pi^b = \pi^{be}$. As an example, consider the system setup in Figure 8.5(a), where $N_b = N_e = 50000$. When $(K^b, K^{be}) = (60, 95)$, following the analysis in Section 8.3.2.2, β should be in the range $[0.5308, 0.5858]$, which gives $0.0454 \le \pi^b \le 0.1084$ and $0 \le \pi^{be} \le 0.0390$. In this example, even though $\mathbb{S}_p \ne \emptyset$, $\max(\pi^{be}) < \min(\pi^b)$ and, therefore, the absolute fairness solution does not exist. In this scenario, colluders should consider other fairness criteria.

Max–Min fairness: To guarantee the minimum payoff that a colluder can receive by participating in collusion, colluders might choose to maximize the minimum utility over all colluders. For our two-player colluder game, the max–min fairness solution is

$$\pi_{maxmin} = \max_{(\pi^b,\pi^{be})\in\mathbb{S}_p} \min\left(\pi^b, \pi^{be}\right). \tag{8.39}$$

Max–sum fairness: With the max–sum fairness criterion, colluders select the β in the Pareto optimal set $\mathbb{S}_p$ that maximizes the summation of all colluders' utilities. Mathematically, the max–sum fairness solution can be formulated as follows:

$$\pi_{maxsum} = \max_{(\pi^b,\pi^{be})\in\mathbb{S}_p} \sum_{i\in SC} \pi^{(i)} = K^b\pi^b + K^{be}\pi^{be}. \tag{8.40}$$

Nash bargaining solution: The Nash bargaining solution is a famous bargaining solution in game theory. Mathematically, the general Nash bargaining solution selects $(\pi^b, \pi^{be}) \in \mathbb{S}_p$ that maximizes

$$g(\pi^b, \pi^{be}) = \left(\pi^b - \underline{\pi}^b\right)^{a_b} \left(\pi^{be} - \underline{\pi}^{be}\right)^{a_{be}}, \tag{8.41}$$

where a_b and a_{be} are the bargaining powers of SC^b, SC^{be}, respectively. Here, as defined in (8.24), $\underline{\pi}^b$ and $\underline{\pi}^{be}$ are the payoffs for SC^b and SC^{be}, respectively, when the two subgroups fail to reach an agreement and do not cooperate with each other. When $a_b = a_{be} = 1$, the Nash bargaining solution divides the additional utility between the two players in a ratio that is equal to the rate at which this utility can be transferred. If $a_b \neq a_{be}$, then the bargaining solution deviates from the proportional fairness solution and favors the player with a higher bargaining power.

8.4.2 Time-restricted bargaining

For time-restricted bargaining with $0 \le \delta_b, \delta_{be} < 1$, the reward that colluder i receives from collusion decays by a constant factor every round that colluders fail to reach an agreement. Thus, the feasible set and the Pareto optimal set change every time colluders must go to the next bargaining round. Figure 8.10 illustrates the feasible regions of this bargaining process. In Figure 8.10, $SC^{be} = 250$, $SC^b = 100$, and $P_{fa} = 10^{-3}$. The thin solid line is the feasible region that the colluders can choose before the bargaining process starts, the wide (circled) line and the dashed line are the feasible regions after the first and the second rounds, respectively. As can be seen from Figure 8.10, when colluders take more rounds to reach an agreement, the utilities that they receive from collusion are smaller. Thus, to maximize their payoffs, colluders should finish the bargaining process as soon as possible, ideally in the first round.

Assume that colluders reach an agreement at the kth stage. Colluders seek the utility pair $(\beta_k \in \mathbb{S}_p, \beta_{k+1} \in \mathbb{S}_p)$ that is stationary, which are the bottom lines of the two players. Here, β_k gives $\left(\pi_k^b, \pi_k^{be}\right)$ in stage k, and β_{k+1} gives $\left(\pi_{k+1}^b, \pi_{k+1}^{be}\right)$ in stage $k+1$. In this game, the stationary strategy for SC^b is to offer π_k^{be} to subgroup SC^{be} whenever it is its turn to act, and to accept any offer larger or equal to π_{k+1}^b. Similarly, the stationary strategy for SC^{be} is to offer π_k^b to SC^b whenever it is its turn to act, and to accept any offer larger or equal to π_{k+1}^{be}.

For an equilibrium, if SC^{be} acts in stage k, its offer, π_k^b should be large enough that SC^b will accept, but not larger. On the other hand, SC^b should accept the offer π_k^b if it

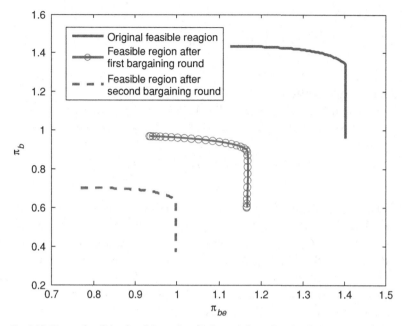

Fig. 8.10 Example of the feasible region for bargaining after the first two rounds

is not smaller than π_{k+1}^b, which is SC^b's received payoff if it rejects SC^{be}'s offer in stage k and if SC^{be} accepts SC^b's counteroffer in stage $k+1$. That is,

$$-P_d^b(\beta_k) * L + \left[1 - P_d^b(\beta_k)\right] (\delta_b)^{k-1} R^b$$
$$= -P_d^b(\beta_{k+1}) * L + \left[1 - P_d^b(\beta_{k+1})\right] (\delta_b)^k R^b. \qquad (8.42)$$

Similarly, if SC^b acts in stage k, then β_k and β_{k+1} should satisfy

$$-P_d^{be}(\beta_k) * L + \left[1 - P_d^{be}(\beta_k)\right] (\delta_{be})^{k-1} R^{be}$$
$$= -P_d^{be}(\beta_{k+1}) * L + \left[1 - P_d^{be}(\beta_{k+1})\right] (\delta_{be})^k R^{be}. \qquad (8.43)$$

8.5   How to collude: examples

In this section, we consider three different scenarios and show how colluders bargain with one another.

8.5.1   Scenario 1: Reward is independent of risk

We first consider the scenario in which the market value of the colluded copy is not time-sensitive; that is, $\delta_b = \delta_{be} = 1$. In addition, $\gamma = 0$ and $D(P_d^{(i)}) = 1$ in the reward function, and the total reward received from collusion is equally distributed among all colluders. In such a scenario, the utility and the payoffs for SC^b and SC^{be} can be simplified as

$$\pi^b = -\left(\frac{\theta}{K} + L^b\right) P_{d,c}^b + \frac{\theta}{K} \quad \text{and} \quad \pi^{be} = -\left(\frac{\theta}{K} + L^{be}\right) P_{d,c}^{be} + \frac{\theta}{K}. \quad (8.44)$$

8.5.1.1   Bargaining solutions

- *Absolute fairness solution:* If all colluders agree to have the same utility, given the utility function in (8.44), the absolute fairness solution is the pair $(\pi^b, \pi^{be}) \in \mathbb{S}_p$ that satisfies

$$\frac{P_{d,c}^{be}(\beta)}{P_{d,c}^b(\beta)} = \frac{L^b + \theta/K}{L^{be} + \theta/K}, \quad (8.45)$$

where $P_{d,c}^{be}(\beta)$ and $P_{d,c}^b(\beta)$ are the SC^b and SC^{be}'s probabilities of being detected as defined in (8.3) and (8.11), respectively, and L^b and L^{be} are the loss terms claimed by SC^b and C^{be}, respectively. From the analysis in Section 8.1.2, for $\beta \in [\underline{\beta}, \min(\bar{\beta}, \beta^+)]$, $P_{d,c}^{be}$ is a monotonically decreasing function of β, and $P_{d,c}^b(\beta)$ is a monotonically increasing function of β. Thus, $P_{d,c}^{be}/P_{d,c}^b$ is a monotonically decreasing function of β, and (8.45) can be solved easily using numerical methods if the absolute fairness solution exists.

- *Max–min solution:* The max–min fairness solution with the payoff function defined in (8.44) always exists and is unique. If the absolute fairness solution exists, then the max–min solution and the absolute fairness solution are the same.

  To prove this statement, note that π^b in (8.44) is a monotonically decreasing function of β and π^{be} in (8.44) is a monotonically increasing function of β. Therefore, there is one and only one max–min solution. Assume that the absolute fairness solution exists, and there is a $\beta' \in [\underline{\beta}, \min(\bar{\beta}, \beta^+)]$ that satisfies $\pi^b(\beta') = \pi^{be}(\beta')$. Then, for all $\beta \leq \beta', \pi^b(\beta) \geq \pi^b(\beta') = \pi^{be}(\beta') \geq \pi^{be}(\beta)$ and $\min(\pi^b(\beta), \pi^{be}(\beta)) = \pi^{be}(\beta) \leq \pi^{be}(\beta')$, whereas for all $\beta \geq \beta', \pi^b(\beta) \leq \pi^b(\beta') = \pi^{be}(\beta') \leq \pi^{be}(\beta)$ and $\min(\pi^b(\beta), \pi^{be}(\beta)) = \pi^b(\beta) \leq \pi^b(\beta')$. Therefore, $\max \min(\pi^b, \pi^{be}) = \pi^b(\beta') = \pi^{be}(\beta')$, and the max–min solution is the same as the absolute fairness solution.

- *Max–sum solution:* From (8.44), the max–sum solution in (8.40) is equivalent to

$$\min_{\underline{\beta} \leq \beta \leq \min(\bar{\beta}, \beta^+)} C_{sum} = P_{d,c}^b K^b \left(\frac{\theta}{K} + L^b\right) + P_{d,c}^{be} K^{be} \left(\frac{\theta}{K} + L^{be}\right). \quad (8.46)$$

The minimum C_{sum} is achieved either when β is at the boundary points or when $\frac{\partial C_{sum}}{\partial \beta} = 0$. Taking the first-order derivative of C_{sum} with respect to β, we have

$$\frac{\partial C_{sum}}{\partial \beta} = \frac{\partial P_{d,c}^b}{\partial \beta} K^b \left(\frac{\theta}{K} + L^b\right) + \frac{\partial P_{d,c}^{be}}{\partial \beta} K^{be} \left(\frac{\theta}{K} + L^{be}\right) = 0,$$

$$\text{where} \quad \frac{\partial P_{d,c}^b}{\partial \beta} = \frac{\sqrt{N_b}\sigma_w}{K^b \sqrt{2\pi}\sigma_n} \exp\left\{-\frac{(h - \beta\sqrt{N_b}/K^b)^2}{2\sigma_n^2}\right\},$$

$$\text{and} \quad \frac{\partial P_{d,c}^{be}}{\partial \beta} = -\frac{N_b\sigma_w}{K^{be}\sqrt{2\pi(N_b + N_e)}\sigma_n} \exp\left\{-\frac{\left[h - \frac{(1-\beta)N_b + N_e}{K^{be}\sqrt{N_b + N_e}}\right]^2}{2\sigma_n^2}\right\}. \quad (8.47)$$

The solution to (8.47) can be found numerically.

- *Nash bargaining solution:* With the utility function defined in (8.44), (8.41) becomes

$$g(\beta) = \left(\pi^{be} - \underline{\pi}^{be}\right)^{a_{be}} \left(\pi^b - \underline{\pi}^b\right)^{a_b},$$

where
$$\pi^{be} = -\left(\frac{\theta}{K} + L^{be}\right) Q \left(\frac{h - \frac{(1-\beta)N_b + N_e \sigma_w}{K^{be}\sqrt{N_b + N_e}}}{\sigma_n}\right) + \frac{\theta}{K},$$

$$\underline{\pi}^{be} = \max\left\{0, -\left(\frac{\theta}{K^{be}} + L^{be}\right) Q \left(\frac{h - \frac{\sqrt{N_b + N_e}\sigma_w}{K^{be}}}{\sigma_n}\right) + \frac{\theta}{K^{be}}\right\},$$

$$\pi^b = -\left(\frac{\theta}{K} + L^b\right) Q \left(\frac{h - \frac{\beta\sqrt{N_b}\sigma_w}{K^b}}{\sigma_n}\right) + \frac{\theta}{K},$$

and
$$\underline{\pi}^b = \max\left\{0, -\left(\frac{\theta}{K^b} + L^b\right) Q \left(\frac{h - \frac{\sqrt{N_b}\sigma_w}{K^b}}{\sigma_n}\right) + \frac{\theta}{K^b}\right\}, \qquad (8.48)$$

The β that maximizes $g(\beta)$ in (8.48) either is on the boundary of $[\underline{\beta}, \min(\bar{\beta}, \beta^+)]$ or satisfies $\partial g / \partial \beta = 0$. If $\partial g(\beta)/\partial \beta = 0$, then

$$0 = a_{be} \left(\pi^{be} - \underline{\pi}^{be}\right)^{a_{be}-1} \left(\pi^b - \underline{\pi}^b\right)^{a_b} \frac{\partial \pi^{be}}{\partial \beta}$$

$$+ a_b \left(\pi^{be} - \underline{\pi}^{be}\right)^{a_{be}} \left(\pi^b - \underline{\pi}^b\right)^{a_b-1} \frac{\partial \pi^b}{\partial \beta},$$

that is, $\quad 0 = a_{be} \left(\pi^b - \underline{\pi}^b\right) \left(\frac{\theta}{K} + L^{be}\right) \frac{\partial P_{d,c}^{be}}{\partial \beta}$

$$+ a_b \left(\pi^{be} - \underline{\pi}^{be}\right) \left(\frac{\theta}{K} + L^b\right) \frac{\partial P_{d,c}^b}{\partial \beta},$$

or equivalently, $\quad \dfrac{a_{be}}{a_b} = \dfrac{\pi^{be} - \underline{\pi}^{be}}{\pi^b - \underline{\pi}^b} \cdot \dfrac{\frac{\partial P_{d,c}^b}{\partial \beta}}{-\frac{\partial P_{d,c}^{be}}{\partial \beta}} \cdot \dfrac{\frac{\theta}{K} + L^b}{\frac{\theta}{K} + L^{be}}. \qquad (8.49)$

Note that both $\frac{\pi^{be} - \underline{\pi}^{be}}{\pi^b - \underline{\pi}^b}$ and $-\frac{\partial P_{d,c}^b}{\partial \beta} / \frac{\partial P_{d,c}^{be}}{\partial \beta}$ are increasing functions of β. Therefore, the solution to (8.49) is a monotonically increasing function of $\frac{a_{be}}{a_b}$, and a larger ratio of $\frac{a_{be}}{a_b}$ results in a larger β and a higher π^{be}. That is, the Nash bargaining solution favors the subgroup of colluders with a larger bargaining power.

8.5.1.2 Simulation setting and results

In our simulations, we first generate independent vectors following Gaussian distribution $\mathcal{N}(0, 1)$, and then apply Gram–Schmidt orthogonalization to generate orthogonal fingerprints. The lengths of the fingerprints embedded in the base layer and the enhancement layer are $N_b = N_e = 50000$, and each layer contains 20 frames. The total number of users is 500, and each subgroup has 250 users. The probability of accusing an innocent user, P_{fa}, is 10^{-3}. Among the $K = 250$ colluders, $K^b = 100$ of them receive the fingerprinted base layer only, and the other $K^{be} = 150$ of the colluders receive fingerprinted

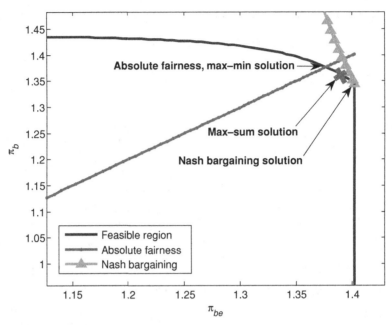

Fig. 8.11 Feasible region and bargaining solutions with utility function as in (8.44), $P_{fa} = 10^{-3}$, $N_b = N_e = 50000$, $K^b = 100$, $K^{be} = 150$, and $|U^b| = |U^{be}| = 250$

copies of high resolution. We consider the scenario in which all colluders have the same loss term $L = 0.1$, and $\theta = 5$.

Figure 8.11 shows the feasible region and the four bargaining solutions in Section 8.4.1 with the utility function as in (8.44), and bargaining powers in (8.41) are $a_b = 2$ and $a_{be} = 3$, which are proportional to K^b and K^{be}, respectively. This matches the real-world scenario, in which the larger group, with more people, has a larger bargaining power. In the example in Figure 8.11, the absolute fairness and the min–max solutions are the same. In addition, compared with the absolute fairness solution, the max–sum solution favors the subgroup with more people, which is SC^{be} in this example. The Nash bargaining solution also favors SC^{be}, as this group has a larger bargaining power.

8.5.2   Scenario 2: Reward is proportional to risk

In this section, we consider the scenario in which colluder i's reward is proportional to his or her risk of being detected; in the reward function in (8.12), we use $\gamma = 0.1$ and $D(P_d^{(i)}) = P_d^{(i)}$ as an example. That is,

$$\pi^g = -P_{d,c}^g L^g + \left(1 - P_{d,c}^g\right) R^g,$$

$$\text{where } R^g = \frac{\theta \left(f^g\right)^{0.1} P_{d,c}^g}{K^b \left(f^b\right)^{0.1} P_{d,c}^b + K^{be} P_{d,c}^{be}}. \tag{8.50}$$

Here, g is the subgroup index, and is either b if $i \in SC^b$ or be for $i \in SC^{be}$.

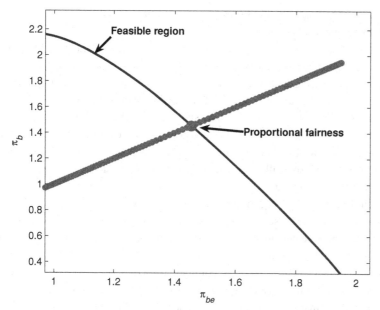

Fig. 8.12 Feasible region and proportional fairness solutions with utility function as in (8.50), $P_{fa} = 10^{-3}$, $N_b = N_e = 50000$, $K^b = 80$, $K^{be} = 170$, and $|U^b| = |U^{be}| = 250$

In this case, the reward each colluder gets is linear to his or her probability of being detected. In addition, colluders who contribute higher-resolution copies also gain more reward. The analysis of the four bargaining solutions is similar to that in Section 8.5.1 and is not repeated here. We also assume that all colluders have the same loss term $L = 0.1$.

To illustrate the feasible set and the bargaining solutions when the reward is proportional to the risk, we run simulations with the same setting as in Figure 8.11. Figure 8.12 shows the feasible region and the absolute fairness solution with the utility function in (8.50). First, different from Figure 8.11, every point in the feasible set with $0 \leq \beta \leq 1$ is Pareto optimal, and there are no non–Pareto optimal feasible points as the straight line segment in Figure 8.11. This is because, although P_d^{be} remains the same for all $\beta > \beta^+ = 1 - \frac{\sqrt{N_e}(\sqrt{N_b+N_e}-\sqrt{N_e})}{N_b}$, P_d^b is a monotonically increasing function of β. Hence, when $\beta > \beta^+$, for all colluders in SC^{be}, the denominator of the reward function (8.50) is an increasing function of β and the numerator is a constant. Consequently, different from Scenario 1, π^{be} is a decreasing function of β when $\beta > \beta^+$ in this scenario, and all points in the feasible set are Pareto optimal.

Furthermore, comparing the feasible region in Figure 8.11 with that in Figure 8.12, it is clear that both the maximum utilities that SC^b and SC^{be} can achieve are much higher if the reward is distributed proportionally ($\pi_{max}^b = 1.441$ and $\pi_{max}^{be} = 1.403$ in Figure 8.11, whereas $\pi_{max}^b = 2.182$ and $\pi_{max}^{be} = 1.947$ in Figure 8.12).

8.5.3 Scenario 3: Time-restricted bargaining

Here we consider the most general utility function in (8.50) to illustrate the time-sensitivity of collusion when colluder i's reward is proportional to the resolution of his or her contributing copy and his or her risk. We apply our analysis to the real video data and verify our results.

Figure 8.13 shows the bargaining equilibrium versus the number of stages that colluders take to make agreement with the utility function defined in (8.50) and different discount factors. Figure 8.13(a) uses $\delta_b = 0.7$ and $\delta_{be} = 0.85$, and Figure 8.13(b) is the result with $\delta_b = 0.7$ and $\delta_{be} = 0.85$. The feasible region and the Pareto optimal set with the same utility function for the first stage of the game are shown in Figure 8.10. It is clear from Figure 8.13 that all colluders' utilities decrease as the negotiation takes more rounds, and they have the incentives to finish the bargaining process as soon as possible. This is especially true for SC^b, whose utility decays faster. Therefore, at the very first bargaining stage, the first mover will make an offer based on the analysis of the equilibrium by solving (8.42) and (8.43). By comparing Figure 8.13(a) with (b), it is clear that a higher discount factor results in a higher payoff. The discount factors δ_b and δ_{be} can also be considered as the bargaining powers of SC^b and SC^{be}, respectively. For instance, let us consider the scenario in which the two groups of colluders cannot make agreement and they decide to collude within their own subgroups and generate two colluded copies with different qualities. Apparently, SC^b members would get much a smaller reward than SC^{be} members because their colluded copy has lower quality. Thus, SC^b has more incentives to cooperate with SC^{be}, which gives them a smaller bargaining power.

8.6   Maximum payoff collusion

The previous sections analyze when colluders will cooperate, how colluders select the collusion parameter β to ensure that cooperation increases all attackers' utilities, and how they bargain with one another. During collusion, in addition to β, attackers can also select with whom to collude and the number of fellow colluders – that is, K^b and K^{be}. In this section, we investigate the impact of (K^b, K^{be}) on colluders' utilities and analyze how attackers choose K^b and K^{be} to maximize their own payoffs. In this section, we consider the scenario in which the market value of the colluded copy is not time-sensitive ($\delta_b = \delta_{be} = 1$) and a colluder's received reward depends only on the resolution of his or her contributed copy but not his or her risk; that is, $D(P_d^{(i)}) = 1$.

8.6.1   Single-resolution collusion

We first start with the simple case in which all colluders receive fingerprinted copies of the same quality. Because all fingerprinted copies have the same resolution, there is no bargaining among colluders, and the risk and the reward are both evenly distributed among colluders, as shown in the utility function (8.17). In this scenario, recall that from Figure 8.2, colluding with more people does not always increase a colluder's payoff, and $\pi^{(i)}$ starts to decrease when the total number of colluders exceeds K_{max}. This section analyzes K_{max}.

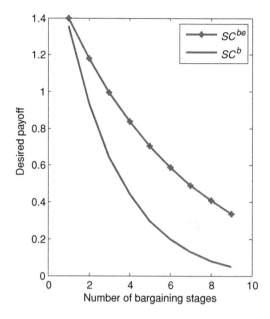

(a) $\delta_b = 0.7$ and $\delta_{be} = 0.85$

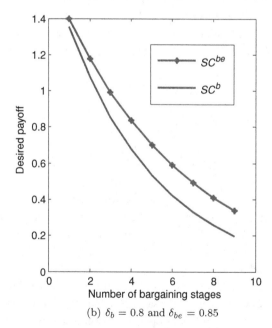

(b) $\delta_b = 0.8$ and $\delta_{be} = 0.85$

Fig. 8.13 Utilities of SC_b and SC^{be} versus the number of bargaining rounds. $P_{fa} = 10^{-3}$, $N_b = N_e = 50000$, $K^b = 100$, $K^{be} = 150$, and $|U^b| = |U^{be}| = 250$ with different discount factors

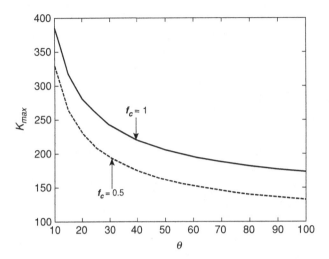

Fig. 8.14 K_{max} versus θ. $N_b = N_e = 50000$, $\sigma_w^2 = \sigma_n^2 = 1$, and $P_{fa} = 10^{-3}$

Given N_b, N_e and θ, to find K_{max}, we solve $\frac{\partial \pi^{(i)}}{\partial K} = 0$, or equivalently, find the root of $\frac{\partial \pi^{(i)}}{\partial x} = 0$, where $\pi^{(i)}$ is in (8.17) and $x = 1/K$ is the inverse of K. As in the previous section, we use $f_c = 1$ as an example; the analysis for other values of f_c is similar and is thus omitted. From (8.17), to find K_{max}, we solve

$$\frac{\partial \pi^{(i)}}{\partial x} = -\frac{\partial P_d^{(i)}}{\partial x}(1 + \theta x) + \left(1 - P_d^{(i)}\right)\theta = 0,$$

$$\text{where} \quad P_d^{(i)} = Q(a - b_s x)$$

$$\text{and} \quad \frac{\partial P_d^{(i)}}{\partial x} = \frac{b_s}{\sqrt{2\pi}}\exp\left\{-\frac{(a - b_s x)^2}{2}\right\}. \tag{8.51}$$

Here, $a = h/\sigma_n$ and $b_s = \sqrt{N_b + N_e}\sigma_w/\sigma_n$. Because of the existence of both the Gaussian tail function and the exponential function, it is difficult to find the analytical solution to (8.51), and we use numerical methods to solve (8.51). Given the solution x to (8.51), we consider the two neighboring integers $\lfloor 1/x \rfloor$ and $\lceil 1/x \rceil$, find the one that gives a larger $\pi^{(i)}$, and let K_{max} equal to that number.

Figure 8.14 shows K_{max} as a function of θ when the colluded copy has high and low resolutions, respectively. The system setup is the same as in Figure 8.3. From Figure 8.14, K_{max} takes a smaller value when the colluded copy has a lower resolution. For example, in Figure 8.14, with $\theta = 50$, $K_{max} = 206$ when the colluded copy has high resolution, and $K_{max} = 162$ when $f_c = 0.5$.

Furthermore, K_{max} is a decreasing function of θ. For example, with $f_c = 1$, $K_{max} = 385$ when $\theta = 10$ and $K_{max} = 173$ when $\theta = 100$. This is because, when θ takes a smaller value and when attackers emphasize risk minimization more, they prefer to collude with more people to lower their risk. Mathematically, it can be proved as follows. After rearranging both sides of (8.51), we have

$$\frac{1 - P_d^{(i)} - \frac{\partial P_d^{(i)}}{\partial x}}{\frac{\partial P_d^{(i)}}{\partial x}} = [1 - Q(a - b_s x)] \frac{\sqrt{2\pi}}{b_s} \exp\left\{\frac{(a - b_s x)^2}{2}\right\} - 1 = \theta^{-1}. \quad (8.52)$$

Assume that $\theta_1 \geq \theta_2$, and x_1 and x_2 are the solutions to (8.52) when $\theta = \theta_1$ and $\theta = \theta_2$, respectively. The left-hand side of (8.52) is a constant of θ. Consequently, $[1 - Q(a - b_s x_1)] \frac{\sqrt{2\pi}}{b_s} \exp\left\{\frac{(a - b_s x_1)^2}{2}\right\} - 1 = \theta_1^{-1} \leq \theta_2^{-1} = [1 - Q(a - b_s x_2)] \frac{\sqrt{2\pi}}{b_s} \times \exp\left\{\frac{(a - b_s x_2)^2}{2}\right\} - 1$. Here, we consider the scenario in which colluders collude only if their probability of being detected is smaller than 0.5, that is, $P_d^{(i)} = Q(a - b_s x) < 0.5$ and $a - b_s x > 0$. In such a scenario, both $[1 - Q(a - b_s x)]$ and $\exp\left\{\frac{(a - b_s x)^2}{2}\right\}$ are decreasing functions of x, thus $x_1 \geq x_2$ and $K_{max}^1 = 1/x_1 \leq K_{max}^2 = 1/x_2$.

8.6.2 Multiresolution collusion

From the previous section, when colluders receive fingerprinted copies of the same resolution, colluding with more people does not always increase an attacker's payoff. This is also true when colluders receive copies of different resolutions. Using SC^{be} as an example, Figure 8.15 shows the impact of the number of colluders on π^{be}. The system setup in Figure 8.15 is the same as that in Figure 8.5(a), and $K^{be} = 150$ is fixed.

In Figure 8.15(a), we consider the scenario in which K^{be} is fixed as 150 and colluders select the Nash bargaining solution with $a_b : a_{be} = K^b : K^{be}$. Figure 8.15(a) plots π^{be} when K^b takes different values. As shown in Figure 8.15(a), in this example, π^{be} achieves the maximum of 0.1681 when $K^b = 76$, and it decreases if K^b continues to increase. When $K^b > 159$, collaborating with colluders in SC^b does not help colluders in SC^{be} further increase their utilities, and colluders will collude only with their fellow attackers in the same subgroup. Figure 8.15(b) compares π^{be} when colluders select different collusion strategies, including the absolute fairness solution, the max–sum solution, and the Nash bargaining solution. It shows that, in this example, colluders in SC^{be} receive the highest payoffs from collusion if they choose the Nash bargaining solution with $a_b : a_{be} = K^b : K^{be}$. To conclude, from Figure 8.15(b), with a fixed $K^{be} = 150$, if colluders in SC^{be} want to maximize their own payoffs, the best strategy is to find another 76 attackers who receive the low-resolution copy and to choose the Nash bargaining solution with $a_b : a_{be} = K^b : K^{be}$. In the example in Figure 8.15, we fix $K^{be} = 150$ and find the optimum K^b to maximize SC^{be}'s utility. In practice, a colluder may wish to select both K^{be} and K^b to maximize his or her payoff. When attackers choose the absolute fairness solution, a colluder in SC^{be} is interested in finding the K^{b*} and K^{be*}

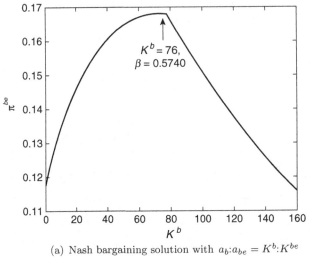

(a) Nash bargaining solution with $a_b{:}a_{be} = K^b{:}K^{be}$

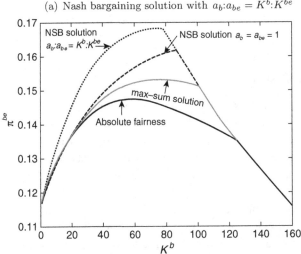

(b) Comparison of different fairness solutions

Fig. 8.15 π^{be} when $K^{be} = 150$ is fixed. $N_b = N_e = 50000$, $\sigma_w^2 = \sigma_n^2 = 1$, $\gamma = 1/3$, $P_{fa} = 10^{-3}$, and $\theta = 50$

that maximize π_{abs}, that is,

$$\pi_{abs}^* = \max_{(K^b, K^{be}) \in \mathbb{K}_p} -P_{d,c}^{be} + \left[1 - P_{d,c}^{be}\right] \frac{\theta}{K^b(f_b)^\gamma + K^{be}},$$

$$\text{s.t.} \quad -P_{d,c}^b + \left[1 - P_{d,c}^b\right] \frac{\theta(f_b)^\gamma}{K^b(f_b)^\gamma + K^{be}} = -P_{d,c}^{be} + \left[1 - P_{d,c}^{be}\right] \frac{\theta}{K^b(f_b)^\gamma + K^{be}}.$$

$$(8.53)$$

The analyses for the max–sum and the Nash bargaining solutions are the same and are omitted here.

Table 8.2 Maximization of π^{be}. $N_b = N_e = 50000$, $\theta = 50$, $\gamma = 1/3$, and $P_{fa} = 10^{-3}$

Fairness criteria	K^{b*}	K^{be*}	$\beta*$	v^{be*}
Absolute fairness	1	182	0.0062	0.1624
Max–sum fairness	16	206	0.1274	0.1682
Proportional fairness, $a_b = a_{be} = 1$	46	188	0.3792	0.1707
Proportional fairness, $a_b : a_{be} = K^b : K^{be}$	52	176	0.4421	**0.1742**

Table 8.3 Maximization of π^b. $N_b = N_e = 50000$, $\theta = 50$, $\gamma = 1/3$, and $P_{fa} = 10^{-3}$

Fairness criteria	K^{b*}	K^{be*}	$\beta*$	v^{b*}
Absolute fairness	1	182	0.0062	0.1624
Max–sum fairness	106	97	0.5858	0.1803
Proportional fairness, $a_b = a_{be} = 1$	1	126	0.0076	**0.2074**
Proportional fairness, $a_b : a_{be} = K^b : K^{be}$	106	97	0.5858	0.1803

For the example in Figure 8.5(a), the optimal (K^{b*}, K^{be*}) with different fairness criteria are shown in Table 8.2. As an example, from Table 8.2, if colluders prefer the Max–sum fairness solution, then for a colluder in SC^{be} to maximize his or her own payoff, he or she should find another 205 attackers who also receive high-resolution copies and 16 attackers who have the base layer only. If we compare the four solutions in Table 8.2, to maximize π^{be}, the optimal collusion strategy for a colluder in SC^{be} is to let $(K^b, K^{be}) = (52, 176)$ and to select the Nash bargaining solution with $a_b : a_{be} = K^b : K^{be}$. It helps colluders in SC^{be} receive a maximum payoff of 0.1742 among all possible payoffs that they could have.

For colluders in SC^b, we can use the same method to find the optimum pair (K^b, K^{be}) that maximizes π^b; Table 8.3 shows the results. For instance, from Table 8.3, if colluders decide to select the Nash bargaining solution with $a_b : a_{be} = K^b : K^{be}$, then to maximize π^b, a colluder in SC^b should find an additional 105 attackers who receive the base layer only and another 97 attackers who have the high-resolution copies. Similarly, by comparing all four collusion strategies, if colluder $i \in SC^b$ hopes to maximize his or her payoff, he or she should let $(K^b, K^{be}) = (1, 126)$ – that is, find another 126 attackers who receive both layers but no more attackers who have the base layer only – and select the Nash bargaining solution with $a_b = a_{be} = 1$. By doing so, π^b achieves the maximum of 0.2047.

8.6.3   Simulation results

In our simulations, we test the first forty frames of the "carphone" sequence in QCIF format, which is a popular test sequence for video processing. The base layer includes all the odd frames, and the enhancement layer contains all the even frames. The length of the fingerprints embedded in each frame is 2500, and the lengths of the fingerprints embedded in the base layer and the enhancement layer are $N_b = 50000$ and $N_e = 50000$, respectively. We use orthogonal fingerprint modulation [61], and use spread spectrum

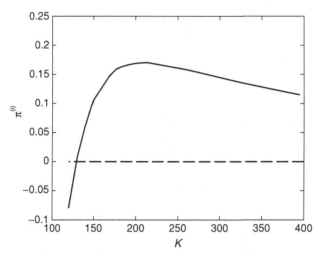

Fig. 8.16 Simulation results of a colluder's utility when all attackers receive fingerprinted copies of high resolution. The system setup is the same as that in Figure 8.2(a). The results are based on 2000 simulation runs

embedding [58] to embed fingerprints into the host signals. During collusion, colluders follow the two-stage collusion, and they adjust the power of the additive noise such that $||\mathbf{n}_j||^2 = ||JND_j\mathbf{W}_j^{(i)}||^2$. In addition, as an example, when defining the utility function, we let colluders select $\theta = 50$ and $\gamma = 1/3$. When identifying colluders, the fingerprint detector uses the self-probing detector in Chapter 6. The fingerprint detector selects the threshold h such that the probability of falsely accusing an innocent is $P_{fa} = 10^{-3}$.

We first consider the scenario in which all colluders receive the high-resolution copies. In such a scenario, they simply average all the fingerprinted copies that they have and then add additive noise to further hinder the detection. The simulation results are shown in Figure 8.16, and it is consistent with our analysis results shown in Figure 8.2(a). A colluder receives a positive payoff when there are more than 125 colluders, and $\pi^{(i)}$ reaches the maximum when K is around 206. As K continues to increase, $\pi^{(i)}$ starts decreasing.

We then consider the scenario in which colluders receive fingerprinted copies of different resolutions. Following the example in Figure 8.15(a), we fix the number of colluders who receive high resolution copies as $K^{be} = 150$. We select K^b such that $(K^b, K^{be} = 150) \in \mathbb{K}_p$ and it is possible for colluders to find at least one β that increases all colluders' payoffs. Furthermore, as in Figure 8.15, we consider the scenario in which colluders select the Nash bargaining solution with $a_b:a_{be} = K^b:K^{be}$. Figure 8.17 shows the simulation results of π^{be}, which is consistent with our analytical results shown in Figure 8.15(a). In the example in Figure 8.17, π^{be} reaches the maximum when $K^b = 76$. If there are more than 76 colluders who have the base layer only, the β that maximizes $g(\beta)$ in (8.41) with $a_b:a_{be} = K^b:K^{be}$ is the upper bound β^+, and π^{be} decreases quickly as K^b continues to increase.

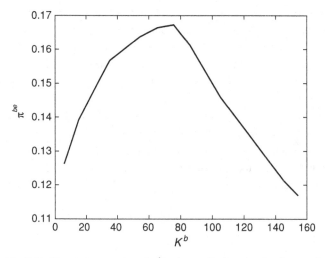

Fig. 8.17 Simulation results of π^{be} when colluders receive fingerprinted copies of different resolutions. The system setup is the same as that in Figure 8.15(a). The results are based on 2000 simulation runs

8.7   Chapter summary and bibliographical notes

A social network exists only if the users gain from cooperation and have the incentives to interact with one another. Therefore, to understand the necessary conditions for the formation of media-sharing social networks, we must investigate when and how users collaborate. We use multimedia fingerprinting system as an example, and study when and how attackers collude.

To study when colluders form a coalition, we first consider a scenario in which all attackers receive fingerprinted copies of the same resolution and the colluded copy is a simple average of all copies with equal weights. In such a scenario, attackers collude with one another if and only if the total number of colluders is larger than or equal to K_0. We then show that colluding with more attackers does not always increase a colluder's payoff, and analyze the optimum number of colluders (K_{max}) that maximizes a colluder's utility.

We then consider the scenario in which attackers receive fingerprinted copies of different resolutions. We first investigate the necessary conditions for colluders to cooperate with one another. We analyze $\mathbb{K}_p$, the set including all pairs of (K^b, K^{be}) where it is possible for all colluders to benefit from cooperation, and explore all possible collusion strategies that increase every attacker's utility for a given $(K^b, K^{be}) \in \mathbb{K}_p$. We then examine how the number of colluders in each subgroup, (K^b, K^{be}), affects colluders' utilities, and analyze the optimum strategy to select fellow attackers if a colluder wants to maximize his or her own payoff.

Furthermore, we study how the colluders reach agreements. We extend our model to address the special time-sensitive property of multimedia contents to analyze the colluders' behavior by modeling collusion as a time-sensitive bargaining process and find the equilibrium of the bargaining game. Our analysis shows that in the colluder social network, the colluders will make an agreement at the first bargaining stage and reach equilibrium; if the market value of the colluded copy is not time-sensitive, colluders choose different points in the feasible set, depending on the colluders' definition of "fairness" and their agreement on how to distribute the risk and the reward among themselves.

Interested readers can refer to references [63,66,68] for the study of equal-risk fairness criteria in different fingerprinting systems. When colluders' goal is only to minimize their probability of being detected, the analysis in references [61,63] shows that in a collusion attack with more attackers, each attacker has a smaller chance of being caught. The exponentially decay model for the reward during bargaining are also discussed in references [96–98].

9 Cooperation stimulation in peer-to-peer video streaming

In the previous chapter, we used colluders in multimedia fingerprinting as an example to study the necessary conditions for colluders to cooperate with one another, and investigated how attackers negotiate with one another and reach an agreement. In this chapter, we consider P2P networks, investigate how users in P2P systems cooperate with one another to form a social network, and study the optimal cooperation strategies.

As introduced in Chapter 3, mesh-pull–based P2P video streaming is one of the largest types of multimedia social networks on the Internet and has enjoyed many successful deployments. However, because of the voluntary participation nature and limited resources, users' full cooperation cannot be guaranteed. In addition, users in P2P live streaming systems are strategic and rational, and they are likely to manipulate any incentive systems (for example, by cheating) to maximize their payoffs [119]. As such, in this chapter, we study the incentives for users in video streaming systems to collaborate with one another and design the optimal cooperation strategies.

Furthermore, with recent developments in wireless communication and networking technologies and the popularity of powerful mobile devices, low-cost and high-quality–service wireless local area networks (WLANs) are becoming rapidly ingrained in our daily lives via public hotspots, access points, digital home networks, and many others. Users in the same WLAN form a wireless social network; such wireless social networks have many unique properties that make cooperation stimulation more challenging. For example, video streaming users in the wireless social network may use different devices – example, laptops PDAs, cell phones, and MP4 video players – and they have different requirements for the streaming quality and the power. For instance, compared with PDA users, laptop users would prefer high-resolution videos, and they are willing to use more transmission power for cooperation. Second, many users in the wireless social networks have high mobility, and the quality of their network connections may be unstable. All these motivate us to investigate the optimal cooperation strategies in wireless video streaming social networks [120].

In this chapter, we study P2P live streaming over the Internet, consider a simple scenario with nonscalable video coding, and study how to stimulate cooperation between any two users. We investigate the Nash equilibriums of the game and derive cheat-proof stimulation strategies. Such an analysis aims to stimulate each pair of users in P2P video streaming to cooperate with each other and achieve better performance. Then, we consider the unique properties of mobile devices and wireless channels and provide

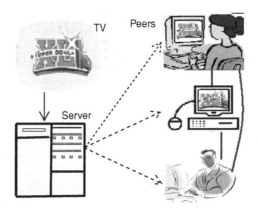

(a) Overall system

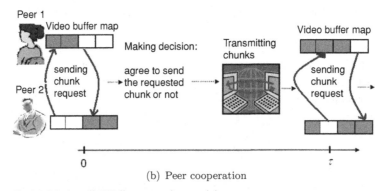

(b) Peer cooperation

Fig. 9.1 Mesh-pull P2P live streaming model

incentives for user cooperation in wireless video streaming social networks. Finally, to provide personalized service to users with different resources, we consider scalable video coding and design a chunk-request algorithm to maximize the quality of users' received videos.

9.1   Incentives for peer cooperation over the Internet

In this section, we consider P2P live streaming over the Internet, introduce a game-theoretic framework to model and analyze user dynamics, and study how to stimulate two users to cooperate with each other.

9.1.1   Mesh-pull P2P live streaming

We first briefly review the basic protocol and streaming mechanisms of mesh-pull P2P live streaming systems as shown in Figure 9.1(a). In a mesh-pull delivery architecture for live video streaming, a compressed video bit stream of bit rate B bps is divided into media chunks of M bits per chunk, and all chunks are available at the original streaming server's side. When a peer first joins the P2P live streaming network, he or she obtains a list of peers who are currently watching the video, and establishes a partnership with several peers. At any instance, a peer buffers up to a few minutes' chunks within a sliding window. Each user keeps a "buffer map," indicating the chunks that he or she has currently buffered and can share with others, and they frequently exchange buffer map information. Time is divided into rounds of τ seconds each; Figure 9.1(b) shows an example of how peers cooperate with each other. At the beginning of each round, every user sends a chunk request either to one of the peers or to the original streaming server. Then, the supplier either replies by providing the requested chunk or rejects the request.

9.1.2 Two-player game model

We assume that there are N users in the live streaming social network and every user buffers L chunks. The video stream is originally stored in the streaming server whose upload bandwidth can only afford to transmit K' chunks in one round (τ seconds) with $K' \ll N$. The server has no information of the network topology, and the P2P system is information-pull, which means that the server sends only chunks that are requested by some users, and it replies the chunk requests in a round-robin fashion. Because of the playback time lags among peers, different users request different chunks in one round, and the server cannot answer all users' requests. In such a scenario, peers have to help each other to receive more chunks and thus better-quality videos [121].

This section investigates the incentive mechanisms for peer cooperation in live streaming. We start with a simple scenario with two cooperating users and nonscalable video coding structure. To simplify the analysis, in this chapter, we consider a simple scenario in which in each round, every peer can request only one chunk from, and uploads at most one chunk to, the other peer. We build a game-theoretic framework to model and analyze the dynamics between these two users, and design an optimal and cheat-proof cooperation stimulation strategy.

We first define the utility (payoff) function of the two-player game. In each round, if player i accepts the other player k's request and sends the requested data chunk to k, we define i's cost c_i as the percentage of i's upload bandwidth used to transmit the requested chunk. That is, $c_i = M/(W_i \tau)$, where W_i is player i's total available upload bandwidth, M is the size of the chunk, and τ is the time duration of the round. If player k forwards the data that i requested and player i receives the chunk correctly, then i receives a gain of g_i, which is a user-defined value between 0 and 1. Every user in the P2P live streaming network defines his or her own value of g_i depending on how much he or she wants to watch the video. A higher value of g_i also implies that user i is willing to cooperate with others to get a better-quality video. For instance, if all the user does is watch the live streaming program and he or she is not distracted by other activities, he or she may select $g_i = 1$. On the other hand, if the user is watching several videos, browsing the Internet, or downloading files simultaneously, he or she will not value the live streaming much, and thus will set a lower value g_i. Here, we consider only the scenario in which all g_is are positive. This is because $g_i = 0$ corresponds to the scenario in which user i does not care about the video quality and thus will not join the P2P live streaming network. We further assume that there exists a minimum upload bandwidth W_{min} for all users, and thus c_i is upper-bounded by $c_{max} = M/W_{min}$. The minimum upload bandwidth constraint is necessary because if a user cannot even completely upload a chunk in one round period, other users have no incentives to cooperate with him or her. Here, W_i and g_i are player i's private information, which is not known to the other player unless player i reports it.

Let $a_i(j) \in \{0, 1\}$ be player i's action in round j, where $a_i(j) = 0$ means that player i does not cooperate in round j, and $a_i(j) = 1$ indicates that player i is willing to cooperate in this round. Let P_{12} denote the probability that the requested chunk is successfully transmitted from user 1 to user 2, and P_{21} is defined as the probability that user 2 successfully transmits the requested chunk to user 1. Then, player 1 and 2's payoffs in round j are

$$\pi_1(a_1(j), a_2(j)) = a_2(j)P_{21}g_1 - a_1(j)c_1 = a_2(j)P_{21}g_1 - a_1(j)\frac{M}{W_1\tau}$$

$$\pi_2(a_1(j), a_2(j)) = a_1(j)P_{12}g_2 - a_2(j)c_2 = a_1(j)P_{12}g_2 - a_2(j)\frac{M}{W_2\tau}. \quad (9.1)$$

This payoff function consists of two terms: the first term in π_i denotes user i's gain with respect to the other user's action, and the second term denotes user i's cost with respect to his or her own action. From (9.1), it is reasonable to assume that $P_{21}g_1 \geq c_1$ and $P_{12}g_2 \geq c_2$, as users will cooperate with each other only if cooperation can benefit both users and give them positive payoffs. Let $\pi(a_1(j), a_2(j)) = (\pi_1(a_1(j), a_2(j)), \pi_2(a_1(j), a_2(j)))$ be the payoff profile.

It is easy to check that, if this game will be played for only one time, the only Nash equilibrium is $(0, 0)$, which means no one will answer the other's request. According to the backward induction principle, this is also true when the repeated game will be played for a finite number of times with the game termination time known to both players. Therefore, in such scenarios, each player's only optimal strategy is to always play non-cooperatively. However, in many live streaming systems, these two players will interact for many rounds and no one can know exactly when the other user will quit the game.

Next, we show that cooperative strategies can be obtained under a more realistic setting. Let $\mathbf{a}_i = (a_i(1), a_i(2), \ldots)$ denote player i's behavior strategy in the infinitely repeated game, and $\mathbf{a} = (\mathbf{a}_1, \mathbf{a}_2)$ is the strategy profile. When the game is played more than one time, the summation of a user's payoffs in all times should be considered as each player's utility. However, in infinite-time game model, the summation usually goes to infinity, and therefore, the averaged payoff is considered instead. Define $x_i \overset{\Delta}{=} \lim_{T\to\infty} \frac{1}{T}\sum_{j=0}^{T} a_i(j) \in [0, 1]$ as user i's averaged action. In this chapter, for infinitely repeated games, we consider the following averaged utilities:

$$v_1(\mathbf{s}) = \lim_{T\to\infty} \frac{1}{T}\sum_{j=0}^{T} \pi_1(a_1(j), a_2(j)) = x_2 P_{21}g_1 - x_1\frac{M}{W_1\tau}$$

$$v_2(\mathbf{s}) = \lim_{T\to\infty} \frac{1}{T}\sum_{j=0}^{T} \pi_2(a_1(j), a_2(j)) = x_1 P_{12}g_2 - x_2\frac{M}{W_2\tau}. \quad (9.2)$$

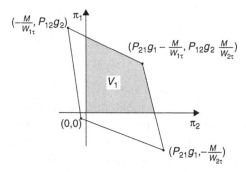

Fig. 9.2 Feasible and enforceable payoff profiles

For two different action profiles $(\mathbf{a}_1, \mathbf{a}_2) \neq (\mathbf{a}_1', \mathbf{a}_2')$, if their averaged actions are the same – that is, $(x_1, x_2) = (x_1', x_2')$ – then their corresponding averaged utilities are the same and $(v_1, v_2) = (v_1', v_2')$.

Let us analyze the Nash equilibriums of the infinitely repeated game with the utility function defined in (9.2). According to the Folk theorem, there exists at least one Nash equilibrium to achieve every feasible and enforceable payoff profile. A feasible payoff profile is the payoff that can be achieved, and an enforceable payoff profile is a feasible payoff in which both users have nonnegative utilities. The set of feasible and enforceable payoff profiles for the above game is:

$$V_1 = \text{convex hull} \{(v_1, v_2) | v_1 = \pi_1(\mathbf{a}_1, \mathbf{a}_2) \geq 0, v_2 = \pi_1(\mathbf{a}_1, \mathbf{a}_2) \geq 0,$$

$$a_1(j), a_2(j) \in \{0, 1\}\}. \tag{9.3}$$

Figure 9.2 illustrates the feasible and the enforceable regions of the previous infinitely repeated game. The feasible region is inside the convex hull of $\left\{(0, 0), (P_{21}g_1, -\frac{M}{W_{2\tau}}),\right.$ $(P_{21}g_1 - \frac{M}{W_{1\tau}}, P_{12}g_2 - \frac{M}{W_{2\tau}}), \left.(-\frac{M}{W_{1\tau}}, P_{12}g_2)\right\}$. V_1 is the gray region shown in Figure 9.2, which is the intersection of the feasible region and the first quadrant. It is clear that there exists an infinite number of Nash equilibriums.

For each feasible and enforceable payoff profile $(v_1, v_2) \in V_1$, let $(\mathbf{a}_1, \mathbf{a}_2)$ be a Nash equilibrium that achieves (v_1, v_2), and let (x_1, x_2) be the corresponding averaged strategy profile of $(\mathbf{a}_1, \mathbf{a}_2)$. If there are two or more Nash equilibriums that can achieve the same payoff profile (v_1, v_2), their averaged strategy profiles are the same, and we focus on the analysis of the average strategy profile here.

9.1.3   Nash equilibrium refinement

From the preceding analysis, one can see that the infinitely repeated game has an infinite number of Nash equilibriums, and apparently, not all of them are acceptable simultaneously. For example, the payoff profile $(0, 0)$ is not acceptable from either player's point of view. Therefore, in this section, we discuss how to refine the equilibriums based on new optimality criteria, analyze how to eliminate the less-rational Nash equilibriums, and find out which equilibrium is cheat-proof. In this section, we consider the most widely used optimality criteria as introduced in Chapter 4: Pareto optimality, proportional fairness, and absolute fairness.

9.1.3.1   Pareto optimality

A payoff profile $(v_1, v_2) \in V_1$ is Pareto optimal if and only if there is no $(v'_1, v'_2) \in V_1$ such that $v'_i \geq v_i$ for both users. Pareto optimality means that no one can increase his or her payoff without degrading the other's, which the rational players will always go to. It is clear from Figure 9.2 that the line segment between $(P_{21}g_1, -\frac{M}{W_{2}\tau})$ and $(P_{21}g_1 - \frac{M}{W_{1}\tau}, P_{12}g_2 - \frac{M}{W_{2}\tau})$ in the first quadrant and the line segment between $(-\frac{M}{W_{1}\tau}, P_{12}g_2)$ and $(P_{21}g_1 - \frac{M}{W_{1}\tau}, P_{12}g_2 - \frac{M}{W_{2}\tau})$ in the first quadrant is the Pareto optimal set.

9.1.3.2   Proportional fairness

Next, we further refine the solution set based on the criterion of proportional fairness. Here, we consider the Nash bargaining solution that maximizes the product $g(x_1, x_2) = v_1 \cdot v_2$; that is

$$\max_{0 \leq x_1, x_2 \leq 1} g(x_1, x_2) = v_1 v_2 = x_1 x_2 (P_{12} P_{21} g_1 g_2 + c_1 c_2)$$

$$- x_1^2 c_1 P_{12} g_2 - x_2^2 c_2 P_{21} g_1. \tag{9.4}$$

It can be easily shown that the objective function $g(x_1, x_2)$ and the constraint functions are continuously differentiable at any feasible points, satisfying the KKT conditions [122]. Thus the optimal (x_1^*, x_2^*) either satisfies $\nabla g(x_1^*, x_2^*) = 0$ or is on the boundary of the feasible region [0, 1]. If $\nabla f(x_1^*, x_2^*) = 0$, then (x_1^*, x_2^*) satisfies

$$\left. \frac{\partial \pi_1(x) \pi_2(x)}{\partial x_1} \right|_{(x_1^*, x_2^*)} = x_2^* (P_{12} P_{21} g_1 g_2 + c_1 c_2) - 2x_1^* P_{12} g_2 c_1 = 0$$

$$\left. \frac{\partial \pi_1(x) \pi_2(x)}{\partial x_2} \right|_{(x_1^*, x_2^*)} = x_1^* (P_{12} P_{21} g_1 g_2 + c_1 c_2) - 2x_2^* P_{21} g_1 c_2 = 0, \tag{9.5}$$

which has only one solution $(x_1^* = 0, x_2^* = 0)$ and $g(0, 0) = 0$. Apparently, this is not a desired solution. If (x_1^*, x_2^*) is on the boundary of the feasible region, then it satisfies

$$x_1^* = 1, \ x_2^* = \min \left\{ 1, \arg\max_{x_2} f(1, x_2) \right\}$$

$$= \min \left\{ 1, \frac{P_{12} P_{21} g_1 g_2 + c_1 c_2}{2c_2 P_{21} g_1} \right\},$$

$$\text{or} \quad x_2^* = 1, \ x_1^* = \min \left\{ 1, \arg\max_{x_1} f(x_1, 1) \right\}$$

$$= \min \left\{ 1, \frac{P_{12} P_{21} g_1 g_2 + c_1 c_2}{2c_1 P_{12} g_2} \right\}. \tag{9.6}$$

Combining (9.5) and (9.6), we can obtain the Nash bargaining solution

$$
x^* = \begin{cases} \left(\frac{P_{12} P_{21} g_1 g_2 + M^2/(W_1 W_2 \tau^2)}{2 P_{12} g_2 M/(W_1 \tau)}, 1 \right) & \text{if } \frac{P_{12} P_{21} g_1 g_2 + M^2/(W_1 W_2 \tau^2)}{2 P_{12} g_2 M/(W_1 \tau)} \leq 1, \\ \left(1, \frac{P_{12} P_{21} g_1 g_2 + M^2/(W_1 W_2 \tau^2)}{2 P_{21} g_1 M/(W_2 \tau)} \right) & \text{if } \frac{P_{12} P_{21} g_1 g_2 + M^2/(W_1 W_2 \tau^2)}{2 P_{21} g_1 M/(W_2 \tau)} \leq 1, \\ (1, 1) & \text{otherwise.} \end{cases}
\tag{9.7}
$$

9.1.3.3 Absolute fairness

Here we consider the absolute fairness solution, in which both players in the game have the same payoff. By solving $v_1(x_1^*, x_2^*) = v_2(x_1^*, x_2^*)$, we can get the unique absolute fairness solution as follows:

$$
x^* = \begin{cases} \left(\frac{P_{21} g_1 + M/(W_2 \tau)}{P_{12} g_2 + M/(W_1 \tau)}, 1 \right) & \text{if } P_{12} g_2 + \frac{M}{W_1 \tau} \geq P_{21} g_1 + \frac{M}{W_2 \tau}, \\ \left(1, \frac{P_{12} g_2 + M/(W_1 \tau)}{P_{21} g_1 + M/(W_2 \tau)} \right) & \text{if } P_{21} g_1 + \frac{M}{W_2 \tau} \geq P_{12} g_2 + \frac{M}{W_1 \tau}. \end{cases}
\tag{9.8}
$$

9.1.4 Optimal and cheat-proof strategies

In the previous section, we obtained several Nash equilibriums with different optimality criteria. However, as shown in (9.7) and (9.8), all these solutions need some private information (g_i, W_i, P_{ji}) reported by each player. Because of players' selfishness, reporting private information honestly cannot be taken for granted and players may cheat whenever they believe cheating can help increase their payoffs.

9.1.4.1 Cheating on private information

One way of cheating is to cheat on the private information (g_i, W_i, P_{ji}). First, let us examine whether the proportional fairness solution in (9.7) is cheat-proof with respect to (g_i, W_i, P_{ij}).

From (9.7), when

$$
\frac{P_{12} P_{21} g_1 g_2 + M^2/(W_1 W_2 \tau^2)}{2 P_{21} g_1 M/(W_2 \tau)} = \frac{P_{12} g_2}{2M/(W_2 \tau)} + \frac{M/(W_1 \tau)}{2 P_{21} g_1} \leq 1,
\tag{9.9}
$$

$x_1^* = 1$ is fixed and

$$
x_2^* = \frac{P_{12} g_2}{2M/(W_2 \tau)} + \frac{M/(W_1 \tau)}{2 P_{21} g_1}.
\tag{9.10}
$$

From (9.10), if user 2 reports false and lower values of the product $P_{12} g_2 W_2$, he or she can lower x_2^* and, therefore, further increase his or her own payoff $\pi_2(1, x_2^*) = P_{12} g_2 - x_2^* \frac{M}{W_2 \tau}$. Similarly, when

$$
\frac{P_{12} P_{21} g_1 g_2 + M^2/(W_1 W_2 \tau^2)}{2 P_{12} g_2 M/(W_1 \tau)} = \frac{P_{21} g_1}{2M/(W_1 \tau)} + \frac{M/(W_2 \tau)}{2 P_{12} g_2} \leq 1,
\tag{9.11}
$$

$x_2^* = 1$ is fixed and

$$
x_1^* = \frac{P_{21} g_1}{2M/(W_1 \tau)} + \frac{M/(W_2 \tau)}{2 P_{12} g_2}.
\tag{9.12}
$$

By falsely reporting lower values of the product $P_{21}g_1W_1$, user 1 can lower x_1^* and thus further increase his or her own payoff $\pi_1(x_1^*, 1) = P_{21}g_1 - x_1^* \frac{M}{W_1\tau}$. Therefore, the proportional fairness solution in (9.7) is not cheat-proof. Applying similar analysis to the absolute fairness solution in (9.8), we can also prove that the absolute fairness solution is also not cheat-proof with respect to private information. Therefore, players have no incentives to honestly report their private information. On the contrary, they will cheat whenever cheating can increase their payoff.

From the preceding analysis, to maximize their own payoffs, both players will report the minimum value of the product $P_{ji}g_iW_i$. Because we have assumed that $P_{ji}g_i \geq c_i = M/(W_i\tau)$ and $W_i \geq W_{min}$, both players will claim the smallest possible value $P_{ji}g_iW_i = M/\tau$, and both solutions in (9.7) and (9.8) become

$$\mathbf{x}^* = (1, 1); \tag{9.13}$$

the corresponding payoff profile is:

$$\mathbf{v}^* = \left(P_{21}g_1 - \frac{M}{W_1\tau}, P_{12}g_2 - \frac{M}{W_2\tau} \right). \tag{9.14}$$

To summarize, both players will report the smallest possible product $P_{ji}g_iW_i = M/\tau$, and they should always cooperate with each other. It is clear that the solution in (9.13) is an Nash equilibrium, Pareto optimal, and cheat-proof with respect to private information g_i, W_i, and P_{ji}.

9.1.4.2   Cheating on buffer map information

In mesh-pull P2P live streaming, at the beginning of each round, players exchange *buffer map information* with each other, and each player knows which chunks the other player has in his or her buffer. Thus, another way of cheating is to cheat on the buffer map information – that is, to hide the availability of some chunks in his or her buffer to the other user. It may help the cheating user reduce the number of requests from its peer and thus reduce the number of chunks that the cheating user needs to upload. As a result, it may help increase the cheating user's payoff.

The only circumstance under which cheating on buffer map information is effective is when all chunks claimed by the cheating user are already in the honest user's buffer, and the honest user has some chunks that the cheating user needs. That is, the honest user can help the cheating user but the cheating user cannot provide any useful chunks to the honest user. To prevent this cheating on buffer map information, a possible strategy is to let both users upload the same number of chunks.

To summarize, the two-player cheat-proof P2P live streaming cooperation strategy is as follows: in the two-player P2P live streaming game, to maximize each user's payoff and be resistant to possible cheating behavior, a player should not send more chunks than the opponent sends. Specifically, each player in each round should always agree to send the requested chunk unless the opponent did not cooperate in the previous round or there is no useful chunk in the opponent's buffer.

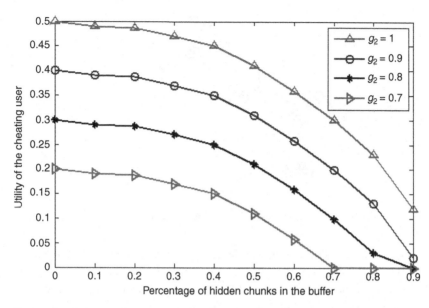

Fig. 9.3 Simulation results on two-person cheat-proof P2P live streaming cooperation strategy

9.1.5 Performance of the cheat-proof cooperation strategy

Here we study the performance of the two-player cheat-proof P2P live streaming cooperation strategy discussed earlier. In our simulations, there are 500 users in the network, and everyone can directly download chunks from the server. Each peer is either a DSL peer with 768 kbps uplink bandwidth, or a cable peer with 300 kbps uplink bandwidth. We fix the ratio between DSL peers and cable peers as 4:6. The video is initially stored at an original server with upload bandwidth of 3 Mbps. The request round is one second and each peer has a buffer that can store 30 seconds of video. We choose the "Foreman" CIF video sequence at frame rate of 30 frames per second, and pad the video to two hours. An MPEG-4 video codec is used to encode the video sequence into a nonscalable bit stream with bit rate 150 kbps. We divide the video into one-second chunks, thus each chunk has $M = 150K$ bits. Among those peers, we randomly choose two who cooperate with each other using the two-player cheat-proof P2P live streaming cooperation strategy. We set $g_1 = 1$, $g_2 = 1, 0.9, 0.8, 0.7$, and every peer claims the lowest bandwidth $W_{min} = 300$ kbps.

In our simulations, user 1 always reports accurate private information to user 2, and user 2 cheats on his or her buffer map information. Among all the chunks that user 2 has, he or she randomly selects p_c percent of them, and changes their status from "available" to "missing." Figure 9.3 shows user 2's utility with different gain g_2, where the x axis is p_c and the y axis is the utility v_2. From Figure 9.3, for a given g_2, a higher value of p_c gives the cheating user a lower payoff. In addition, when g_2 is small (for example, when $g_2 = 0.7$), if the cheating user selects a larger p_c, then he or she receives a zero payoff. That is, cheating cannot help a user increase his or her payoff, but rather decreases his or her utilities. It clearly demonstrates the cheat-proof property of the proposed cooperation

Fig. 9.4 Illustration of a wireless live-streaming social network

strategy. In addition, from our simulations, by cooperating with each other, both peers double the number of chunks that they receive: each user receives an average of 278 chunks without cooperation and an average of 542 chunks with cooperation. Therefore, cooperation enables users to reconstruct a better-quality video.

9.2   Wireless peer-to-peer video streaming

In this section, we first describe the wireless video streaming system model and study how two users in a wireless video streaming social network cooperate with each other. We then define the payoff function and introduce the game-theoretic framework to analyze user dynamics in wireless P2P video streaming.

9.2.1   Wireless video streaming model

Figure 9.4 shows the architecture of a wireless video live streaming social network. The wireless network service is provided by an access point connected to the Internet. The video bit stream is divided into media chunks of M' bits in the original server, each of which is channel-coded to M bits and transmitted in t seconds with data rate M/t bits per second. All chunks are available at the streaming server in the Internet. Here we assume that there is a dedicated channel of bandwidth B Hz for user cooperation, and that this channel is different from the channels between the access point and the users. We assume that the dedicated channel for user cooperation is symmetric and is a slow-fading channel with additive white complex Gaussian noise with variance σ_n^2. Here we use the wireless signal model

$$Y_i = Z_i + \frac{A_{ji}}{\sqrt{d_{ij}}} X_i, \qquad (9.15)$$

where X_i is the signal transmitted from user j to user i, Y_i is the signal that user i receives, Z_i is the additive Gaussian noise, A_{ji} is the channel fading factor, and d_{ij} is

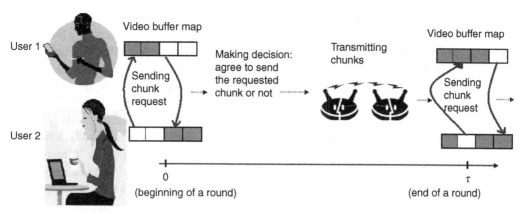

Fig. 9.5 Cooperation model for users in the P2P video streaming social network

the distance between user i and user j. Because of the symmetric channel assumption, $A_{ij} = A_{ji}$ and we will not differentiate between these two in the following section.

We assume that two users, u_1 and u_2, try to cooperate with each other and exchange chunks. Each user has a buffer of length L, which stores L_f chunks to be played and $L - L_f$ chunks that have been played. First, u_1 and u_2 exchange their buffer map information and the transmission power P_1 and P_2 that u_1 and u_2 use to transmit the chunks, respectively. To ensure that the packet can be successfully received, intuitively, users will cooperate only with people whose transmission power is above the minimum required transmission power P_{min}. Hence we assume that $P_1 \geq P_{min}$ and $P_2 \geq P_{min}$. The chunk exchange is done on a round-by-round basis, and each round is of duration τ seconds. At the beginning of each round, users send requests to each other, and at the same time keep downloading from the original server. Each user is allowed to send multiple requests in each round, and he or she can also answer multiple requests. Figure 9.5 shows how two users cooperate with each other: at the beginning of each round, every user sends chunk requests to the other. Then, the supplier either replies with the requested chunks or rejects the request. This request-answering process is repeated for all rounds.

9.2.2 Two-player game model

Assume that in the original structure, every user in the wireless live-streaming social network asks only the original server in the Internet for the media chunks, and two of them, u_1 and u_2, want to see if they can cooperate with each other to get a better-quality video. We use the following game to model the interaction between u_1 and u_2.

- **Players and player types:** There are two players, u_1 and u_2, in this game. Each player u_i has a type $\theta_i \in \{\text{laptop, PDA, PDA2}\}$. Users with different types have different costs to share chunks with each other and receive different gains if they receive chunks correctly. We assume that the battery used in the type PDA2 has a shorter life than that in PDA, and thus the cost per unit energy for PDA2 is higher than that for PDA. Also, the cost per unit energy for a laptop user is the smallest among the three.

- **Strategies:** In each round, the two players first exchange their buffer map information, and then send the chunk requests to each other. Upon receiving the chunk requests, each player u_i decides how many chunks he or she will send to the other user in this round. We define the number of chunks u_i agrees to send in round k as his or her strategy $a_i(k) \in \mathbb{Z}$. The two users use time division multiplexing access (TDMA) to share the cooperation channel with bandwidth B, and the bits to be transmitted within a round cannot exceed the channel capacity. Assume that in round k, u_1 and u_2 take turns to transmit, and are allocated τ_1 and τ_2 seconds, respectively, with $\tau_1 + \tau_2 \le \tau$. For u_1, the transmission data rate should be upper-bounded by the channel capacity; that is, $a_1(k)M \le B\tau_1 \log\left(1 + \frac{P_1 A_{12}}{d_{12}\sigma_n^2}\right)$. Similarly, for u_2, we have $a_2(k)M \le B\tau_2 \log\left(1 + \frac{P_2 A_{21}}{d_{12}\sigma_n^2}\right)$. Therefore, the constraint on the strategy profile $(a_1(k), a_2(k))$ at round k is

$$\frac{a_1(k)}{\log(1 + P_1 A_{12}^2/(d_{12}\sigma_n^2))} + \frac{a_2(k)}{\log(1 + P_2 A_{21}^2/(d_{12}\sigma_n^2))} \le \frac{\tau B}{M}. \qquad (9.16)$$

If (9.16) is not satisfied and users are transmitting above the channel capacity, the probability of erroneous transmission would be high and users will not receive any chunks successfully.

- **Utility function:** u_i's utility function π_i is the gain if he or she successfully receives the chunks minus the cost to forward chunks to the other user. Because users in the wireless live-streaming social network use mobile devices, the battery energy is the most limited resource. Hence we consider the transmission energy as the cost of cooperation, and let c_i be the cost per unit transmission energy for u_i. Different types of players would give different weights to the energy cost. For example, clients running on tight energy budgets would have a higher cost than those with powerful batteries. The total cost of user i transmitting M bits will be c_i times transmission power times transmission time, which is M divided by the channel capacity.

Let g_i be u_i's gain if he or she successfully receives one chunk. As with the P2P live streaming over the Internet in the previous section, every user in the P2P wireless video streaming social network defines his or her own value of g_i depending on how much he or she wants to watch the video. A higher value of g_i implies that user i is willing to cooperate with others to get a better-quality video.

Based on the preceding discussion, given the strategy profile $(a_1(k), a_2(k))$ in round k, the players' payoffs in the kth round are

$$\pi_1(a_1(k), a_2(k)) = a_2(k)g_1 - a_1(k)c_1 \frac{M P_1}{B \log(1 + \frac{P_1 A_{12}^2}{d_{12}\sigma_n^2})}$$

$$\pi_2(a_1(k), a_2(k)) = a_1(k)g_2 - a_2(k)c_2 \frac{M P_2}{B \log(1 + \frac{P_2 A_{21}^2}{d_{12}\sigma_n^2})}. \qquad (9.17)$$

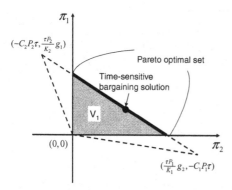

Fig. 9.6 Feasible and enforceable payoff profiles

Let $\pi(a_1(k), a_2(k)) = (\pi_1(a_1(k), a_2(k)), \pi_2(a_1(k), a_2(k)))$ be the payoff profile for round k. Define $K_1 = MP_1/\left[B\log\left(1 + \frac{P_1 A_{12}^2}{d_{12}\sigma_n^2}\right)\right]$, and $K_2 = MP_2/\left[B\log\left(1 + \frac{P_2 A_{21}^2}{d_{12}\sigma_n^2}\right)\right]$. K_i can be considered as the energy that user i spends on transmitting a chunk. It is reasonable to assume that $g_i \geq c_i K_i$; that is, cooperation always benefits participating users and brings them positive gain. In addition, we assume that there exists a C_{max}, where $c_i K_i \leq C_{max}$. Here c_i and g_i are user i's private information depending on user i's type, and are not known to others. We assume that users do not exchange their private information – that is, their types. Thus this is a game with incomplete information. We assume that users have the belief of the probability of the other users' type, which is independent of their own type. Users believe that another user's probabilities of being a laptop, PDA, and PDA2 user are P_1, P_2, and P_3, respectively, and we assume that this belief is independent of their own type.

9.3 Optimal cooperation strategies for wireless video streaming

Similar to the analysis in Section 9.1.2, in this section, we first extend the one-stage game model in Section 9.2.2 into an infinitely repeated game. We then apply several optimization criteria, such as Pareto optimality and time-restricted bargaining solution, to refine the Bayesian–Nash equilibriums (BNEs) of the game. Furthermore, we discuss the possible cheating behavior, and design cheat-proof cooperation strategy to stimulate cooperation between two users.

Follow the same discussion as in the previous section, the one-stage game model can be extended to an infinitely repeated game. Figure 9.6 illustrates both the feasible region and the enforceable region: the feasible region is inside the triangle bounded by dashed lines, and the enforceable feasible set V_1 is the shaded region shown in Figure 9.6. It is clear that there exists an infinite number of BNEs. To simplify our equations, in this chapter, we use $\mathbf{x} = (x_1, x_2)$ to denote the averaged BNE strategies corresponding to an enforceable payoff profile $(x_2 g_1 - x_1 c_1 K_1, x_1 g_2 - x_2 c_2 K_2)$ in V_1.

From the preceding analysis, we can see that the infinitely repeated game has an infinite number of equilibriums, and apparently, not all of them are simultaneously acceptable. In this section, we discuss how to refine the equilibriums based on new optimality criteria to eliminate those less-rational solutions and find cheat-proof equilibriums.

9.3.1   Nash equilibrium refinement

We use Pareto optimality and time-restricted bargaining to further refine Nash equilibriums.

Pareto optimality: It is clear from Figure 9.6 that the solid segment between $(-C_2 P_2 \tau,$ $g_1 \tau P_2 / K_2)$ and $(g_2 \tau P_1 / K_1, -C_1 P_1 \tau)$ in the first quadrant is the Pareto optimal set.

Time-sensitive bargaining solution: Because the players' action pair (a_1, a_2) must satisfy (9.16), and both players are rational and greedy, they will try to maximize the quality of their video streaming by requesting as many chunks as possible in each round. Every user will request all the chunks that are in his or her opponent's buffer and that he or she needs. However, the maximum number of bits transmitted in a round is limited by the channel capacity to ensure that the transmitted chunks are received correctly. Thus, users must bargain for their chunk-request quotas for every round to ensure that the total number of bits to be transmitted is not larger than the channel capacity. Also, the gain of receiving a chunk is time-sensitive. For instance, if users cannot reach an agreement before a chunk's playback time, users gain nothing even if they receive that chunk later. Therefore, users want chunks with earlier playback time to arrive as early as possible, and for chunks with later playback time, the worry of not receiving them on time to keep their value is less. To model such a phenomenon, the reward of receiving chunks is decreasing as the time spent on bargaining process increases.

We model the time-sensitive bargaining process for round k as follows: One user offers an action pair $(a_1^{(1)}, a_2^{(1)})$ first, and the other user can decide whether to accept this offer or to reject and offer back another action pair $(a_1^{(2)}, a_2^{(2)})$. This process continues until both players agree on the offer. If users reach an agreement at the jth action pair, then g_i decreases to $\delta_i^{j-1}(LC_{k,i})g_i$ for $i = 1$ or 2, where $\delta_i(LC_{k,i})$ is the discount factor for u_i, $LC_{k,i} = \{I_1, \ldots, I_q\}$ denotes the indexes of the chunks u_i wants to ask in the kth round, and $I(k)$ denotes the index of the chunk playing at the beginning of the kth round. Let t be the length of a chunk (in seconds).

Depending on how early the chunks must be received to preserve their value, we separate the chunks into two groups with different discount factors $\delta_i(LC_{k,i})$. The first group contains the chunks whose playback time is within the current bargaining round. The second group includes the chunks that are currently missing in the buffer and whose playback time is later than the end of the current bargaining round. At time t, for the $L - L_f$ chunks whose playback time has already passed, the reward of receiving

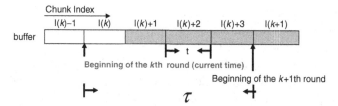

Fig. 9.7 Example of a user's buffer with six chunks

these $L - L_f$ chunks will remain 0 after time t. Therefore, the reward of these $L - L_f$ chunks will not decrease as time goes by, which means they should not be taken into consideration when calculating $\delta_i(LC_{k,i})$.

For the first group of chunks, assume that there are q' of them, and they are the first q' terms in $LC_{k,i}$. That is, among all the chunks that user u_i needs, there are q' of them whose playback time is the current (kth) round. Therefore, for these q' chunks, if users cannot reach agreement in the kth round, user u_i gains nothing by receiving them, as their playback time has already passed. For each of these q' chunks whose playback time is within the kth round, the later its playback time, the higher chance that the gain of receiving it can be preserved. We use such a chunk index difference to model how fast the value of a chunk decreases with time.

For the second group of chunks, which are the left $q - q'$ chunks in $LC_{k,i}$ that would be played after the kth round, user u_i still gains by receiving them if the bargaining process does not end within a round duration. However, if the bargaining process in round k takes more time, the number of chunks that can be transmitted in the kth round would decrease. Consequently, a smaller portion of the $q - q'$ chunks can be received in the kth round; thus, users receive a small gain. Therefore, even for the chunks that would be played after the kth round, their value would have a higher risk to be dropped if the bargaining time in the kth round is longer. That is, the value of these $q - q'$ chunks is also decreasing with time and should be counted in δ. Because chunks in the second group are not as urgently needed as the first group of chunks, we multiply the discounter factor δ of chunks in the second group by a "reduced constant" d.

According to this analysis, we define the discount factor of gain for user i at round k as follows:

$$\delta_i(LC_{k,i}) = 1 - \frac{\sum_{i=1}^{q'} \left[\frac{\tau}{t} - (I_i - I(k)) \right] + (q - q')d}{\frac{\tau}{t}(\frac{\tau}{t} + 1)/2 + (L_f - \frac{\tau}{t})d}, \qquad (9.18)$$

The first term in the numerator of (9.18) is the sum of the index difference between the requested chunks and the last chunk that can be played in the kth round.

Figure 9.7 gives an example to illustrate the time-sensitive property for the live-streaming scenario. The white blocks are the chunks that u_1 has in the buffer, the gray ones are the chunks he or she needs, and the buffer contains $L = 6$ chunks; $L - L_f = 1$ chunk's playback time has already passed, and this chunk does not play a role in the calculation of the discount factor δ. In this example, the number of chunks that u_1

would request is $q = 4$; $q' = 3$, and $\tau/t = 4$. Therefore, $\sum_{i=1}^{q'} \left[\frac{\tau}{t} - (I_i - I(k)) \right] = (4-1) + (4-2) + (4-3) = 6$, and $q - q' = 1$. Let $d = 0.8$, then the discount factor of gain for user i at round k is $\delta_i(LC_{k,i}) = 0.37$.

Because both players' payoffs decrease as the time for bargaining increases, the first mover would seek the equilibrium and offer at the first bargaining round for his or her maximum payoff. Let δ_1 and δ_2 be the averaged discount factor for u_1 and u_2 over all rounds. Here, we are discussing the equilibrium of the infinite game, which is the outcome when the game's end time goes to infinity. Thus, at each round, users do not need to predict the discount factor averaged over all rounds (including the future). Instead, for each round, users can calculate the averaged δ_i until the previous round, and find the equilibrium. Such a mechanism will result in the equilibrium as follows. The Pareto optimal equilibrium pair $((x_1^{(1)}, x_2^{(1)}), (x_1^{(2)}, x_2^{(2)}))$ for the infinitely repeated game happens when

$$x_2^{(2)} g_1 - x_1^{(2)} c_1 K_1 = \delta_1 x_2^{(1)} g_1 - x_1^{(1)} c_1 K_1$$

$$x_1^{(1)} g_2 - x_2^{(1)} c_2 K_2 = \delta_2 x_1^{(2)} g_2 - x_2^{(2)} c_2 K_2,$$

$$\text{where} \quad x_1 \frac{K_1}{P_1} + x_2 \frac{K_2}{P_2} = \tau. \tag{9.19}$$

Because two users take turns to make the first offer, the time-sensitive bargaining strategy (x_1^*, x_2^*) is

$$x_1 = \frac{1+m}{2} \times \frac{(1-\delta_1)\frac{P_2}{K_2} g_1 \tau}{(m-1)K_1 c_1 + (m-\delta_1)\frac{K_1 P_2}{K_2 P_1} g_1}$$

$$x_2 = P_2 \frac{\tau - x_1 \frac{K_1}{P_1}}{K_2}, \quad \text{where} \quad m = \frac{g_2 + c_2 K_2 \frac{P_2}{P_1}}{\delta_2 g_2 + c_2 K_2 \frac{P_2}{P_1}}. \tag{9.20}$$

It is clear that the bargaining solution in (9.20) depends on the knowledge of both users' types – that is, the private information, which is unavailable. Both players know the discount factors because the discount factors depend only on the chunks to be requested that the two users must exchange every round. Although at the beginning, users do not know each other's type, they can probe it during the bargaining process using the following mechanism. Let T_1 be u_1's type, which is only known to u_1; let T_2 be u_2's type, and $T(j)$ is the jth type. At the first bargaining stage, without loss of generality, let u_1 be the first mover. User u_1 calculates all the bargaining equilibriums $(a_1^{(1)}(T_1, T(j)), a_2^{(1)}(T_1, T(j)))$ for $j = 1, 2, 3$, corresponding to the three possible types of u_2. Then u_1 chooses the the equilibrium j' that gives highest $p_{j'} \pi_1(a_1^{(1)}(T_1, T(j')), a_2^{(1)}(T_1, T(j')))$. User u_2 will accept the offer if $\pi_2(a_1^{(1)}(T_1, T(j)), a_2^{(1)}(T_1, T(j)))$ is larger than or equal to $\pi_2(a_1^{(1)}(T_1, T_2), a_2^{(1)}(T_1, T_2))$. If not, u_2 will offer back $(a_1^{(2)}(T_1, T_2), a_2^{(2)}(T_1, T_2))$, and the two users reach an agreement. Because u_1 calculates the offer based on the equilibrium in (9.20), which depends on u_1's own type, u_2 can probe u_1's type based on the offer he or she made. Thus after the first bargaining stage in the first chunk-requesting round, u_2 knows u_1's type, and because u_2 will make the first move in the next round, after two rounds, each user has the information about the other's type.

9.3.2   Cheat-proof cooperation strategy

Users in peer-to-peer wireless video streaming social networks will try to maximize their own utility even by cheating. Therefore, to ensure fairness and to give incentives to users, it is crucial that the cooperation strategy be cheat-proof. In this subsection, we first discuss possible cheating methods, and then design the two-person cheat-proof cooperation strategy in peer-to-peer wireless video streaming social networks.

9.3.2.1   Cheating on private information

Because users know each other's private information (g_i, c_i) from the offers they make, users can cheat by making different offers. First, let us examine whether the time-sensitive bargaining solution in (9.20) is cheat-proof with respect to (g_i, c_i) – that is, determine whether a smaller x_2 can help increase x_1 or decrease P_2 and thus increase π_2.

x_1 is a function of m and

$$\frac{\partial x_1}{\partial m} = -\frac{(1+m)\left(K_1 c_1 + \frac{K_1 P_2}{K_2 P_1} g_1\right)(1-\delta_1)\frac{P_2}{K_2} g_1 \tau}{2\left[(m-1)K_1 c_1 + (m-\delta_1)\frac{K_1 P_2}{K_2 P_1} g_1\right]^2}, \tag{9.21}$$

which is always smaller than 0 because $m \geq 1 \geq \delta_1$. Thus, x_1 is a monotonically decreasing function of m if $\delta_1 < 1$.

Furthermore,

$$\frac{\partial m}{\partial g_2} = \frac{(\delta_2 - 1)c_2 K_2 \frac{P_2}{P_1}}{\left(\delta_2 g_2 + c_2 K_2 \frac{P_2}{P_1}\right)^2} \leq 0 \quad \text{and}$$

$$\frac{\partial m}{\partial c_2} = \frac{(1-\delta_2)K_2 \frac{P_2}{P_1}}{\left(\delta_2 g_2 + c_2 K_2 \frac{P_2}{P_1}\right)^2} \geq 0. \tag{9.22}$$

Therefore, m is a monotonically decreasing function of g_2 and is a monotonically increasing function of c_2 if $\delta_2 < 0$. Thus, u_2 can have a higher payoff by making the bargain offer by using a smaller g_2, a larger c_2, and a smaller P_2. Similarly, u_1 can also achieve higher utility by offering the equilibrium based on a smaller g_1, a larger c_1, and a smaller P_1.

From this analysis, both players will bargain based on the minimum value of g_i and maximum value of c_i. As we have assumed that $g_i \geq c_i K_i$, and $P_i \geq P_{min}$, both players will make the offer based on $g_i = c_i K_i = C_{max}$, and $P_i = P_{min}$. Thus the solution (9.20) becomes:

$$x_1^* = \frac{(\delta_2 + 3)(1-\delta_1)}{2(4-(1+\delta_1)(1+\delta_2))} \times \frac{\tau}{M/\left[B\log(1+\frac{P_1 A_{12}^2}{d_{12}\sigma_n^2})\right]},$$

$$x_2^* = \frac{\tau}{M/\left[B\log(1+\frac{P_1 A_{21}^2}{d_{12}\sigma_n^2})\right]} - x_1^*, \tag{9.23}$$

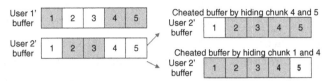

Fig. 9.8 Example of how to cheat on buffer information

which implies that both players should always cooperate with each other. It is clear that the solution in (9.23) forms a Nash equilibrium, is Pareto optimal, and is cheat-proof with respect to private information g_i and c_i. The user whose discount factor is closer to 1 has an advantage, and if $\delta_1 = \delta_2$, then $x_1^* = x_2^*$ and is equal to half the number of chunks that can be transmitted in τ seconds.

9.3.2.2   Cheating on buffer map information

The other way of cheating is to cheat on buffer information – that is, although player i has chunk k in the buffer, he or she does not report it to the opponent in an effort to reduce the number of requests from the opponent. However, from (9.18), hiding chunks that the other user needs might increase the other user's discount factor.

Take Figure 9.8 as an example. The white blocks are chunks in the buffer, and the gray blocks are the chunks that the user needs. Assume that user 1 always honestly reports his or her buffer information, and the time-sensitive bargaining solution gives a two-chunk request quota for user 1 and a two-chunk request quota for user 2. Apparently, user 1 will ask for two of chunks 1, 4, and 5 from user 2, and user 2 will ask for chunks 2 and 3 from user 1. Now assume that user 2 wants to hide chunks in his or her buffer from user 1, so the number of chunk requests from user 1 to user 2 will decrease, resulting in an increase of user 2's payoff in this round. It is clear that user 2 must hide at least two chunks to increase his or her payoff, because if user 2 hides only one chunk, there are still two chunks in user 2's buffer that user 1 needs. User 2 can choose two among chunks 1, 4, and 5 to hide; hiding different chunks will lead to different utilities. For instance, if user 2 hides chunks 1 and 4, this means chunk 5 is the only chunk that user 1 needs. However, user 2 would request chunks 2 and 3 from user 1. Because chunk 4 has a later playback time than those of chunks 2 and 3, the discount factor of user 1's gain will be larger than that of user 2. Thus, user 1 will have more advantage in the time-sensitive bargaining process, and the bargaining solution might be changed to a three-chunk request quota for user 1 and a one-chunk request quota for user 2. As a result, user 2's utility decreases because now he or she can request only one chunk from user 1. Therefore, user 2 has no incentive to cheat on buffer information by hiding chunks 1 and 4.

Although user 2's cheating on buffer information will always increase the discount factor of user 1's gain (δ_1), it does not necessarily always decrease the chunk-request quota. The reason is that the chunk-request quota is always an integer, as a partial chunk gives no gain for either user and the users would like to round the time-sensitive solution to the nearest integer. For instance, if before cheating, the time-sensitive bargaining solution is $(1.8, 2.2)$, and the solution changes to $(2.4, 1.6)$ after cheating, both solutions round to $(2, 2)$, which means that if user 2 hides the chunks properly to keep δ_1 low so that the chunk-request quota does not change after cheating, cheating on buffer information will increase user 2's utility because user 2 can still request two chunks from user 1, and there is only one chunk in user 2's buffer that user 1 needs.

Therefore, to prevent selfish users from gaining higher utility by cheating on buffer information, each player should not send more chunks than what the other user has sent to him or her.

9.3.2.3 Cheating on transmission power

The transmission power that user 1 and user 2 use for cooperation, P_1 and P_2, are declared in the beginning of the game, and they directly influence the feasible region, as in Figure 9.6 and the bargaining solution (9.23). As discussed in Section 9.3.2.1, user i can increase his or her payoff by decreasing P_i; thus, both users will declare that they use the minimum power P_{min}. However, if the user declares that he or she transmits the chunks using P_{min} but the actual power used for transmission is smaller than P_{min}, he or she can have higher utility by paying less cost for cooperation.

Given the signal model in (9.15), the receiver must estimate the attenuation term $A_{ij}/\sqrt{d_{ij}}$ before estimating the transmitted power. Suppose user i wants to estimate $A_{ij}/\sqrt{d_{ij}}$. If user j is honest, user i can simply ask user j to transmit a probing signal using P_2 to estimate the attenuation. However, in the fully distributed system, user j might be cheating and may transmit the probing signal with power smaller than P_2; thus the attenuation that user i estimated will be more serious than the real attenuation. To solve this problem, user i sends to user j a probing signal that user j cannot decode and asks user j to transmit back the received signal; user i can then investigate the attenuation from the replied signal.

If user i sends the probing signal X to user j, then the signal Y_j that user j receives is $Z_j + A_{ij}/\sqrt{d_{ij}}X$. Suppose the selfish user j wants to manipulate the signal; he or she can secretly amplify Y_j with a constant $\alpha < 1$ and then send αY_j back to user i. Then the replied signal Y_i that user i receives will be

$$Y_i = Z_i + \alpha \frac{A_{ij}}{\sqrt{d_{ij}}} Z_j + \alpha \frac{A_{ij}^2}{d_{ij}} X. \tag{9.24}$$

Because user i knows X and the noise power σ_n^2, he or she can easily extract $\alpha \frac{A_{ij}^2}{d_{ij}} X$ from Y_i, divide the energy of the residue by σ_n^2, and get the estimation of $1 + \alpha^2$. Given α, the attenuation term $\frac{A_{ij}^2}{d_{ij}}$ can easily be estimated. From this analysis, such probing procedure is cheat-proof, because no matter how user j manipulates the signal, the estimation of the attenuation term is independent of α.

After estimating $\frac{A_{ij}^2}{d_{ij}}$, the transmission power can easily be estimated by calculating the averaged power of the signal at the receiver's side. Therefore, user i can compute the estimated transmission power $P_j'(k)$ by user j at the kth round by

$$P_j'(k) = \frac{d_{ij}}{A_{ij}^2} \frac{1}{\tau_j} \int_{t=t_k}^{t=t_k+\tau_j} \left[y^2(t) - \sigma_n^2 \right], \tag{9.25}$$

where $y(t)$ is the received signal, t_k is the beginning of user j's transmission in the kth round, and τ_j is the duration of user j's transmission in the kth round.

Thus we design a mechanism to prevent cheating on transmitted power based on $P'_j(k)$ in (9.25):

- Each user i at each round k uses (9.25) to estimate user j's transmission power. If $P'_j(k)$ is smaller than P_{min}, then at the $(k+1)$th round, user i transmits the chunks using $P'_j(k)$. If $P'_j(k) \geq P_{min}$, user i uses P_{min} for cooperation.
- Each user estimates the transmission power at every round and follows the preceding procedure.

Using this mechanism, if user i decides to cheat by transmitting chunks with power $P'_i \leq P_{min}$, then the other user j can estimate P'_i and use P'_i to transmit the chunks for user i in the next round. Therefore, although user i increases his or her payoff in the current round, the payoff will be decreased in the next round; thus, the actual channel capacity is less than the user's estimation using P_{min}. Therefore, the probability for both users of successfully receiving the requested chunks would decrease and lead to no gain, as they cannot receive the extra chunks by cooperation. Thus, no user has the incentive to cheat on the transmission power if both follow this mechanism.

9.3.2.4   Two-player cheat-proof cooperation strategy

Based on the preceding analysis, we can conclude that, in the two-player wireless live streaming game, to maximize each user's own payoff and resist possible cheating behavior, each player in each round should always agree to send the requested chunks up to the bargained chunk-request quota as in (9.20) and should not send more chunks than his or her opponent has sent to him or her. Also, each user should estimate the other user's transmission power in every round, and use the estimated power for transmission if the estimation result is smaller than P_{min}. We refer to this strategy as the *two-player cheat-proof wireless video streaming cooperation strategy*.

9.3.3   Simulation results for two-user case

To demonstrate the cheat-proof property of the two-user cooperation strategy, we use a WiFi network as an example. The link from the wireless router to the Internet is a DSL link with 1.5 Mbits download bandwidth. There are ten users in the network using live-streaming service without cooperation, and another five users using Internet resources at the same time. We fix the ratio among laptop, PDA, and PDA2 users as 1:2:2. The video is initially stored at an original server with an upload bandwidth of 3 Mbps, and there are another 400 users in the Internet watching the same video stream. The request round is one second and the buffer length is ten seconds. We choose the "Foreman" CIF video sequence with frame rate thirty frames per second. We encode the video into a single-layer bit stream with 150 kbps, and divide the video into chunks of 0.1 second; thus the chunk size before channel coding is $M' = 15$ kilobits. For the WiFi network, we apply the BCH code with rate 15/31. Hence the chunk size for wireless live-streaming users is $M = 31$ kilobits. Among the ten live streaming users in the WiFi network, we

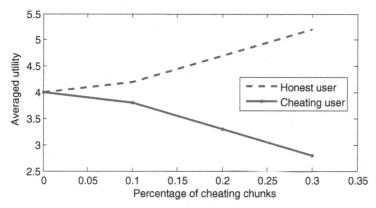

Fig. 9.9 Utilities of cheating player and honest player versus number of cheating chunks in buffer

randomly choose two users to cooperate using the two-player cheat-proof wireless live streaming cooperation strategy. We run the simulation forty times with forty different pairs of users. We set $g_i = c_{max} = 1, c_{PDA2} = 0.8c_{max}, c_{PDA2} : c_{PDA} : c_{laptop} = 1 : 0.9 : 0.4$. $P_{min} = 100$ mW, the noise power is 20 mW, and the bandwidth is $B = 200$ kHz. δ_1 and δ_2 in (9.18) are set to be the same, and equal to 0.1. We assume that one user is always honest and the other user cheats on the buffer map information.

Figure 9.9 shows the averaged utilities of the cheating player versus the number of hidden chunks in the buffer. The utility is averaged over all the simulation runs and all the rounds. It is clear that the cheating user receives a lower payoff if he or she hides more chunks in his or her buffer. Thus the best policy is to be honest, which demonstrates the cheat-proof property of our cooperation strategy. In addition, from the simulation, a user receives an averaged number of 3.7 chunks per second without cooperation; this number is increased to 6.2 chunks per second per peer with cooperation. Thus, cooperation can help significantly improve the received videos' quality in wireless live streaming social networks.

9.4   Optimal chunk request algorithm for P2P video streaming with scalable coding

The previous sections discussed the cheat-proof cooperation stimulation strategies for P2P live streaming over the Internet and wireless networks when nonscalable video coding is used. In this section, we extend the cooperation strategy to a scenario with layered video coding, in which different chunks may belong to different layers and thus have different gains. In this scenario, an important issue is to schedule the chunk requests to maximize each peer's utility. We investigate the chunk-request algorithm for a two-person P2P live streaming social network that optimizes three different video quality measures in Section 9.4.2. We then design a chunk request algorithm considering the tradeoff between different quality measures.

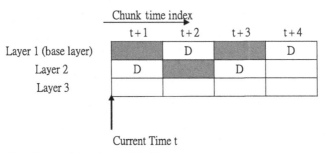

Fig. 9.10 Example of a user's buffer map

9.4.1   P2P video streaming with scalable video coding

In P2P video streaming social networks, peers belong to different domains with different uplad/download bandwidths, and scalable video coding is often widely adopted to accommodate heterogenous networks. Scalable video coding decomposes the video sequence into different layers of different priority. The base layer contains the most important information in the video and is received by all users, and the enhancement layers gradually refine the reconstructed sequence at the decoder's side. Although scalable video coding provides service depending on peers' bandwidth capacity, it also has unique challenges when used in P2P live streaming social networks. The importance of different layers is unequal, as higher layers cannot be decoded without successful decoding of all the lower layers. Therefore, in a P2P live streaming social network using scalable video coding, lower layers should be assigned higher priorities in the chunk-request algorithm.

In this chapter, we encode a video into L layers, and assume that every layer has the same bit rate B_s bps. We further divide each layer into layer chunks of τ seconds each. Figure 9.10 shows an example of the buffer map at one user's end. The gray blocks represent chunks that are available in the buffer, the white blocks denote the missing chunks, and D stands for missing chunks that are directly decodable after arriving. A chunk is *decodable* if and only if all the lower layers with the same time index have been decoded correctly.

For user j, we define $a_t^{(j)}$ as the number of decodable chunks with time index t that are available in j's buffer, and define the decodable chunk number vector as $A^{(j)} = \left[a_1^{(j)}, a_2^{(j)}, \dots\right]$. For instance, in the example in Figure 9.10, $a_{t+1}^{(j)} = 1$, $a_{t+2}^{(j)} = 0$, $a_{t+3}^{(j)} = 1$, and $a_{t+4}^{(j)} = 0$. $N^{(j)}$ is the total number of chunks that peer j receives during his or her stay in the P2P live streaming social network.

9.4.2   Video quality measure

With scalable video coding, the chunk-request algorithm should maximize the received video quality. Here, we use the following three popular criteria to measure the video quality.

Chunk decodable rate: Every user in the P2P live streaming social network has limited bandwidth available, and every peer wants to use it as efficiently as possible. The chunk decodable rate $R^{(j)}$ measures the bandwidth efficiency of the chunk-request algorithm, and it is defined as

$$R^{(j)} \triangleq \frac{\sum_i a_i^{(j)}}{N^{(j)}}. \tag{9.26}$$

It calculates the percentage of decodable chunks among all those that user j receives.

Video smoothness: A video stream with nearly constant quality gives a better user experience than one with large variations in quality. The video smoothness measure S is defined as

$$S^{(j)}(A) \triangleq \sum_i |a_i^{(j)} - a_{i-1}^{(j)}|. \tag{9.27}$$

where $|\cdot|$ is the absolute value operator. $S^{(j)}(A)$ takes a larger value when the variance of $\{a_t^{(j)}\}$ increases, and decreases when the difference between adjacent $\{a_t^{(j)}\}$ is smaller. To improve the quality and maximize the smoothness of the received video, user j should request the chunks to minimize $S^{(j)}(A)$.

Video discontinuity ratio: The discontinuity ratio $\alpha^{(j)}$ of peer j is defined as the percentage of instances that a video is undecodable and unplayable. With scalable video coding, if all chunks in the base layer are available, then the video is decodable and playable. Thus, the discontinuity ratio considers only the availability of chunks in the base layer, and peer j's video is unplayable at chunk time i if $a_t^{(j)} = 0$. So $\alpha^{(j)}$ is defined as

$$\alpha^{(j)} \triangleq \frac{\sum_i I[a_i^{(j)} > 0]}{T^{(j)}}, \tag{9.28}$$

where $I[\cdot]$ is the indicator function. $T^{(j)}$ is the duration for which player j stays in the P2P live streaming network.

9.4.3 Optimal chunk-request algorithms

In this section, we first design three chunk-request algorithms that maximize the chunk decodable rate, maximize the smoothness, and minimize the discontinuity ratio, respectively. We then design the optimal chunk-request algorithm for P2P live streaming that jointly considers the above three video quality measurements. Here, we consider the simple scenario in which each user request at most one chunk per round; for the scenario in which a user can ask for multiple chunks in one round, the analysis is similar and is thus omitted.

Assume that player j's current decodable chunk number vector is $A^{(j)} = \left[a_1^{(j)}, a_2^{(j)}, \ldots\right]$, and player j requests the chunk $LC(t', l)$ with time index t' and layer index l. Once j receives chunk $LC(t', l)$, his or her new decodable chunk number vector is $A'^{(j)} = \left[a_1'^{(j)}, a_2'^{(j)}, \ldots \right]$ where $a_i'^{(j)} = a_i^{(j)}$ for all $i \neq t'$. $a_{t'}'^{(j)} > a_{t'}^{(j)}$ if $LC(t', l)$ is decodable, and $A'^{(j)} = A^{(j)}$ if $LC(t', l)$ is not decodable.

- **Maximizing chunk decodable rate:** We first design a chunk-request algorithm that maximizes the chunk decodable rate. According to the definition in (9.26), chunks that are not decodable do not give any immediate quality improvement. Thus, if peer j requests and receives a chunk $LC(t', l)$ with time index t' and layer index l, the gain that j receives is

$$
g_j = \begin{cases} g & \text{if } LC(t', l) \text{ is decodable,} \\ 0 & \text{otherwise,} \end{cases} \tag{9.29}
$$

where $g > 0$ is a constant. Therefore, maximizing the payoff function in (9.1) is equivalent to always requesting chunks that are directly decodable after arriving and thus $g_j = g$. In the example in Figure 9.10, at the current state, requesting any one of the D chunks, $LC(t + 1, 2)$, $LC(t + 2, 1)$, $LC(t + 3, 2)$, and $LC(t + 4, 1)$, will maximize the player's payoff.

- **Maximizing video smoothness:** We now consider the scenario in which the player places more emphasis on the smoothness of the received video. If player j requests and successfully receives the chunk $LC(t', l)$, player j's increment in the smoothness is

$$
g_j = \begin{cases} S^{(j)}(A) - S^{(j)}(A') & \text{if } LC(t', l) \text{ is decodable,} \\ 0 & \text{otherwise.} \end{cases} \tag{9.30}
$$

For a decodable chunk $LC(t', l)$, we have $S^{(j)}(A) - S^{(j)}(A') = \sum_{i=t'}^{t'+1} \times \left(|a_i^{(j)} - a_{i-1}^{(j)}| - |a_i'^{(j)} - a_{i-1}'^{(j)}| \right)$. Therefore, to maximize the video smoothness, player j should choose the decodable chunk $LC(t', l)$ that maximizes the difference $\sum_{i=t'}^{t'+1} \left(|a_i^{(j)} - a_{i-1}^{(j)}| - |a_i'^{(j)} - a_{i-1}'^{(j)}| \right)$. Using the buffer map in Figure 9.10 as an example, the peer should request $LC(t + 4, 1)$.

- **Minimizing video discontinuity ratio:** For the scenario in which peer j wants to minimize the video discontinuity ratio, if player j receives a chunk $LC(t', l)$, his or her gain is

$$
g_j = \begin{cases} g & \text{if } l = 1, \\ 0 & \text{otherwise.} \end{cases} \tag{9.31}
$$

To maximize g_j, the peer should request chunks in the base layer. For the example in Figure 9.10, requesting either $LC(t+2, 1)$ or $LC(t+4, 1)$ will maximize g_j.

The preceding three algorithms use different video quality measurements and select different chunks to maximize each individual criteria. To address the tradeoff between different video quality measurements, we combine the above three chunk-request algorithms as follows.

For user j, for each chunk $LC(t', l)$ that is missing in j's buffer but available at other peers' buffers, user j assigns a score $SC(t', l)$ as follows:

- User j first assigns an original score $SC(t', l) = ((t + L) - t')/L$ to the chunk $LC(t', l)$, where t is the current time and L is user j's buffer size. It addresses the stringent time constraint in video streaming, and gives higher priority to chunks that are closer to their playback.
- If $LC(t', l)$ is decodable once it is received, then the score is updated as $SC(t', l) = SC(t', l) + w_1$.
- If $g_2 = \sum_{i=t'}^{i=t'+1} \left(|a_i^{(j)} - a_{i-1}^{(j)}| - |a_i'^{(j)} - a_{i-1}'^{(j)}| \right) > 0$, then j updates $SC(t', l) = SC(t', l) + w_2 g_2$.
- If $l = 1$, then $SC(t', l) = SC(t', l) + w_3$.

Here, $0 \leq w_1, w_2, w_3 \leq 1$ with $w_1 + w_2 + w_3 = 1$ are the weights that the peer can adjust depending on the importance of different video quality criteria. Then player j chooses the chunk with the highest score and requests it from a peer who has this chunk in his or her buffer.

9.5   Chapter summary and bibliographical notes

The P2P video streaming social network is one of the largest social networks over the Internet; the performance of the streaming service depends heavily on user cooperation. In this chapter, we investigate the incentives for cooperation in P2P video streaming social networks and stimulate cooperation between any two users. Also, because some users might try to cheat to increase their own payoff during cooperation, the optimal cooperation strategies discussed in this chapter are also cheat-proof.

Furthermore, as mobile devices with large computational capacity have become very popular nowadays, we also study the optimal and cheat-proof cooperation strategies in wireless P2P video streaming networks. To accommodate heterogenous networks and mobile devices and to provide personalized service, we also investigate the optimal chunk request algorithm for layered video coding under various video quality measures.

The P2P service model [79,81] was first studied to reduce the heavy load of servers. Recently, several industrial large-scale P2P video streaming systems have been developed, including Coolstreaming [81], PPLive [79], PPStream, UUSee, and Sopcast. In these P2P streaming systems, scalable video coding is widely adopted to provide higher-quality service [94,123]. In the research on cooperation in P2P video streaming, a rank-based peer-selection mechanism was introduced by Habib and Chuang [124], and Tan and Jarvis [93] proposed a payment-based incentive mechanism, in which peers pay points to receive data and earn points by forwarding data to others. Both works use reputation or micropayment-based mechanisms, which often demand a centralized archi-tecture and thus hinder their scalability. The work of Liu *et al.* [94] proposed a distributed incentive mechanism on mesh-pull P2P live video streaming systems without consider-ing the cheating behavior. Interested readers can also refer to references [125–128] for the development of media streaming over wireless networks.

10 Optimal pricing for mobile video streaming

Mobile phones are among the most popular consumer devices; the recent developments of 3G networks and smart phones enable users to watch video programs by subscribing to data plans from service providers. Because of the ubiquity of mobile phones and phone-to-phone communication technologies, subscribers can redistribute the video content to nonsubscribers. Such a redistribution mechanism is a potential competitor for the service provider and is very difficult to trace, given users' high mobility. The service provider must set a reasonable price for the data plan to prevent such unauthorized redistribution behavior and to protect the provider's own profit. In this chapter, we analyze the optimal price setting for the service provider by investigating the equilibrium between the subscribers and the secondary buyers in the content redistribution network. We model the behavior between the subscribers and the secondary buyers as a noncooperative game and find the optimal price and quantity for both groups of users. Such an analysis can help the service provider preserve the profit under the threat of the redistribution networks and can improve the quality of service for end users.

10.1 Introduction

The explosive advance of multimedia processing technologies is creating dramatic shifts in the ways that video content is delivered to and consumed by end users. Also, the increased popularity of wireless networks and mobile devices has drawn a great deal of attention in the past decade about ubiquitous multimedia access in the multimedia community. Network service providers and researchers are focusing on developing efficient solutions to ubiquitous access to multimedia data, especially videos, from everywhere using mobile devices (laptops, PDAs, or smart phones that can access 3G networks) [129]. Mobile phone users can watch video programs on their devices by subscribing to the data plans from network service providers [130,131], and they can easily use their programmable hand devices to retrieve and reproduce the video content. Therefore, it is important to understand end users' possible actions in order to provide better ubiquitous video access services.

According to a survey on the popularity of mobile devices [132], almost every person in developed countries has at least one cell phone. With such a high popularity and the convenient phone-to-phone communication technologies, it is very possible for a data-plan subscriber to redistribute video content without authorization. For example, some users who do not subscribe to a data plan may wish to watch TV programs while waiting for public transportation, and some of them might want to check news from time to time. Hence, these users have incentives to buy the desired video content from neighboring data subscribers if the cost is lower than the subscription fee charged by the service provider. Unlike generic data, multimedia data can be easily retrieved and modified, which facilitates the redistribution of video content. In addition, subscribers also have incentives to redistribute the content with a price higher than their transmission cost, as long as such an action will not be detected by the content owner. Because of the high mobility, time-sensitivity, and small transmission range characteristics of mobile devices, each redistribution action exists for only a short period of time and is very difficult to track. Consequently, a better way to prevent copyright infringement is to set a price at which no subscriber will have the incentive to redistribute the video.

The content subscribers and the secondary buyers who are interested in the video data interact with each other and influence each other's decisions and performance. They will reach an agreement at the equilibrium price from which all users have no incentives to deviate. Hence, such an equilibrium price will serve as the upper bound for the price set by the network service provider to prevent copyright infringement. Because of the small coverage area and limited power of each mobile device, a subscriber can sell the content only to secondary buyers within his or her transmission range, and the distance between users and the channel conditions dominates users' decisions. In this chapter, we consider the simple scenario in which one subscriber can sell the content to only one secondary buyer, and propose a multiuser game-theoretic framework to analyze the hybrid user dynamics in the live-video marketing social network.

The rest of the chapter is organized as follows. We introduce the system model in Section 10.2. We then analyze the optimal strategies for all users in Section 10.3 and prove the existence and the uniqueness of the equilibrium when there is only one secondary buyer. We analyze the mixed-strategy equilibrium for a scenario with multiple secondary buyers in Section 10.4. In Section 10.5, the content owner sets the price to maximize its payoff, but not to prevent the video redistribution among users.

10.2   System model

In this section, we introduce the channel, transmission, and rate–distortion models for live video transmission over wireless networks.

The system diagram is shown in Figure 10.1. There are N_s subscribers in the network, trying to sell the video content to N_b secondary buyers. At the beginning, each subscriber sends his or her own price per unit transmission power, as well as the probing signal, to secondary buyers. Because the price information contains only a few bits, we assume that it can be received immediately and perfectly. The probing signal enables secondary buyers to estimate the maximal achievable transmission rate. A secondary buyer must decide how much power he or she wants to buy from each subscriber. As scalable video coding is used widely in mobile video streaming [133], secondary buyers can purchase

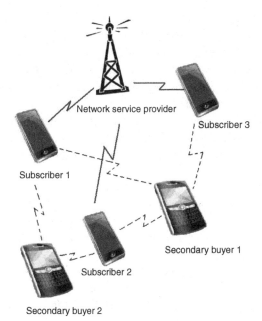

Network service provider

Subscriber 3

Subscriber 1

Secondary buyer 1

Subscriber 2

Secondary buyer 2

Fig. 10.1 Example of a mobile video stream redistribution network

different coding layers of the video from different subscribers and combine these streams during the decoding process.

Assume that the jth secondary buyer purchases part of the video stream from subscriber S_i with transmission power $P_i^{(j)}$. The channel between them is a slow fading channel with channel gain H_{ij}, the distance between them is d_{ij}, and the variance of the $\triangle$WGN at the receiver's side is σ^2. Assume that the total bandwidth available for the video redistribution network is W, which will be evenly allocated to all N' subscribers from whom secondary buyers purchase the video stream. The SNR and the maximal achievable bit rate of the video stream between S_i and B_j are

$$SNR_{ij} = \frac{P_i^{(j)} H_{ij}}{\sqrt{d_{ij}}\sigma^2},$$

$$\text{and} \quad R_{ij} = \frac{W}{N'+1} \log_2\left(1 + \frac{SNR_{ij}}{\gamma}\right), \tag{10.1}$$

where γ is the capacity gap.

For video streaming services, two commonly used objective quality measurements are the video's peak signal-to-noise-ratio (PSNR) and the streaming delay. Here, we adopt the polynomial delay model, as developed by Aniba and Aissa [134]. The overall delay D_B at the secondary buyers' end is the network delay between the subscribers and the service provider plus the maximal processing time of the subscribers. Therefore,

$$D_B = D_q\left(\frac{N'+K}{M}\right) + \max_i D_p(i), \tag{10.2}$$

where N' is the number of subscribers from whom the secondary buyers purchase the video stream, and K is the number of other subscribers within the coverage of the same base station who cannot establish a direct link to secondary buyers. M is the maximal number of users that the network service provider can afford simultaneously. $D_q((N'+K)/M)$ is the network delay between subscribers and the service provider, and $D_p(i)$ is subscriber i's processing time.

Without loss of generality, in this chapter, we use the two-parameter rate–distortion model [76], which is widely employed in a medium to high bit rate situation, and the analysis for other models is similar. The two-parameter rate–distortion model is as follows:

$$\text{Distortion} = \alpha e^{-\beta R}, \tag{10.3}$$

where α and β are two positive parameters determined by the characteristics of the video content.

A secondary buyer is able to purchase the video from different subscribers in two different ways. Because the log function and the exponential function are convex over $\mathbb{R}^+$ and the exponential function is nondecreasing over $\mathbb{R}^+$, it is easy to prove that buying different layers from different subscribers is a better choice, as buying all layers from one subscriber is just a special case. Given the bit rate in (10.1), the mean square error (MSE) of the video stream reconstructed by the secondary buyer B_j is

$$MSE_j = \alpha \exp\left(-\beta \sum_i R_{ij}\right) = \alpha \exp\left\{-\beta \frac{W}{N'+1} \sum_i \log_2\left(1 + \frac{SNR_{ij}}{\gamma}\right)\right\}.$$
$$\tag{10.4}$$

10.3   Optimal strategies for single secondary buyer

In this section, we focus on the scenario in which there is only one secondary buyer; that is, $N_B = 1$. We model the behavior of the subscribers and the secondary buyer as a Stackelberg game, and then analyze and prove the existence of the equilibrium that leads to the optimal strategies for all users. When there is only one secondary buyer, we can remove the superscript j for the secondary buyer index and have

$$SNR_i = \frac{P_i H_i}{\sqrt{d_i}\sigma^2}, \quad R_i = \frac{W}{N'+1} \log_2\left(1 + \frac{SNR_i}{\gamma}\right),$$

$$\text{and} \quad MSE = \alpha \exp\left(-\beta \sum_i R_i\right)$$

$$= \alpha \exp\left\{-\beta \frac{W}{N'+1} \sum_i \log_2\left(1 + \frac{SNR_i}{\gamma}\right)\right\}. \tag{10.5}$$

10.3.1 Video stream redistribution game definition

Because the video stream redistribution network is a dynamic system in which all users have high mobility and join and leave at any time, it is very difficult to have a central authority to control users' behavior. In addition, because this redistribution is unauthorized and illegal, to minimize their risk of being detected by the service provider, the participating users (subscribers and secondary buyers) have no incentives to trust one extra person, the central authority, and a distributed strategy is preferred.

Given the fact that there is only one secondary buyer, we propose a Stackelberg game model to analyze how the secondary buyer provides incentives for subscribers to redistribute the video stream, and find the optimal price and quantity that the secondary buyer should offer. The ultimate goal of this analysis is to help the content owner set an appropriate subscription fee such that the equilibrium of the game between subscribers and the secondary buyers leads to negative payoffs. Thus, subscribers will have no incentive to redistribute the video.

Before the game starts, each user, either a subscriber or the secondary buyer, will declare his or her presence to all other users within his or her transmission range. The game between subscribers and the second buyer is modeled as a Stackelberg game.

- **Game Stages:** The first stage of the game is the subscribers' (leaders') move. Each subscriber i will set his or her unit price p_i per unit transmission power as well as his or her maximal transmission power $P_i^{(max)}$.

  Then, in the second stage of the game, the secondary buyer (follower) will decide from whom to buy the video and how much power he or she wants the subscriber to transmit. The secondary buyer then pays each subscriber accordingly at the price that the subscriber sets in stage 1.
- **Utility function of the secondary buyer/follower:** We first define the secondary buyer's utility function and study his or her optimal action. The secondary buyer B gains rewards by successfully receiving the video with a certain quality. On the other hand, B must pay for the power that the subscribers use for transmission. Let P_i be the power that the secondary buyer B decides to purchase from the ith subscriber S_i, the channel gain between S_i and B is H_i, and the distance between them is d_i. Therefore, given the video rate–distortion model, the utility function of the secondary buyer B_i can be defined as

$$\pi_B = g_Q \left(PSNR_B - PSNR_{max}\right) - g_D \left(D_B - D_q \left(\frac{K+1}{M}\right)\right) - \left(\sum_i p_i P_i - p_o\right),$$

(10.6)

where g_Q is a user-defined constant measuring the received reward if the PSNR of the reconstructed video is improved by one dB, and g_Q is a constant measuring the user's loss if the video stream is further delayed by one second. $PSNR_{max}$ is the maximal PSNR of the video that can be obtained by subscribing to the service, and p_o is the price set by the content owner.

The preceding utility definition can be viewed as the difference between the utility if the secondary buyer buys the video stream from the subscribers and the utility if he or she subscribes to the data plan. The first term in (10.6) reflects the visual quality difference between the subscriber's video stream and the service provider's video stream. The second term considers the delay difference between the subscriber's video stream and the service provider's video stream. D_B was defined in (10.2), and $D_q((K + 1)/M)$ is the delay profile if the secondary buyer subscribes to the data plan and becomes an extra subscriber in the network. The third term indicates the price difference. The two constants g_Q and g_D control the balance between the gain and the loss of the secondary buyer.

- *Utility functions of the subscribers:* Each subscriber S_i can be viewed as a seller who aims to earn the payment that covers his or her transmission cost and also to gain as much extra reward as possible. We introduce a parameter c_i, the cost of power for relaying data, which is determined by the characteristics of the device that subscriber S_i uses. Hence, the utility of S_i can be defined as

$$\pi_{S_i} = (p_i - c_i) P_i, \tag{10.7}$$

where P_i is the power that subscriber i uses to transmit to the secondary buyer. Thus, subscriber S_i will choose the price p_i that maximizes his or her utility π_{S_i}.

The choice of the optimal price p_i is affected not only by the subscriber's own channel condition but also by other subscribers' prices, because different subscribers play noncooperatively and they compete to be selected by the secondary buyer. Thus, a higher price may not help a subscriber improve his or her payoff.

10.3.2   Equilibrium analysis

This video stream redistribution game is a game with perfect information; the secondary buyer has perfect information about each subscriber's action (the selected price). According to backward induction [97], a game with perfect information has at least one equilibrium. Therefore, the optimal strategies for both the secondary buyer and subscribers exist and can be obtained by solving the optimal decision for each stage using backward induction.

10.3.2.1   Secondary buyer's optimal strategy

We analyze the game using backward induction, and first study the secondary buyer's optimal strategy for a given price list from the subscribers. The secondary buyer B aims to determine the optimal power P_i that B should buy from each subscriber to maximize his or her own utility defined in (10.6).

Let L be the set including all subscribers who want to sell the video to the secondary buyer. Given that the secondary buyer purchases transmission power P_i from subscriber S_i, the secondary buyer's received video rate is

$$R_B = \frac{W}{\sum_{i \in L} I[P_i > 0] + 1} \sum_{i \in L} \log_2\left(1 + \frac{P_i H_i}{\gamma \sqrt{d_i} \sigma^2}\right), \tag{10.8}$$

where $I[\cdot]$ is the indicator function. Following the rate–distortion model and the transmission rate given in (10.5), the first term in (10.6) can be rewritten as a function of the transmission rate and equals to

$$g_Q \left(PSNR_B - PSNR_{max} \right)$$

$$= g_Q \left[10 \log_{10} \frac{255^2}{\alpha \exp\left(-\beta R_B\right)} - 10 \log_{10} \frac{255^2}{\alpha \exp\left(-\beta R_{max}\right)} \right] = g'_Q \left(R_B - R_{max} \right),$$

$$(10.9)$$

where $g'_Q = 10 \beta g_Q \log_{10} e$, and R_{max} is the video rate provided by the service provider.

Combining (10.5) and (10.6) with the preceding equation, we can formulate the utility function of B as a function of $\{P_i, \forall i \in L\}$. According to Aniba and Aissa [134], the network delay of the 3G network is reciprocal to the network utilization percentage. Hence, the optimal strategy for the secondary buyer is

$$\max_{\{P_i\}} \pi_b = g'_Q \left(R_B - R_{max} \right) - g_D \left(\max_{P_i > 0} D_p(i) + D_B \right) - \left(\sum_{i \in L} p_i P_i - p_o \right),$$

$$\text{s.t. } R_B \le R_{max}, \ P_i \le P_i^{(max)}, \ \forall i \in L,$$

$$\text{where } D_B = \frac{MC}{M - K - \sum_{i \in L} I[P_i > 0]} - \frac{MC}{M - K - 1}, \quad (10.10)$$

and C is the network constant [134].

In (10.10) and (10.8), $\sum_{i \in L} I[P_i > 0]$ and $\max_{i \in L} D_p(i) I[P_i > 0]$ are piecewise continuous functions and are not necessarily continuous cross different sets of $S = \{i \,|\, P_i > 0\}$. Therefore the optimization problem cannot be solved at once for the whole feasible set and must be divided into subsets. Let $\mathbf{P} = [P_1, P_2, \ldots, P_{N_s}]$ be the power vector, where P_i is the power that the secondary user purchases from subscriber i. The subset $\mathbb{S}_{N'}^{(k)}$ includes all possible scenarios in which the secondary buyer purchases the video from N' subscribers $\left(\sum_{i \in L} I[P_i > 0] = N' \right)$, and where among these N' subscriber k has the largest processing delay $\left(k = \arg\max_{i \in L} D_p(i) I[P_i > 0] \right)$.

We can find the optimal power vector $\mathbf{P}_{N'}^{(k)}$ for subset $\mathbb{S}_{N'}^{(k)}$ by making the first-order derivative of π_B with respect to P_i be zero:

$$\frac{\partial \pi_B}{\partial P_i} = g'_Q \frac{W \ln 2}{N' + 1} \frac{A_i}{1 + A_i P_i} - p_i = 0, \ \forall S_i \in L, \quad (10.11)$$

where $A_i = \sqrt{d_i} \sigma^2 \gamma / H_i$. Therefore, if the secondary buyer purchases from any N' subscribers with the same maximal processing delay,

$$P_i(\mathbb{S}_{N'}^{(k)}) = \frac{g'_Q W \ln 2}{p_i(N' + 1)} - \frac{1}{A_i}, \quad \forall S_i \in L \quad (10.12)$$

is the optimal solution. Note that (10.12) can be proved to be the unique maximizer for the subscriber set $\mathbb{S}_{N'}^{(k)}$ by finding the maximizer on the boundary. From (10.12), given the same maximal processing delay and the same number of subscribers from whom the secondary buyer is going to purchase, the second buyer purchases less from subscribers with higher prices. Also, the secondary buyer tends to purchase more from subscribers with better channels.

After the maximizer over each feasible subset is obtained, the secondary buyer should choose the one that gives himself or herself the largest utility. Let $\mathbf{P}^*$ be the optimal decision of the secondary user; then $\mathbf{P}_i^* = \arg\max_{0 \le N', k \le N_s} \pi_B \left(\mathbf{P}(\mathbb{S}_{N'}^{(k)}) \right)$.

10.3.2.2 Subscribers' best strategies

Given the preceding optimal strategy $\mathbf{P}^*$ for the secondary buyer, each subscriber selects the optimal price p_i that maximizes his or her utility. The optimal price $p_i^*(\mathbf{H}_i, \mathbf{d}_i)$ should satisfy

$$\frac{\partial \pi_{S_i}}{\partial p_i} = P_i^* + (p_i - c_i)\frac{\partial P_i^*}{\partial p_i} = 0$$

$$\text{s.t.} \quad p_i \le c_i \quad \forall i \in L, \tag{10.13}$$

or be on the boundary, which means $p_i^* = c_i$. Given a set of subscribers $\mathbb{S}_{N'}^{(k)}$, this problem is a convex optimization problem and the solutions can be found numerically. The subscriber is willing to redistribute the video stream only if he or she can profit from the redistribution action. Therefore, a subscriber's claimed price should be higher than his or her cost.

10.3.3 Existence of the equilibrium

In this subsection, we prove that the optimal strategies of the subscribers p_i^* in (10.13) and that of the secondary buyers $\mathbf{P}^*$ in (10.12) form an equilibrium. By definition, if $(\{p_i^*\}, \mathbf{P}^*)$ is an equilibrium, then p_i^* is the best response of the subscriber i if other subscribers choose $\{p_j^*\}_{j \ne i}$ and the secondary buyer chooses $\mathbf{P}^*$, and $\mathbf{P}^*$ is the secondary buyer's best response if subscribers choose the prices $\{p_i^*\}$.

The optimization problem in (10.10) can be solved only by dividing the problem into subproblems with different sets of subscribers from whom the secondary buyer actually purchases the video stream; that is, $\mathbb{S}_{N'}^{(k)}$. Therefore, here we first prove that given any $\mathbb{S}_{N'}^{(k)}$, (10.12) and (10.13) form an equilibrium for the secondary buyer and all subscribers in $\mathbb{S}_{N'}^{(k)}$. Then the actual equilibrium is the one that maximizes the secondary buyer's utility among these solutions.

For any given $\mathbb{S}_{N'}^{(k)}$, the optimization problem in (10.10) is equivalent to

$$\max_{P_i} \ \pi_B' = g_Q' R_B - \sum_{i \in \mathbb{S}_{N'}^{(k)}} p_i P_i,$$

$$\text{s.t.} \ R_B \le R_{max}, \ P_i \le P_i^{(max)}, \ \forall i \in \mathbb{S}_{N'}^{(k)},$$

$$\text{where} \ R_B = \frac{W}{k+1} \sum_{i \in \mathbb{S}_{N'}^{(k)}} \log_2\left(1 + \frac{P_i H_i}{\gamma \sqrt{d_i}\sigma^2}\right). \tag{10.14}$$

We first show that the solution $\mathbf{P}^*$ in (10.12) is the global optimum of (10.14) by showing the objective function in (10.14) being a concave function in P. The second-order derivatives of π_B' in (10.14) are

$$\frac{\partial^2 \pi_B'}{\partial P_i^2} = -\frac{g_Q' W \ln 2}{k+1} \frac{A_i^2}{(1 + A_i P_i)^2} < 0,$$

$$\frac{\partial^2 \pi_B'}{\partial P_i \partial P_j} = 0, \quad \text{and}$$

$$\frac{\partial^2 \pi_B'}{\partial P_i^2} \frac{\partial^2 \pi_B'}{\partial P_j^2} - \left(\frac{\partial^2 \pi_B'}{\partial P_i \partial P_j}\right)^2 = \left(\frac{g_Q' W \ln 2}{k+1}\right)^2 \frac{A_i^2 A_j^2}{(1 + A_i P_i)^2 (1 + A_j P_j)^2} > 0.$$

$$(10.15)$$

Moreover, π_B' is a continuous function of P_i. So for $0 \le P_i \le P_{max}$, π_B' is strictly concave in P_i, and jointly concave over **P** as well. Therefore, the solution $\mathbf{P}^*$ in (10.12) is the global optimum that maximizes the secondary buyer's utility π_B. Furthermore, in the real scenario, the secondary buyer can gradually increase the power P_i for each subscriber to reach the optimal solution $\mathbf{P}^*$ if there are information mismatches. For example, the knowledge of channel coefficients may change slowly, and the secondary buyer needs to adjust the strategy accordingly.

Now we show that when other subscribers' prices are fixed, subscriber S_i cannot arbitrarily increase the price p_i to get higher payoff. Given $\mathbb{S}_{N'}^{(k)}$, we take the first-order derivative of the optimal $P_i^*(\mathbb{S}_{N'}^{(k)})$ in (10.12) with respect to the price p_i

$$\frac{\partial P_i^*(\mathbb{S}_{N'}^{(k)})}{\partial p_i} = -\frac{g_Q' W \ln 2}{k+1} \frac{1}{p_i^2} < 0, \tag{10.16}$$

which means $P_i^*(\mathbb{S}_{N'}^{(k)})$ is a decreasing function of p_i. Such a phenomenon is reasonable because the secondary buyer tends to purchase less from subscribers with higher prices. Furthermore, when other subscribers' prices and the power that the secondary buyer purchases from each subscriber are fixed, the utility of subscriber i is a concave function of the price p_i. The first-order derivative of subscriber i's utility π_{S_i} with respect to the price p_i, is

$$\frac{\partial \pi_{S_i}}{\partial p_i} = \frac{g_Q' W \ln 2}{k+1} \frac{c_i}{p_i^2} > 0, \tag{10.17}$$

and we can also derive the second-order derivative of subscriber i's utility π_{S_i} with respect to the price p_i,

$$\frac{\partial^2 \pi_{S_i}}{\partial p_i^2} = -\frac{2 c_i g_Q' W \ln 2}{k+1} \frac{1}{p_i^3} < 0. \tag{10.18}$$

Therefore, π_{S_i} is concave with respect to the price p_i. Owing to the concavity of π_{S_i}, subscriber S_i can always find its optimal price p_i^*. As a result, $(\{p_i^*\}, \mathbf{P}^*)$ form an equilibrium.

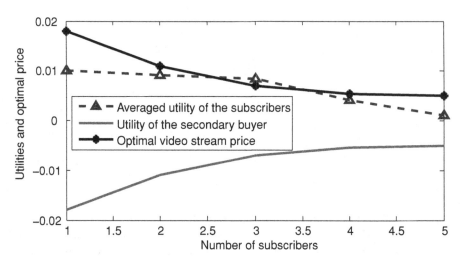

Fig. 10.2 Utilities of the users and the optimal video stream price versus different number of subscribers

10.3.4   Simulation results

In this section, we show the equilibrium of the video stream redistribution game under different scenarios as well as the optimal price for the content owner.

In our simulations, the secondary buyer is located at the origin (0,0), and the subscribers are initially uniformly distributed in a rectangle of size 100 meters by 100 meters centered around the origin. The pricing game is played 100 times, and each subscriber changes its location each time the game restarts. For each subscriber, the location change is normally distributed, with zero mean and unit variance. The direction of each subscriber's location change follows the uniform distribution. For all users, the maximal transmit power P_{max} is 100 mW, and the noise level is 10^{-8} W. The capacity gap is $\gamma = 1$, the total available bandwidth is W $= 1$ MHz, $g_Q = 0.1$ and $g_D = 0.1$ per millisecond, and for subscriber S_i, the cost per unit transmission power c_i is a random variable following uniform distribution in the range [0.05, 0.15]. The processing delay of each subscriber $D_p(i)$ is also a uniformly distributed random variable in [0.1, 10.1] millisecond. We use the video sequence "Akiyo" in QCIF format encoded using the H.264 video codec. The resulted rate–distortion parameter $\beta = 0.0416$, and $\alpha = 6.8449$. We set the maximal PSNR, which is provided by the original content owner, as 35 dB, and the corresponding maximal bit rate for Akiyo is $R_{max} = 84$ kbps. For simplicity and without loss of generality, the subscription price p_o for the video sequence is set to be 0 so the optimal price for the content owner can be simply viewed as $-\pi_B$. This implies that if the secondary buyer's utility is negative, he or she has incentives to purchase the video stream from the content owner.

First we let $M - K = 50$, and the number of subscribers K varies from 1 to 5; that is, the network is not crowded and the number of subscribers is small when compared with the maximal number that the network can afford. In Figure 10.2, we observe that as the number of available subscribers increases, the competition among subscribers

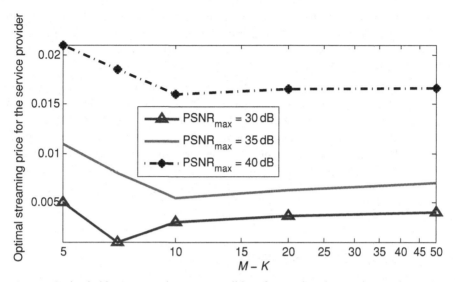

Fig. 10.3 Optimal video stream price versus qualities of network and streaming service

becomes more severe, so the optimal price for the content owner decreases. When there are no more than three subscribers, the averaged utility of the subscribers does not vary much, because in such cases, the secondary buyer is trying to purchase the maximum video rate R_{max} from all subscribers to increase the reconstructed video's quality, and these subscribers are not competing with one another. However, when there are more subscribers, the secondary buyer can easily get the video quality close to $PSNR_{max}$, and subscribers compete with one another to motivate the secondary buyer to purchase from himself or herself. Such a phenomenon is the nature of a free market with more sellers.

Next we examine the impact of network quality on the optimal price of the video stream. From Figure 10.2 we can see that the competition among subscribers dominates their own utilities, and the optimal price for the video stream (at the content owner's side) does not vary much when there are more than three subscribers. Therefore, here we set the total number of subscribers as three; Figure 10.3 shows how the video bit rate R_{max} and network usage influence the optimal video stream price. In Figure 10.3, $M - K$ varies from 5 to 50, and a smaller $M - K$ means that the current number of users in the network is approaching the network capacity and the network is more crowded. We select $R_{max} = 26.23$, 32.91, and 36.34 kbps; the resulting maximum PSNRs are 30 dB, 35 dB, and 40 dB, respectively. From Figure 10.3, we can see that if the service provider can offer a good video quality that the redistribution network cannot achieve, the provider can charge more for the streaming service. Also, when the network is very busy with a smaller $M - K$, the video delay dominates the video quality and, therefore, the secondary buyer tends to purchase from a smaller number of subscribers, but each subscriber can provide only limited video quality. Hence for the content owner, providing a better-quality streaming service is critical when the network is busy and the content owner must maintain the network quality at a level at which the quality of the video stream is not sacrificed. Furthermore, for a fixed R_{max}, when the network delay is small enough – for example, $M - K$ is less than 10 for PSNR $= 40$ dB – the content price starts to degrade as $M - K$ increases, because the secondary buyer purchases the video from the redistribution network. However, when the network delay is relatively small, buying from more subscribers will not introduce too much delay, and secondary buyers will be willing to purchase from enough number of subscribers to reach the highest data rate R_{max} and the maximum possible $PSNR_{max}$. But compared with purchasing from the content owner, although the video rate is the same, delay will be slightly larger. For instance, when $M - K$ is larger than 20 and PSNR $= 30$ dB, the optimal content price starts to get slightly higher because buying from subscribers introduces a slightly larger delay.

10.4   Multiple secondary buyers

In this section, we will extend the optimal strategy for the single secondary buyer case to the scenario with multiple secondary buyers.

10.4.1   Game model

Assume that there are N_s subscribers and $N_b > 1$ secondary buyers. The first two stages of the game are the same as in the single secondary buyer scenario: each subscriber declares the price per unit energy p_i, and then each secondary buyer B_j chooses the transmission power vector $\mathbf{P}^{(j)} = [P_1^{(j)}, P_2^{(j)}, \ldots, P_{N_s}^{(j)}]$, where $P_i^{(j)}$ is the power that secondary buyer j plans to purchase from subscriber i. With multiple secondary buyers, each subscriber i may receive several power purchase orders from different secondary buyers. Here, we let one subscriber transmit to one secondary buyer only. Thus, in the multiple secondary buyer scenario, the game model has an additional stage in which each subscriber i chooses the secondary buyer B_j who purchases the largest $P_i^{(j)}$ among all the N_b secondary buyers. Thus, for buyer B_j, the set of subscribers who will transmit video data to him or her is

$$N_j' = \{x \mid P_x^{(m)} < P_x^{(j)}, \ \forall m \neq j, \ m \in N_b\}. \tag{10.19}$$

Because each subscriber will answer to the secondary buyer who purchases the highest power, subscriber S_i's utility function is

$$\pi_{S_i} = (p_i - c_i) \max_{j \in N_b} P_i^{(j)}. \tag{10.20}$$

For a secondary buyer B_j, the utility becomes

$$\pi_{B_j} = g_Q'(R_B - R_{max}) - g_D \left(D_B + \max_{i \in N_j', P_i^{(j)} > 0} D_p(i) \right) - \left(\sum_{i \in N_j'} p_i P_i^{(j)} - p_o \right),$$

$$\tag{10.21}$$

where

$$R_B = \frac{W}{\sum_{i \in N_s, j \in N_b} I[P_i^{(j)} > 0] + 1} \sum_{i \in N_j'} \log_2 \left(1 + \frac{P_i^{(j)} H_{ij}}{\gamma \sqrt{d_{ij}} \sigma^2} \right). \quad (10.22)$$

10.4.2 Mixed-strategy equilibrium

Given the preceding definition of the utility functions, our next step is to find the subscribers and the secondary buyers' optimal decisions $(\{\mathbf{P}^{(j)*}\}_{j=1}^{N_b}, \{p_i^*\}_{i=1}^{N_s})$ from which no one in the system has an incentive to deviate.

Given the subscribers' price list $\{p_i\}$, for a secondary buyer B_j, the choice of the optimal power quantity $\mathbf{P}^{(j)*}$ is not only influenced by the channel conditions and the distances between subscribers and the secondary buyer B_j, but also depends on the number of subscribers from whom B_j can purchase the video stream. For instance, if B_j is the only secondary buyer within the transmission range of S_i, B_j would always tend to use the optimal power in (10.12). If B_j must compete with other secondary buyers, he or she might need to increase the offer $P_i^{(j)}$ or switch to other subscribers.

The deterministic way to find the optimal strategy is to model the competition among secondary buyers as an auction problem. However, in a fast-changing mobile network, secondary buyers may not have enough interactions to learn how much others value the transmission power and the video bit stream. Also, without a central authorization, the final bid price may not be revealed to all subscribers. Instead, we focus on finding the optimal probability distribution over possible strategies and find the mixed-strategy equilibrium of the game. We use backward induction to find the equilibrium.

When using backward induction, given the subscribers' price list for unit transmission power $\{p_i\}_{i=1}^{N_s}$, as introduced in Chapter 4, the secondary buyer B_j chooses the probability function $f_j(\mathbf{P}^{(j)})$ that maximizes his or her own payoff; that is,

$$\max_{f_j(\mathbf{P}^{(j)})} E[\pi_{B_j} | \{f_m(\mathbf{P}^{(m)})\}_{m \in N_b, m \neq j}], \quad (10.23)$$

where π_{B_j} is as defined in (10.21). Or, equivalently, each secondary buyer j seeks $f_j(\mathbf{P}^{(j)})$ that satisfies

$$E\left[\pi_{B_j}(\mathbf{P}^{(j)}) | \{f_m(\mathbf{P}^{(m)})\}_{m \in N_b, m \neq j}\right] = C_j \quad \forall j \in N_b, \quad (10.24)$$

where C_j is a constant [97].

We use an iterative best response algorithm for the secondary buyers to find the probability distribution $\{f_m(\mathbf{P}^{(m)})\}_{m \in N_b}$ as follows.

- First calculate the equilibrium power $\mathbf{P}^{(j)*}$ of each secondary buyer B_j based on (10.12) as a single secondary buyer. Also, let $f_j(\mathbf{P}^{(j)}) = \delta(\mathbf{P}^{(j)} - \mathbf{P}^{(j)*})$ for all $\mathbf{P}^{(j)} \neq \mathbf{P}^{(j)*}$.
- For each $j \in N_b$, given $\{f_m(\mathbf{P}^{(m)})\}_{m \in N_b, m \neq j}$, solve (10.24) and update $f_j(\mathbf{P}^{(j)})$.
- Repeat the preceding step until the solutions converge.

In the next subsection, we show the convergence of this algorithm by simulations. After solving $\mathbf{P}^{(j)*}$ for a given the price vector $\{p_i\}_{i=1}^{N_s}$, the optimal pricing $\{p_i^*\}_{i=1}^{N_s}$ can be calculated similarly by exhaustive search.

10.4.3   Simulation results

We consider the system setup in which secondary users are uniformly distributed in a 100-meter by 100-meter square centered around the origin. There are three subscribers located at $(25, 10)$, $(25, -10)$ and $(0, -30)$, respectively. In our system setup, we fix $M - K = 50$ and $K = 3$. The game is played 100 times, and each secondary buyer changes location each time the game restarts. For each subscriber, the location change is normally distributed with zero mean and unit variance. The direction of each secondary buyer's location change follows the uniform distribution. The maximal transmit power P_{max} is 100 mW, and the noise level is 10^{-8} W. The capacity gap is $\gamma = 1$, the total available bandwidth is W $= 1$ MHz, $g_Q = 0.1$ and $g_D = 0.1$ per millisecond, and the cost per unit transmission power for each subscriber c_i is a uniformly distributed in the range $[0.05, 0.15]$. The processing delay of each subscriber, $D_p(i)$, is a random variable uniformly distributed in $[0.1, 10.1]$ millisecond. We use the video sequence "Akiyo" in QCIF format as in the single secondary buyer scenario. The subscription price p_o for the video sequence is set to be 0 so the content owner's optimal price can be simply viewed as the negative of the secondary buyers' average payoff.

In Figure 10.4, we observe that as the total number of secondary buyers increases, the competition among the secondary buyers becomes more severe, and the optimal price for the content owner increases. When there are fewer than three secondary buyers, the averaged utility of the secondary buyers does not vary much, as each secondary buyer has a high probability to receive the video from at least one subscriber. Comparing the utilities of the subscribers and that of the secondary buyers when there are more than three secondary buyers, it is clear that the increment in the subscribers' utilities is much smaller than the decrease in the secondary buyers' utilities. Such a phenomenon occurs because secondary buyers compete with one another and some secondary buyers may not even receive anything from the subscribers.

Figure 10.5 shows the convergence speed of the iterated algorithm to find the mixed strategy equilibriums. It is clear that the algorithm converges. With more users in the network, the algorithm takes more iterations to find the equilibriums.

10.5   Optimal pricing for the content owner

In the previous sections, we discussed the equilibriums and the optimal pricing strategy in the video redistribution network. Our assumption there is that the content owner would like to set the price p_o smaller than the equilibrium price in the redistribution network. By doing so, the secondary buyers would have no incentives to purchase the video content from the subscribers and will always subscribe to the data plan from the service provider. However, such a strategy may not always maximize the total

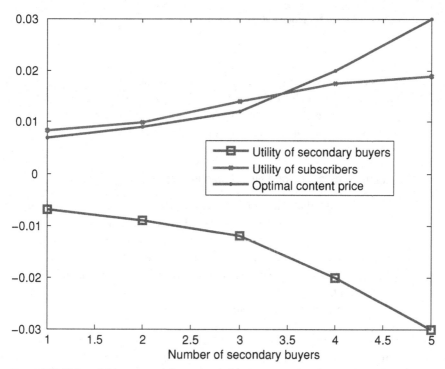

Fig. 10.4 Utilities of the users and the optimal video stream price versus different number of secondary buyers with three subscribers

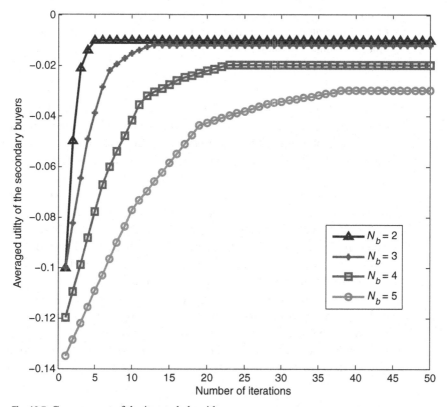

Fig. 10.5 Convergence of the iterated algorithm

income – that is, the price times the number of subscribers. In this section, we consider the scenario in which the service provider's goal is not prevention of video redistribution but rather maximization of his or her own income. We include the service provider as a player in the game and find his or her optimal strategies.

10.5.1   Pricing game model and evolution dynamics

Here, we model the video pricing problem for the content onwer as a non-cooperative game, which can be played several times. For example, in practical scenarios, the service provider can always change the price if the total income is below the expectation. Also, even when the price is fixed, mobile users can change their minds on whether to subscribe to the data plan or purchase from other subscribers. Such natural repetitions help the players find the equilibrium.

The basic elements of the game are listed as follows.

- **Game stages:** In the video pricing game, the first mover is the service provider, who first sets the price of the video content p_o. Then N_a mobile users who are interested in the video content decide whether to subscribe to the video streaming service. Because, based on the analysis in Sections 10.3 and 10.4, redistribution of the video content is possible, mobile users also take into consideration the possible payoffs that they can get in the redistribution network when making the decision.
- **Utility function of the service provider:** Obviously, the content owner's utility is price times the number of subscribers,

$$\pi_c = p_o \times N_s, \tag{10.25}$$

where N_s is the number of subscribers. With a higher price, there will be fewer subscribers and a smaller N_s, especially when it is possible for mobile users to receive the video content from the redistribution network. Therefore, the service provider cannot arbitrarily increase the content price p_o and must consider mobile users' utilities.

- **Utility function of the mobile users:** Each mobile user in N_a has the choice to pay p_o to subscribe to the data plan, or to purchase the content from other subscribers instead. Assume that among the N_a mobile users, $N_s \geq 0$ of them subscribe to the data plan, and the remaining $N_b = N_a - N_s \geq 0$ users decide not to. Let $\pi_s(N_s, N_b)$ and $\pi_B(N_s, N_b)$ be the utilities that a subscriber and a secondary buyer can get from the redistribution network as in Section 10.4, respectively.

  If user i decides to subscribe to the data plan, then his or her utility contains two parts. The first part is from the subscription to the streaming service, wherein he or she enjoys the video content with higher quality and shorter delay at a cost of the subscription fee. The second part is from the redistribution of the video to secondary buyers. Hence, if user i chooses to become a subscriber, his or her utility is

$$\pi_i(s, N_s, N_b) = \pi_s(N_s, N_b) + g_{Qi} * PSNR_{max} - g_{Di} D_q \left(\frac{K+1}{M}\right) - p_o, \tag{10.26}$$

where g_{Qi} is user i's gain per dB improvement in PSNR of the reconstructed video, and g_{Di} is user i's cost-per-second delay in receiving the bit stream. The first input parameter s in $\pi_i(s, N_s, N_b)$ denotes the action "subscribe." Note that $\pi_s(N_s, N_b) = 0$ if N_s or N_b equals to 0.

If user i chooses not to subscribe to the data plan, his or her utility comes only from the redistribution network by purchasing the video content from subscribers; that is, the user gains nothing from the service provider, but also pays nothing to the service provider. Hence, the

$$\pi_i(n, N_s, N_b) = g_Q PSNR_B - g_D D_B - \sum_{j \in N_i'} p_j P_j^{(i)}$$

$$= \pi_B(N_s, N_b) + g_{Qi} * PSNR_{max} - g_{Di} D_q \left(\frac{K+1}{M} \right) - p_o, \quad (10.27)$$

where the first input n in $\pi_i(n, N_s, N_b)$ denotes the action "do not subscribe." Note that $\pi_i(n, N_s, N_b) = 0$ if N_s or N_b equals to 0.

To analyze this game, we first investigate the equilibrium strategy of the mobile users given the content price p_o. As mentioned earlier, this pricing game can be played repeatedly and mobile users may use their previous experience to adjust their strategies accordingly. Therefore, a stable strategy for all mobile users that is robust to mutants of users' strategies is preferred in the pricing game. To find the stable equilibrium, we will use the evolutionary game theory introduced in Chapter 4 to analyze the evolution of the mobile users' behavior and derive the evolutionarily stable equilibrium, which leads to the optimal price of the video content.

Because each mobile user is not certain of other users' decisions, he or she may try different strategies in every play and learn from the interactions. For example, a mobile user may try to change from "subscribe" to "do not subscribe" and observe whether his or her utility received from the redistribution network is satisfactory. During such a learning process, the percentage – that is, the population share – of players using a certain pure strategy ("subscribe" or "do not subscribe") may change. The stable percentage of mobile users that chooses to subscribe to the data plan is what we are interested in.

The population evolution can be characterized by replicator dynamics as follows: At time t, let $N_s(t)$ denotes the number of mobile users that subscribe to the data plan. Then the subscribers' population state $x_s(t)$ is defined as

$$x_s(t) = \frac{N_s(t)}{N_a}, \quad (10.28)$$

and $x_b(t) = N_b(t)/N_a = 1 - x_s(t)$ is the secondary buyers' population state. By replicator dynamics, the evolution dynamics of $x_s(t)$ at time t is given by the differential equation

$$\dot{x}_s = \eta[\bar{\pi}_s(x_s) - \bar{\pi}(x_s)]x_s, \quad (10.29)$$

where $\dot{x}_s$ is the first-order derivative of $x_s(t)$ with respect to time t, $\bar{\pi}_s(x_s)$ is the average payoff of mobile users who subscribe to the data plan, and $\bar{\pi}(x_s)$ is the average payoff of all mobile users; η is a positive scale factor. We can see that if subscribing to the data plan can lead to a higher payoff than the average level, the probability of a user switching to "subscribe" will grow and the growth rate $\dot{x}_s/x_s$ is proportional to the difference between the average payoff of subscribers $\bar{\pi}_s(x_s)$ and the average payoff of all mobile users $\bar{\pi}(x_s)$. The other intuition behind x_s is that x_s can be viewed as the probability that one mobile user adopts the pure strategy "subscribe," and the population state vector $\mathbf{x} = \{x_s(t), x_b(t)\}$ is equivalent to a mixed strategy for that player. If subscribing to the data plan results in a higher payoff than the mixed strategy, then the probability of subscribing to the data plan should be higher, and x_s will increase. The rate of the increment is proportional to the difference between the payoff of adopting the pure strategy "subscribe" and the payoff achieved by using the mixed strategy $\mathbf{x} = \{x_s, x_b\}$.

Given the evolution dynamics formulated earlier, in the following sections we derive the evolutionarily stable equilibrium among mobile users in different scenarios.

10.5.2   Analysis of pricing game with homogeneous mobile users

A strategy is an evolutionarily stable strategy (ESS) if and only if it is asymptotically stable to the replicator dynamics. In the pricing game, when time goes to infinity, if (10.29) equals zero, then $\mathbf{x}$ is the evolutionarily stable equilibrium. In this subsection, we first focus on the scenario in which all mobile users value the video equality in the same way and $g_{Qi} = g_{Qj} = g_Q$ and $g_{Di} = g_{Dj} = g_D$ for all $i, j \in N_a$. The scenario in which different mobile users have different values of the video quality is analyzed in the next subsection.

Let $Q = g_Q * PSNR_{max} - g_D D_q(\frac{K+1}{M}) - p_o$; then in the homogeneous case, the utilities of the subscribers and the secondary buyers are

$$\pi(s, N_s, N_b) = \pi_s(N_s, N_b) + Q, \text{ and } \pi(n, N_s, N_b) = \pi_B(N_s, N_b) + Q, \quad (10.30)$$

respectively. Mobile users are homogeneous and they will have the same evolution dynamics and equilibrium strategy. Given that x_s is the probability that a mobile user decides to subscribe to the data plan, the averaged utilities of the subscribers and the secondary users are

$$\bar{\pi}_s(x_s) = \sum_{i=0}^{N_a} \binom{N_a}{i} x_s^i (1 - x_s)^{N_a - i} \pi(s, i, N_a - i),$$

$$\text{and} \quad \bar{\pi}_b(x_s) = \sum_{i=0}^{N_a} \binom{N_a}{i} x_s^i (1 - x_s)^{N_a - i} \pi(n, i, N_a - i), \quad (10.31)$$

respectively. The average utility of all mobile users is

$$\bar{\pi}(x_s) = x_s \cdot \bar{\pi}_s(x_s) + x_b \cdot \bar{\pi}_b(x_s). \quad (10.32)$$

Given the above, (10.29) can be rewritten as

$$\dot{x}_s = \eta[\bar{\pi}_s(x_s) - \bar{\pi}(x_s)]x_s$$

$$= \eta[\bar{\pi}_s(x_s) - \bar{\pi}_b(x_s)]x_b x_s. \quad (10.33)$$

In the equilibrium x_s^*, no player will deviate from the optimal strategy, indicating $\dot{x}_s = 0$ in (10.33). We can then obtain the equilibria, which are $x_s = 0$, $x_s = 1$ or $x_s = \bar{\pi}_s(x_s) - \bar{\pi}_b(x_s)$. To verify that they are indeed ESSs, we will show that these three equilibria are asymptotically stable; that is, the replicator dynamics (10.29) converge to these equilibrium points.

The first step is to guarantee that $x_s(t) + x_n(t) = 1$ for all t, which means that the sum of the probabilities of a mobile user subscribing to the data plan or not equal to one. We can verify it by summing up (10.29) with the reciprocal dynamic function of x_n, which is

$$\dot{x}_b = \eta[\bar{\pi}_b(x_b) - \bar{\pi}(x_s)]x_b. \tag{10.34}$$

Combing (10.29) and (10.34), we have

$$\dot{x}_b + \dot{x}_s = \eta[x_s\bar{\pi}_s(x_s) + x_b\bar{\pi}_b(x_b) - (x_s + x_b)\bar{\pi}(x_s)] = 0. \tag{10.35}$$

Recall that $\bar{\pi}(x_s) = x_s \times \bar{\pi}_s(x_s) + (x_n) \times \bar{\pi}_b(x_s)$, and $x_s(0) + x_n(0) = 1$; the above equation is equivalent to $\dot{x}_n + \dot{x}_s = 0$. As a result, $x_s(t) + x_n(t) = x_s(0) + x_n(0) = 1$ for all t in the evolution process.

Next we need to show that all the non-equilibrium strategies of the pricing game will be eliminated during the evolution process. If the replicator dynamic is a myopic adjustment dynamic, then all non-equilibrium strategies will be eliminated during the process. A dynamic is a myopic adjustment if and only if

$$\sum_{a \in A} \dot{x}_a \bar{\pi}(x_a, x_{-a}) \geq 0, \tag{10.36}$$

where A is the strategy space, x_a is the population of users adopting pure strategy a, and $\bar{\pi}(x_a, x_{-a})$ is the average payoff of users adopting pure strategy a. For our optimal pricing game, the strategy space is $A = \{s, b\}$, where s means "subscribe" and b means "do not subscribe" and be a secondary buyer. Combining (10.29) with (10.34) and (10.32), we have

$$\sum_{a \in \{s,b\}} \dot{x}_a \bar{\pi}_a(x_a) = \sum_{a \in \{s,b\}} \eta[\bar{\pi}_a(x_a) - \bar{\pi}(x_s)]x_a \bar{\pi}_a(x_a)$$

$$= \sum_{a \in \{s,b\}} \eta \left(\bar{\pi}_a(x_a) - \sum_{a' \in \{s,b\}} x_{a'} \bar{\pi}_{a'}(x_{a'}) \right) x_a \bar{\pi}_a(x_a)$$

$$= \eta \left\{ \sum_{a \in \{s,b\}} x_a \bar{\pi}_a^2(x_a) - \left[\sum_{a \in \{s,b\}} x_a \bar{\pi}_a(x_a) \right]^2 \right\} \geq 0. \tag{10.37}$$

In (10.37), the last inequality is from Jensen's inequality, which says $(a_1 x_1 + a_2 x_2)^2 \leq a_1 x_1^2 + a_2 x_2^2$ with $a_1 + a_2 = 1$ and x^2 being a concave function of x. Therefore, the reciprocal dynamic of the pricing game in (10.29) is a myopic adjustment and will eliminate all non-equilibrium strategies.

From (10.33), $\dot{x}_s$ has the same sign as $\bar{\pi}_s(x_s) - \bar{\pi}_b(x_s)$. According to the discussions in Sections 10.3 and 10.4, $\bar{\pi}_s(x_s)$ is a decreasing function of x_s and $\bar{\pi}_b(x_s)$ is an increasing function of x_s. Therefore, when x_s goes from 0 to 1, the sign of $\dot{x}_s$ either does not change or changes only once.

- When $\bar{\pi}_s(x_s) > \bar{\pi}_b(x_s)$ for all $x_s \in [0, 1]$, in the evolution process, $\dot{x}_s = dx_s(t)/dt > 0$ for all t, and (10.29) converges to $x_s = 1$, which is an ESS.
- If $\bar{\pi}_s(x_s) < \bar{\pi}_b(x_s)$ for all $x_s \in [0, 1]$, in the evolution process, $\dot{x}_s = dx_s(t)/dt < 0$ for all t, and (10.29) converges to $x_s = 0$, which is the ESS in this scenario.
- When $\bar{\pi}_s(x_s) - \bar{\pi}_b(x_s) = 0$ has one and only one root, x_s^*, and (10.29) converges to the ESS x_s^*.

Therefore, for each price p_o set by the content owner, we can find the stable number of subscribers $N_s = N_a \cdot x_s^*$, from which we can calculate the service provider's utility. Hence, given the ESS of the mobile users, by backward induction, the service provider can easily choose the optimal content price to maximize his or her own payoff.

10.5.3   Analysis of pricing game with heterogeneous mobile users

In the heterogeneous scenario in which different mobile users value video quality differently, it is very difficult to represent the average payoff of the subscribers and that of the secondary buyers in a compact form. Hence, we start with the simple two-person game and find its ESS. We then extend the ESS into the scenario with multiple heterogeneous mobile users.

We first start with the two-player game. Assume that there are two mobile users with different $\{g_{Qi}, g_{Di}, c_i\}$. If both of them decide not to subscribe to the data plan, then they pay nothing and gain nothing from the service provider. Also, because there are no subscribers, the redistribution network does not exist, and both players' utilities are 0. If both decide to subscribe to the data plan, then the redistribution network does not exist either, as there is no secondary buyers. In this scenario, the utility of player i is $Q_i = g_{Qi} * PSN\, R_{max} - g_{Di} D_q \left(\frac{K+1}{M}\right) - p_o$. If player 1 becomes a subscriber but player 2 decides not to subscribe, then player 1's utility is $Q_1 + \pi_{1s}$, and player 2's utility is π_{2b}. Here, π_{1s} and π_{2b} are the utilities that user 1 and 2 get from the redistribution network as a seller and a buyer, respectively, and their calculation is the same as that in Section 10.3 and Section 10.5. When only player 2 subscribes to the streaming service, the analysis is similar, and we can obtain the matrix form of the game shown in Table 10.1. In Table 10.1, each row represents user 1's decision, and each column represents user 2's decision. For each entry in the table, the first term is user 1's payoff, and the second term is user 2's payoff.

Let x_1 and x_2 be player 1 and 2's probability of adopting the pure strategy "subscribe," respectively. Then the expected payoff $\bar{\pi}_1(s)$ of user 1 by always playing "subscribe" is

$$\bar{\pi}_1(s, x_2) = Q_1 x_2 + [Q_1 + \pi_{1s}](1 - x_2), \tag{10.38}$$

and the expected payoff of player 1 when he or she plays the mixed strategy x_1 is

$$\bar{\pi}_1(\mathbf{x}) = Q_1 x_1 x_2 + [Q_1 + \pi_{1s}]x_1(1 - x_2) + \pi_{1b}(1 - x_1)x_2. \tag{10.39}$$

Table 10.1 Matrix form of the pricing game with two heterogeneous mobile users

	Subscribe	Do not subscribe
Subscribe	(Q_1, Q_2)	$(Q_1 + \pi_{1s}, \pi_{2b})$
Do not subscribe	$(\pi_{1b}, Q_2 + \pi_{2s})$	$(0,0)$

Then, we can write the reciprocal dynamics of x_1 as

$$\dot{x}_1 = x_1(1 - x_1)[(Q_1 + \pi_{1s}) - (\pi_{1s} + \pi_{1b})x_2], \tag{10.40}$$

and similarly,

$$\dot{x}_2 = x_2(1 - x_2)[(Q_2 + \pi_{2s}) - (\pi_{2s} + \pi_{2b})x_1]. \tag{10.41}$$

An equilibrium point must satisfy $\dot{x}_1 = 0$ and $\dot{x}_2 = 0$; then from (10.40) and (10.41), we get five equilibria $(0,0)$, $(0,1)$, $(1,0)$, $(1,1)$, and $((Q_2 + \pi_{2s})/(\pi_{2s} + \pi_{2b})$, $(Q_1 + \pi_{1s})/(\pi_{1s} + \pi_{1b}))$.

If we view (10.41) and (10.40) as a nonlinear dynamic system, then the above five equilibria are ESSs if they are locally asymptotically stable. The asymptotic stability requires that the determination of the Jacobian matrix J be positive and the trace of J be negative. The Jacobian matrix J can be derived by taking the first-order partial derivatives of (10.41) and (10.40) with respect to x_1 and x_2, and

$$J = \begin{bmatrix} (1 - 2x_1)D_1 & -x_1(1 - x_1)(\pi_{1s} + \pi_{1b}) \\ -x_2(1 - x_2)(\pi_{2s} + \pi_{2b}) & (1 - 2x_2)D_2 \end{bmatrix}, \tag{10.42}$$

where $D_1 = (Q_1 + \pi_{1s}) - (\pi_{1s} + \pi_{1b})x_2$ and $D_2 = (Q_2 + \pi_{2s}) - (\pi_{2s} + \pi_{2b})x_1$. By jointly solving $\det(J) > 0$ and $trace(J) < 0$, we can have the following optimal subscription strategies for mobile users under different scenarios:

- When $Q_1 + \pi_{1s} < 0$ and $Q_2 + \pi_{2s} < 0$, there is one ESS, $(0,0)$, and both users tend to not subscribe to the data plan.
- When $Q_1 - \pi_{1b} < -(Q_2 + \pi_{2s})$ and $(Q_2 + \pi_{2s})(Q_1 - \pi_{1b}) < 0$, there is one ESS, $(0,1)$, and the strategy profile user 1 and user 2 adopt converges to (not subscribe, subscribe).
- When $Q_1 - \pi_{1b} < -(Q_1 + \pi_{1s})$ and $(Q_1 + \pi_{1s})(Q_2 - \pi_{2b}) < 0$, there is one ESS, $(1,0)$, and user 1 tends to subscribe while user 2 tends to not subscribe to the data plan.
- When $Q_1 - \pi_{1b} > 0$ and $Q_2 - \pi_{2b} > 0$, there is one ESS, $(1,1)$, and both users tend to subscribe to the data plan.

We can see that when Q_1 is higher with larger g_{Q1} and g_{D1}, user 1 tends to subscribe to the data plan.

Based on this discussion on the ESSs of the two-player game, we can infer that the users who value the video quality more (with higher g_{Qi} and g_{Di}) would intend to subscribe to the data plan. Users with smaller g_{Qi} and g_{Di} would tend to choose "do not subscribe" and become secondary buyers. However, if the content price p_o is too high, so the subscription gives all users a negative payoff, no player would subscribe to the service.

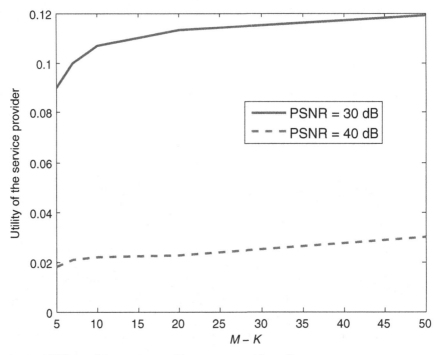

Fig. 10.6 Utilities of the service provider versus network quality

10.5.4   Simulation results

Here, we verify the derived ESS and show by simulation results the optimal price for the content owner if he or she wants to maximize his or her utility. We test with the homogeneous scenario that there are six mobile users who are initially uniformly located in a 100-meter by 100-meter square centered around the origin. All six mobile users have the same gain weighting factors $g_Q = 0.1$ and $g_D = 0.1$ per millisecond. The pricing game is played 100 times, and each secondary buyer changes location after the game restarts. The distance between each secondary buyer's locations in two consecutive games is normally distributed with zero mean and unit variance. The direction of each secondary buyer's location change follows the uniform distribution. Other simulation settings are the same as in Section 10.4. We use the video sequence "Akiyo" in QCIF format, as in the single secondary buyer scenario. The mobile users changes their strategies and evolute according to (10.29).

Figure 10.6 shows the content owner's utility when the PSNRs of the video stream are 30 dB and 40 dB, respectively. $M - K$ reflects how crowded the mobile network is. It is clear that if the content owner provides better-quality network or video, its payoff can be increased. Also, for lower-quality videos, the content owner's utility saturates earlier than high-quality videos with respect to network quality. This means that if the content owner decides to offer low-quality videos, to maximize its utility, it tends to offer low-quality network as well.

10.6   Chapter summary and bibliographical notes

In this chapter, we investigate the cooperation among mobile phone users to balance the price of mobile data plan service. We first analyze the equilibrium price of the video stream redistributed by the subscribers, given the number of subscribers and secondary buyers. Consequently, the results provide a guideline for the content owner to prevent the redistribution behavior. The redistribution behavior is modeled as a Stackelberg game, and we analyze the optimal strategies of both subscribers and secondary buyers. From the simulation results, a secondary buyer will tend to buy more power from subscribers with better channels to maximize his or her utility. If the total number of the subscribers increases, a secondary buyer can obtain a larger utility value, and the payment to each subscriber is reduced owing to more severe competition among the subscribers. Furthermore, when the mobile phone network is crowded, a secondary buyer tends to purchase the video stream from fewer subscribers, and the price for the streaming service can be higher. Nevertheless, the service provider should always offer a high-quality video stream to prevent the illegal redistribution of video via such redistribution networks.

Next, we extend the model by including the content owner in the game and letting the mobile phone users decide whether to subscribe to the data plan. In the extended model, we model the dynamics between the content owner and users who are interested in the video content, and study how the content onwer (the service provider) sets the price for the data plan to maximize his or her overall income. We use the evolutionary game theory to analyze the evolution of the mobile users' behavior and derive the evolutionarily stable equilibrium, which leads to the optimal price for the content owner to maximize his or her total income.

To our knowledge, there is no related work on this topic to date. Interested readers can refer to the following articles for the technologies of video streaming service over mobile devices. Research has been focused on developing efficient solutions for the ubiquitous access to multimedia data and in particular videos, from everywhere with mobile devices (laptops, PDAs, or the smart cellular phones that can access 3G networks) [129]. One of the most popular multimedia content over mobile phones is the video streaming service [130,131]. To accommodate heterogeneous network conditions and devices, scalable video coding is also widely used in mobile video streaming [133].

Part IV

Misbehaving user identification

11 Cheating behavior in colluder social networks

Until now we have discussed cooperation stimulation and how side information changes the behavior dynamics in various types of media-sharing social networks. For colluder social networks, as discussed in Chapters 5 and 8, before collusion, colluders need to reach an agreement regarding the fair distribution of the risk of being detected and the reward from illegal usage of multimedia. Chapters 5 and 8 analyze how colluders bargain with one another to achieve fairness of collusion, assuming all colluders report their private information (their received fingerprinted copies) honestly.

In reality, some colluders might break their fair-play agreement. They still wish to participate in and receive reward from collusion, but they do not want to take any risk of being detected by the digital rights enforcer. To achieve this goal, they may lie to other attackers about their fingerprinted copies [135]. For example, they may process their fingerprinted signals before multiuser collusion and use the processed copies instead of the originally received ones during collusion. The cheating colluders' goal is to minimize their own risk while still receiving reward from collusion. Therefore, they select the most effective precollusion processing strategy to reduce their risk.

Precollusion processing reduces the cheating colluders' risk, and makes other attackers have a higher probability than the cheating colluders of being detected. It is obviously a selfish behavior. In some scenarios, precollusion processing can also increase other attackers' probability of being detected; this is not only selfish, but also malicious. Therefore, to protect their own interest, colluders must examine all fingerprinted copies before collusion, detect and identify cheating colluders, and exclude them from collusion.

From a traitor-tracing perspective, it is important to study this problem of *traitors within traitors* in multimedia fingerprinting and understand the attackers' behavior during collusion to minimize their own risk and protect their own interest. This investigation helps us have a better understanding of multiuser collusion and enables to offer stronger protection of multimedia. In this chapter, we use the equal-risk collusion and the simple collective fingerprint detector in Chapter 5 as an example, investigate possible strategies that cheating colluders may use to minimize their own risk, and evaluate their performance.

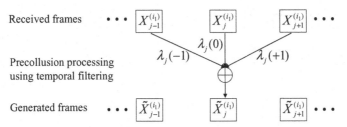

Received frames $\bullet \bullet \bullet$ $X_{j-1}^{(i_1)}$ $X_j^{(i_1)}$ $X_{j+1}^{(i_1)}$ $\bullet \bullet \bullet$

$\lambda_j(0)$

Precollusion processing $\lambda_j(-1)$ $\lambda_j(+1)$
using temporal filtering

Generated frames $\bullet \bullet \bullet$ $\widetilde{X}_{j-1}^{(i_1)}$ $\widetilde{X}_j^{(i_1)}$ $\widetilde{X}_{j+1}^{(i_1)}$ $\bullet \bullet \bullet$

Fig. 11.1 Applying temporal frame averaging during precollusion processing

11.1   Traitors within traitors via temporal filtering

For a cheating colluder to further reduce his or her own probability of being detected, one possible solution is to attenuate the energy of the embedded fingerprints before multiuser collusion. An example is to replace each segment of the fingerprinted signal with another, seemingly similar, segment from different regions of the content, such as averaging or swapping consecutive frames of similar content.

In this section, we take temporal filtering of adjacent frames as an example, and analyze its effects on the cheating colluder's probability of being detected, as well as on the perceptual quality of the fingerprinted copies. We consider a simple scenario in which all colluders receive fingerprinted copies of the same quality. When different colluders receive copies of different quality, the analysis is similar and is this not repeated.

11.1.1   Precollusion processing using temporal filtering

In this chapter, we assume that the cheating colluder uses a simple linear interpolation-based frame average during precollusion processing. A cheating colluder can also apply more complicated motion-based interpolation, and the analysis will be similar. For a cheating colluder i_1, assume that $\{\mathbf{X}_j^{(i_1)}\}_{j=1,2,\cdots}$ are the fingerprinted frames that he or she received from the content owner, and $\mathbf{X}_{j-1}^{(i_1)}$, $\mathbf{X}_j^{(i_1)}$ and $\mathbf{X}_{j+1}^{(i_1)}$ are three consecutive frames. As shown in Figure 11.1, for each frame j, colluder i_1 linearly combines the current frame $\mathbf{X}_j^{(i_1)}$, the previous frame $\mathbf{X}_{j-1}^{(i_1)}$, and the next frame $\mathbf{X}_{j+1}^{(i_1)}$ with weights $\lambda_j(0)$, $\lambda_j(-1)$, and $\lambda_j(+1)$, respectively, and generates a new frame $\widetilde{\mathbf{X}}_j^{(i_1)}$ where

$$\widetilde{\mathbf{X}}_j^{(i_1)} = \lambda_j(-1) \cdot \mathbf{X}_{j-1}^{(i_1)} + \lambda_j(0) \cdot \mathbf{X}_j^{(i_1)} + \lambda_j(+1) \cdot \mathbf{X}_{j+1}^{(i_1)}. \qquad (11.1)$$

In (11.1), $0 \le \lambda_j(-1), \lambda_j(0), \lambda_j(+1) \le 1$ and $\lambda_j(-1) + \lambda_j(0) + \lambda_j(+1) = 1$. For simplicity, we let $\lambda_j(-1) = \lambda_j(+1) = (1 - \lambda_j(0))/2$, and give equal weights to the two neighboring frames. Colluder i_1 repeats this process for every frame in the sequence and generates $\{\widetilde{\mathbf{X}}_j^{(i_1)}\}_{j=1,2,\cdots}$. When $\lambda_j(0) = 1$, $\widetilde{\mathbf{X}}_j^{(i_1)} = \mathbf{X}_j^{(i_1)}$ and it corresponds to the scenario in which colluder i_1 does not process his or her copy before collusion.

We assume that there is only one cheating colluder and other colluders do not discover his or her precollusion processing actions. The analysis is similar when there are multiple cheating colluders. In this scenario, under the averaging collusion, the jth frame in the colluded copy is

$$V'_j = \frac{\sum_{i \in SC, i \neq i_1} X_j^{(i)}}{K} + \frac{\lambda_j(-1) \cdot X_{j-1}^{(i_1)} + \lambda_j(0) \cdot X_j^{(i_1)} + \lambda_j(+1) \cdot X_{j+1}^{(i_1)}}{K} + n_j, \quad (11.2)$$

where n_j is additive noise.

11.1.2   Risk minimization and tradeoff analysis

During precollusion processing, the cheating colluder wishes to generate a new copy of high quality and minimize his or her own risk of being detected. In this section, we first analyze the quality of the newly generated frames $\{\widetilde{X}_j^{(i_1)}\}$ and calculate the cheating colluder's probability of being detected, and then study the selection of the optimal weight vector $[\lambda_1(0), \lambda_2(0), \ldots]$.

We first analyze the perceptual quality of the processed copy. If $\widetilde{X}_j^{(i_1)}$ is generated as in (11.1), then the MSE between $\widetilde{X}_j^{(i_1)}$ and $X_j^{(i_1)}$ is

$$MSE_j = ||\widetilde{X}_j^{(i_1)} - X_j^{(i_1)}||^2 = \left(\frac{1 - \lambda_j(0)}{2} \right)^2 \cdot \phi_j,$$

$$\text{where} \quad \phi_j = 4||X_j^{(i_1)}||^2 + ||X_{j-1}^{(i_1)}||^2 + ||X_{j+1}^{(i_1)}||^2 - 4\langle X_{j-1}^{(i_1)}, X_j^{(i_1)} \rangle$$

$$- 4\langle X_j^{(i_1)}, X_{j+1}^{(i_1)} \rangle + 2\langle X_{j-1}^{(i_1)}, X_{j+1}^{(i_1)} \rangle. \quad (11.3)$$

In (11.3), $||X_j^{(i_1)}||$ is the Euclidean norm of $X_j^{(i_1)}$, and $\langle X_{j-1}^{(i_1)}, X_j^{(i_1)} \rangle$ is the correlation between $X_{j-1}^{(i_1)}$ and $X_j^{(i_1)}$. From (11.3), a larger $\lambda_j(0)$ implies a smaller MSE_j. Consequently, from the perceptual quality's point of view, colluder i_1 should choose a larger $\lambda_j(0)$. Compared with $X_j^{(i_1)}$, $\widetilde{X}_j^{(i_1)}$ has the best possible quality when $\lambda_j(0) = 1$ and colluder i_1 does not apply precollusion processing.

Now, we analyze the cheating colluder's probability of being detected. Given the colluded copy V'_j as in (11.2), the fingerprint extracted from the jth frame is

$$Y_j = \frac{\sum_{i \in SC, i \neq i_1} W_j^{(i_1)}}{K} + \frac{\lambda_j(-1) \cdot W_{j-1}^{(i_1)} + \lambda_j(0) \cdot W_j^{(i)} + \lambda_j(+1) \cdot W_{j+1}^{(i_1)}}{K} + d_j,$$

$$(11.4)$$

where d_j contains terms that are independent of the embedded fingerprints $\{W_j^{(i)}\}$.

With orthogonal fingerprint modulation, given the colluder set SC and the index of the cheating colluder i_1, if the detection noise d_j is i.i.d. and follows Gaussian distribution $\mathcal{N}\left(0, \sigma_n^2\right)$, it is straightforward to show that the detection statistics $T_N^{(i)}$ in (2.2) are independent Gaussian with marginal distribution $N(\mu^{(i)}, \sigma_n^2)$. The detection statistics have zero mean for an innocent user and positive mean for a guilty colluder. Consequently, given a user i, the probability of accusing this user if he or she is innocent is $P_{fa} = Q(h/\sigma_n)$, and the probability of capturing user i if he or she is guilty is $P_d^{(i)} = Q((h - \mu^{(i)})/\sigma_n)$. $Q(\cdot)$ here is the Gaussian tail function and h is the predetermined threshold. For a fixed P_{fa}, the cheating colluder i_1 has a smaller probability of being detected when $\mu^{(i_1)}$ is smaller, and minimizing his or her probability of being detected is equivalent to minimizing the mean of his or her detection statistics.

For the cheating colluder i_1,

$$\mu^{(i_1)} = \sum_j \mu_j^{(i_1)},$$

where $\quad \mu_j^{(i_1)} = \dfrac{\langle \mathbf{W}_{j-1}^{(i_1)}, \mathbf{W}_j^{(i_1)} \rangle + \langle \mathbf{W}_j^{(i_1)}, \mathbf{W}_{j+1}^{(i_1)} \rangle}{2K\sqrt{\sum_l \|\mathbf{W}_l^{(i_1)}\|^2}}$

$$+ \lambda_j(0) \times \dfrac{2\|\mathbf{W}_j^{(i_1)}\|^2 - \langle \mathbf{W}_{j-1}^{(i_1)}, \mathbf{W}_j^{(i_1)} \rangle - \langle \mathbf{W}_j^{(i_1)}, \mathbf{W}_{j+1}^{(i_1)} \rangle}{2K\sqrt{\sum_l \|\mathbf{W}_l^{(i_1)}\|^2}}. \tag{11.5}$$

In (11.5), $\|\mathbf{W}_j^{(i_1)}\|$ is the Euclidean norm of $\mathbf{W}_j^{(i_1)}$, $\langle \mathbf{W}_{j-1}^{(i_1)}, \mathbf{W}_j^{(i_1)} \rangle$ is the correlation between $\mathbf{W}_{j-1}^{(i_1)}$ and $\mathbf{W}_j^{(i_1)}$, and $\langle \mathbf{W}_j^{(i_1)}, \mathbf{W}_{j+1}^{(i_1)} \rangle$ is the correlation between $\mathbf{W}_j^{(i_1)}$ and $\mathbf{W}_{j+1}^{(i_1)}$. From the fingerprint design in Section 2.3, $\langle \mathbf{W}_{j-1}^{(i_1)}, \mathbf{W}_j^{(i_1)} \rangle \leq \langle \mathbf{W}_j^{(i_1)}, \mathbf{W}_j^{(i)} \rangle = \|\mathbf{W}_j^{(i_1)}\|^2$ and $\langle \mathbf{W}_j^{(i_1)}, \mathbf{W}_{j+1}^{(i_1)} \rangle \leq \|\mathbf{W}_j^{(i_1)}\|^2$. Thus, if $\lambda_1(0), \dots, \lambda_{j-1}(0), \lambda_{j+1}(0), \dots$ are fixed, $\mu^{(i_1)}$ is a nondecreasing function of $\lambda_j(0)$ and is minimized when $\lambda_j(0) = 0$. Consequently, from the point of view of risk minimization, a smaller $\lambda_j(0)$ is preferred.

From this analysis, we have seen that during precollusion processing, a cheating colluder should choose larger weights $\{\lambda_j(0)\}$ to minimize the perceptual distortion introduced into his or her fingerprinted copy, whereas smaller weights $\{\lambda_j(0)\}$ are preferred to minimize the risk of being captured. A cheating colluder wishes to minimize his or her probability of being detected while still maintaining good quality of the fingerprinted copies. Thus, for a cheating colluder i_1, the selection of the weight vector $[\lambda_1(0), \lambda_2(0), \dots]$ can be modeled as

$$\min_{\{\lambda_j(0)\}} \left\{ \mu^{(i_1)} = \sum_j \mu_j^{(i_1)} \right\}$$

$$s.t. \ MSE_j \leq \varepsilon, \ 0 \leq \lambda_j(0) \leq 1, \ j = 1, 2, \dots, \tag{11.6}$$

where ε is the constraint on perceptual distortion. In our model of temporal filtering, $\{\lambda_j(0)\}$ for different frames are selected independently. Thus, minimizing $\mu^{(i_1)}$ over the entire video sequence is equivalent to minimizing $\mu_j^{(i_1)}$ in (11.5) for each frame j independently. Therefore, the optimization problem in (11.6) is equivalent to, for each frame j,

$$\min_{\lambda_j(0)} \mu_j^{(i_1)}$$

$$s.t. \ MSE_j \leq \varepsilon, \ 0 \leq \lambda_j(0) \leq 1. \tag{11.7}$$

Given ϕ_j as defined in (11.3), we can show that the solution to (11.7) is

$$\lambda_j^* = \max \left\{ 0, \ 1 - 2\sqrt{\varepsilon/\phi_j} \right\}. \tag{11.8}$$

By using $\{\lambda_j^*\}$ as in (11.8) during temporal filtering, a cheating colluder minimizes his or her own probability of being detected and ensures that the newly generated frames have small perceptual distortion ($MSE \leq \varepsilon$) when compared with the originally received ones.

11.1.3   Simulation results

In our simulations, we use the first forty frames in the sequence "carphone" as an example. At the content owner's side, we adopt the human visual model-based spread spectrum embedding, and embed fingerprints in the DCT domain. The length of the embedded fingerprints is 159608. We generate independent vectors from Gaussian distribution $\mathcal{N}(0, 1/9)$, and then apply Gram–Schmidt orthogonalization. In each fingerprinted copy, fingerprints embedded in adjacent frames are correlated with each other, and the correlation depends on the similarity between the two host frames.

On the colluders' side, we assume that there are a total of 150 colluders. For simplicity, we assume that there is only one cheating colluder and he or she applies temporal filtering to his or her received copy as in (11.1) during precollusion processing. In our simulations, we adjust the power of the noise term $\mathbf{d}_j$ in (11.4) such that $||\mathbf{d}_j||^2 = 2||\mathbf{W}_j^{(i)}||^2$. Other values will give the same trend.

Figure 11.2 shows the simulation results. For each frame j, $PSNR_j$ is defined as PSNR of $\widetilde{\mathbf{X}}_j^{(i_1)}$ compared with $\mathbf{X}_j^{(i_1)}$. In Figure 11.2, $\{\lambda_j^*\}$ are the solution of (11.8) and ε is chosen to satisfy $PSNR_j \geq 40\,\mathrm{dB}$ for all frames. In our simulations, we consider four different scenarios, in which $\lambda_j(0) = 1$, $\lambda_j(0) = 0.8$, $\lambda_j(0) = \lambda_j^*$, and $\lambda_j(0) = 0$, respectively. Note that $\lambda_j(0) = 1$ corresponds to the scenario in which the cheating colluder i_1 does not process his or her copy before multiuser collusion.

Figure 11.2(a) compares the perceptual quality of $\{\widetilde{\mathbf{X}}_j^{(i_1)}\}$, and Figure 11.2(b) plots the cheating colluder i_1's probability of being detected when $\{\lambda_j(0)\}$ take different values. A cheating colluder can reduce his or her own probability of being detected by temporally filtering his or her fingerprinted copy before multiuser collusion. By choosing $\{\lambda_j(0)\}$ of smaller values, the cheating colluder has a smaller probability of being detected, while sacrificing the quality of the newly generated copy. Therefore, during precollusion processing, the cheating colluder must consider the tradeoff between the risk and the perceptual quality.

In Figure 11.2(c), we consider two colluders, the cheating colluder i_1 and an honest colluder i_2 who does not process his or her copy before collusion, and compare their probabilities of being detected. From Figure 11.2(c), precollusion processing makes colluder i_2 take a higher risk of being detected than colluder i_1 and increases the relative risk taken by the honest colluders when compared with that of the cheating colluder.

To address the tradeoff between perceptual quality and the risk, a cheating colluder should choose $\{\lambda_j(0)\}$ as in (11.8). We compare the solution of $\{\lambda_j^*\}$ in (11.8) for different sequences. We choose four representative video sequences: "miss america," which has large smooth regions and slow motion, "carphone" and "foreman," which are moderately complicated, and "flower," whose high frequency band has a large energy and the camera moves quickly. We choose the threshold ε in (11.8) such that $PSNR_j \geq 40\,\mathrm{dB}$ for all

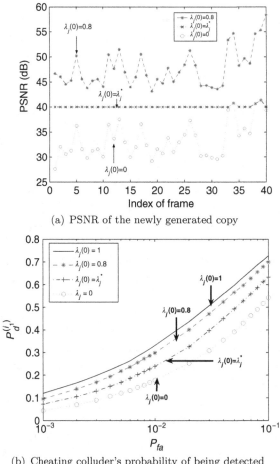

(a) PSNR of the newly generated copy

(b) Cheating colluder's probability of being detected

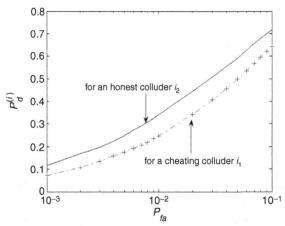

(c) Comparison of an honest colluder's risk with that of the cheating colluder

Fig. 11.2 Simulation results of temporal filtering on sequence "carphone"

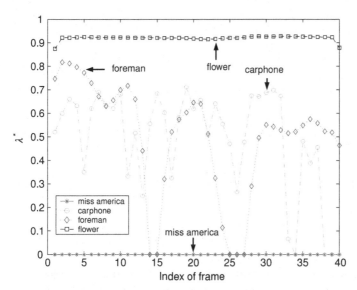

Fig. 11.3 λ_j^* in (11.8) for different sequences

frames in $\{\widetilde{\mathbf{X}}_j^{(i_1)}\}$. Figure 11.3 shows the solutions of (11.8) for various sequences. From Figure 11.3, for sequences that have slow motion ("miss america"), a cheating colluder can choose $\{\lambda_j(0)\}$ with small values, around 0, without significant quality degradation; for sequences that have moderate motion ("carphone" and "foreman"), λ_j^* is around 0.5; and for sequences with fast movement ("flower"), a cheating colluder must choose large $\{\lambda_j(0)\}$, larger than 0.9, to ensure the high quality of the newly generated frames.

11.2 Traitors within traitors in scalable fingerprinting systems

As illustrated in Chapter 5, in scalable multimedia coding systems, for the same multimedia content, different users receive copies of different resolutions and quality, depending on each user's available bandwidth and computation constraints. In scalable multimedia fingerprinting systems, in addition to applying temporal filtering to the received frames, a cheating colluder can also change the resolution of his or her fingerprinted copy before multiuser collusion. In this section, we investigate how cheating colluders behave before multiuser collusion in scalable multimedia fingerprinting systems, and analyze their performance.

11.2.1 Changing the resolution of the fingerprinted copies before collusion

Assume that $F^{(i_1)}$ contains the indices of the frames that a cheating colluder i_1 subscribed to, and $\{\mathbf{X}_j^{(i)}\}_{j\in F^{(i_1)}}$ are the fingerprinted frames that he or she received from the content owner. Before collusion, colluder i_1 processes his or her received copy and generates another copy $\{\widetilde{\mathbf{X}}_j^{(i_1)}\}$, whose temporal resolution is different from that of $\{\mathbf{X}_j^{(i_1)}\}$. Assume

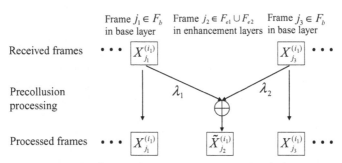

Fig. 11.4 Example of increasing the temporal resolution during precollusion processing. $F^{(i_1)} = F_b$ and $\widetilde{F}^{(i_1)} = F_b \cup F_{e1} \cup F_{e2}$

that $\widetilde{F}^{(i_1)}$ contains the indices of the frames in $\{\widetilde{\mathbf{X}}_j^{(i_1)}\}$ and $\widetilde{F}^{(i_1)} \neq F^{(i_1)}$. During collusion, colluder i_1 uses the newly generated copy $\{\widetilde{\mathbf{X}}_j^{(i_1)}\}_{j \in \widetilde{F}^{(i_1)}}$, instead of $\{\mathbf{X}_j^{(i_1)}\}_{j \in F^{(i_1)}}$. For simplicity, in this section, we assume that the cheating colluders only change the resolution of their received copies and do not further apply temporal filtering during precollusion processing.

We consider a simple scenario in which $\widetilde{F}^{(i_1)} \in \{F_b, \ F_b \cup F_{e1}, \ F_b \cup F_{e1} \cup F_{e2}\}$. We assume that there is only one cheating colluder i_1 who changes the frame rate of his or her copy before multiuser collusion; our analysis can be extended to complicated scenarios where there are multiple cheating colluders.

For a cheating colluder i_1 who changes the temporal resolution of his or her copy during precollusion processing, we define the processing parameter as $CP^{(i_1)} \triangleq \left(F^{(i_1)}, \widetilde{F}^{(i_1)} \right)$, where $F^{(i_1)}$ contains the indices of the frames that colluder i_1 received from the content owner and $\widetilde{F}^{(i_1)}$ contains the indices of the frames in the newly generated copy $\{\widetilde{\mathbf{X}}_j^{(i_1)}\}$. If $\widetilde{F}^{(i_1)} \supset F^{(i_1)}$, the cheating colluder i_1 subscribes to a lower-quality version and increases the frame rate during precollusion processing. If $\widetilde{F}^{(i_1)} \subset F^{(i_1)}$, colluder i_1 subscribes to a higher-quality version and reduces the temporal resolution before multiuser collusion.

11.2.1.1   Increasing resolution before multiuser collusion

In this type of precollusion processing, a cheating colluder i_1 subscribes to a copy of lower frame rate and generates a copy of higher resolution before collusion. Without loss of generality, in this section, we consider the example in Figure 11.4, in which the processing parameter is $CP^{(i_1)} = \left(F^{(i_1)} = F_b, \widetilde{F}^{(i_1)} = F_b \cup F_{e1} \cup F_{e2} \right)$. In this example, the cheating colluder receives the fingerprinted base layer only, and generates a copy $\{\widetilde{\mathbf{X}}_j^{(i_1)}\}$ with all three layers before collusion. He or she then tells the other colluders that $\{\widetilde{\mathbf{X}}_j^{(i_1)}\}_{j \in F_b \cup F_{e1} \cup F_{e2}}$ is the copy that he or she received.

Precollusion processing of the fingerprinted copy: We first study how the cheating colluder i_1 increases the temporal resolution of his or her fingerprinted copy. We assume that for every frame $j \in F^{(i_1)} = F_b$ in the base layer that colluder i_1 received, the cheating colluder simply duplicates $\mathbf{X}_j^{(i_1)}$ in the newly generated copy and we let $\widetilde{\mathbf{X}}_j^{(i_1)} = \mathbf{X}_j^{(i_1)}$.

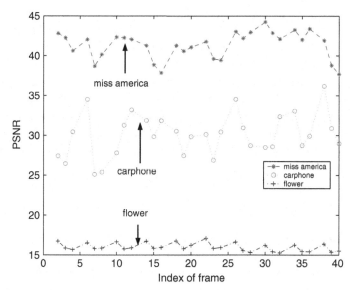

Fig. 11.5 Quality of the enhancement layers that are forged by the cheating colluder during precollusion processing

Colluder i_1 also needs to forge frames $\widetilde{\mathbf{X}}^{(i_1)}_{j \in F_{e1} \cup F_{e2}}$ in the enhancement layers that he or she did not receive. Assume that $\mathbf{X}^{(i_1)}_{j_1}$ and $\mathbf{X}^{(i_1)}_{j_3}$ are two adjacent frames in the base layer that colluder i_1 received. To forge a frame $j_2 \in F_{e1} \cup F_{e2}$ in the enhancement layers where $j_1 < j_2 < j_3$, we consider a simple linear interpolation based method and let $\widetilde{\mathbf{X}}^{(i_1)}_{j_2} = \lambda_1 \cdot \mathbf{X}^{(i_1)}_{j_1} + \lambda_2 \cdot \mathbf{X}^{(i_1)}_{j_3}$, where $\lambda_1 = \frac{j_3 - j_2}{j_3 - j_1}$ and $\lambda_2 = \frac{j_2 - j_1}{j_3 - j_1}$. Other complicated algorithms, such as motion-based interpolation [136], can be used to improve the quality of the forged frames, and the analysis will be similar.

Perceptual quality constraints: Now, we examine the perceptual quality of the forged enhancement layers and study the quality constraints. To increase the frame rate of the fingerprinted copy, the cheating colluder must generate frames in the enhancement layers that he or she did not receive from the content owner. To cover up the fact that he or she processed the copy before collusion and to make other colluders believe him or her, the cheating colluder must ensure that the forged enhancement layers have high quality.

We consider the example in Figure 11.4 with processing parameter $CP^{(i_1)} = (F_b, F_b \cup F_{e1} \cup F_{e2})$ with $F_b = \{1, 5, 9, \ldots\}$, $F_{e1} = \{3, 7, 11, \ldots\}$ and $F_{e2} = \{2, 4, 6, 8, \ldots\}$, and use the previous linear interpolation-based method.

For a cheating colluder i_1 in subgroup SC^b and for a frame $j \in F_{e1} \cup F_{e2}$ in the enhancement layers, define $\mathbf{X}^{(i_1)}_j$ as the fingerprinted frame j that colluder i_1 would have received if he or she had subscribed to frame j. In our simulations, we choose $\mathbf{X}^{(i_1)}_j$ as the ground truth and use the PSNR of $\widetilde{\mathbf{X}}^{(i_1)}_j$ when compared with $\mathbf{X}^{(i_1)}_j$ to measure the perceptual quality of the forged frames in the enhancement layers.

Figure 11.5 shows the results for the first forty frames of sequences "miss america," "carphone," and "flower." From Figure 11.5, for the sequence "miss america" with flat regions and slow motion, the cheating colluder can forge enhancement layers of high quality. For the sequence "flower," which has fast movement, the cheating colluder can generate only low-quality and blurred enhancement layers. Therefore, owing to the quality constraints, for complicated sequences with fast movement, the cheating colluder might not be able to apply this type of precollusion processing and increase the temporal resolution before multiuser collusion. Motion-based interpolation [136] can be used to improve the quality. However, for some sequences with fast movement and complex scene composition, such as "football" and "flower," even with motion-based interpolation, the cheating colluder still may not be able to forge enhancement layers of good enough quality to use. Therefore, for those complicated sequences, the cheating colluders may not be able to increase the resolution of their fingerprinted copies before multiuser collusion.

Cheating colluder's probability of being detected: To analyze the effectiveness of this precollusion processing in reducing a cheating colluder's risk, we compare the cheating colluder's probability of being detected when he or she increases the temporal resolution with that when he or she does not process the fingerprinted copy before collusion. Without loss of generality, we assume that the cheating colluder processes his or her copy as in Figure 11.4 with parameter $CP^{(i_1)} = (F_b, F_b \cup F_{e1} \cup F_{e2})$, and use this example to analyze the impact of resolution change on the cheating colluder's probability of being detected.

Scenario 1: without precollusion processing
We first consider the scenario in which colluder i_1 does not apply precollusion processing, and we assume that $SC^b = \{i \in SC : F^{(i)} = F_b\}$ contains the indices of the colluders who subscribe to copies of the lowest resolution and receive only the base layer from the content owner; $SC^{b,e1} = \{i \in SC : F^{(i)} = F_b \cup F_{e1}\}$ contains the indices of colluders who receive both the base layer and enhancement layer 1 from the content owner; and $SC^{all} = \{i \in SC : F^{(i)} = F_b \cup F_{e1} \cup F_{e2}\}$ contains the indices of colluders who receive all three layers from the content owner. K^b, $K^{b,e1}$, and K^{all} are the number of colluders in SC^b, $SC^{b,e1}$, and SC^{all}, respectively.

Given $(K^b, K^{b,e1}, K^{all})$ and (N_b, N_{e1}, N_{e2}), colluders first check the constraints in Table 5.1, and then choose the collusion parameters $\{\beta_k\}_{k=1,2,3}$ and $\{\alpha_l\}_{l=1,2}$ according to Table 5.1. In this scenario, for each frame $j \in F_b$ in the base layer, the extracted fingerprint is

$$\mathbf{Y}_j = \frac{\beta_1 \cdot \mathbf{W}_j^{(i_1)}}{K^b} + \sum_{i \in SC^b, i \neq i_1} \frac{\beta_1 \cdot \mathbf{W}_j^{(i)}}{K^b} + \sum_{i \in SC^{b,e1}} \frac{\beta_2 \cdot \mathbf{W}_j^{(i)}}{K^{b,e1}} + \sum_{i \in SC^{all}} \frac{\beta_3 \cdot \mathbf{W}_j^{(i)}}{K^{all}} + \mathbf{n}_j,$$

(11.9)

where $\mathbf{n}_j$ is additive noise.

Following the detection procedure in Section 2.3, the detector observes that colluder i_1 received only the fingerprinted base layer from the content owner and, therefore, the detector will use only fingerprints extracted from the base layer to decide whether colluder i_1 is involved in collusion. The detector calculates

$T_N^{(i_1)} = \left(\sum_{j \in F_b} \langle \mathbf{Y}_j, \mathbf{W}_j^{(i_1)} \rangle \right) / \sqrt{\sum_{l \in F_b} ||\mathbf{W}_l^{(i_1)}||^2}$, compares it with the predetermined threshold h, and decides whether colluder i_1 is a colluder. From the analysis of Zhao and Liu [137], with orthogonal modulation, given the colluder set SC and the extracted fingerprint as in (11.9), if the detection noise $\mathbf{n}_j$ is i.i.d. Gaussian $\mathcal{N}(0, \sigma_n^2)$, the detection statistics follow the distribution

$$p\left(T_N^{(i_1)}|SC\right) \sim \mathcal{N}(\mu^{(i_1)}, \sigma_n^2),$$

$$\text{where } \mu^{(i_1)} = \frac{\beta_1}{K^b}\sqrt{\sum_{j \in F_h}||\mathbf{W}_j^{(i_1)}||^2} = \frac{\beta_1 \sqrt{N_b}}{K^b}\sigma_w. \tag{11.10}$$

Colluder i_1's probability of being detected is $P_d^{(i_1)} = Q\left(\frac{h - \mu^{(i_1)}}{\sigma_n}\right)$, where $Q(\cdot)$ is the Gaussian tail function. In this scenario, all colluders share the same risk, and their probability of being detected equals $P_d^{(i_1)}$.

Scenario 2: with precollusion processing
We then consider the scenario in which colluder i_1 increases the frame rate before multiuser collusion and assume that $\widetilde{SC}^b = \{i \in SC : \widetilde{F}^{(i)}\}$ contains the indices of colluders who tell others that they received the base layer only; $\widetilde{SC}^{b,e1} = \{i \in SC : \widetilde{F}^{(i)} = F_b \cup F_{e1}\}$ is the set containing the indices of colluders who tell others that they received both the base layer and enhancement layer 1; and $\widetilde{SC}^{all} = \{i \in SC : \widetilde{F}^{(i)} = F_b \cup F_{e1} \cup F_{e2}\}$ is the set containing the indices of colluders who tell others that they received all three layers. Define $\widetilde{K}^b$, $\widetilde{K}^{b,e1}$, and $\widetilde{K}^{all}$ as the number of colluders in $\widetilde{SC}^b$, $\widetilde{SC}^{b,e1}$, and $\widetilde{SC}^{all}$, respectively.

If colluder i_1 is the only cheating colluder and the processing parameter is $CP^{(i_1)} = (F_b, F_b \cup F_{e1} \cup F_{e2})$, then we have $\widetilde{SC}^b = SC^b \setminus \{i_1\}$, $\widetilde{SC}^{b,e1} = SC^{b,e1}$ and $\widetilde{SC}^{all} = SC^{all} \cup \{i_1\}$. Consequently, $\widetilde{K}^b = K^b - 1$, $\widetilde{K}^{b,e1} = K^{b,e1}$, and $\widetilde{K}^{all} = K^{all} + 1$. If other colluders do not discover i_1's precollusion processing, they assume that the extracted fingerprints from all three layers will be used by the detector to determine whether i_1 is a colluder. Under this assumption, colluders analyze each attacker's detection statistics and follow Table 5.1 to choose the collusion parameters.

As an example, assume that colluders decide to generate a colluded copy including all frames in the base layer and enhancement layer 1, and $(\widetilde{K}^b, \widetilde{K}^{b,e1}, \widetilde{K}^{all})$ and (N_b, N_{e1}, N_{e2}) satisfy the constraint

$$\frac{\widetilde{K}^b \sqrt{N_b}}{\widetilde{K}^b \sqrt{N_b} + (\widetilde{K}^{b,e1} + \widetilde{K}^{all})\sqrt{N_b + N_{e1}}} \leq \frac{N_b}{N_b + N_{e1}} \tag{11.11}$$

listed in Table 5.1. Under the assumption that fingerprints extracted from both layers would be used by the detector to identify colluder i_1, other colluders estimate that i_1's detection statistics have the mean

$$\bar{\mu}^{(i_1)} = \frac{\widetilde{\beta}_3 N_b + \widetilde{\alpha}_2 N_{e1}}{\widetilde{K}^{all}\sqrt{N_b + N_{e1}}}\sigma_w, \tag{11.12}$$

where N_b and N_{e1} are the lengths of the fingerprints embedded in the base layer and enhancement layer 1, respectively. They choose the collusion parameters such that $\bar{\mu}^{(i_1)}$

equals the means of other colluders' detection statistics. From Table 5.1, the selected parameters are

$$\tilde{\beta}_1 = \frac{N_b + N_{e1}}{N_b} \cdot \frac{\tilde{K}_b \sqrt{N_b}}{\tilde{K}^b \sqrt{N_b} + (\tilde{K}^{b,e1} + \tilde{K}^{all}) \sqrt{N_b + N_{e1}}},$$

$$\tilde{\beta}_2 = \frac{\tilde{K}^{b,e1}}{\tilde{K}^{b,e1} + \tilde{K}^{all}} \left(1 - \tilde{\beta}_1\right), \quad \tilde{\beta}_3 = 1 - \tilde{\beta}_1 - \tilde{\beta}_2,$$

$$\tilde{\alpha}_1 = \frac{\tilde{K}^{b,e1}}{\tilde{K}^{b,e1} + \tilde{K}^{all}}, \quad \text{and } \tilde{\alpha}_2 = 1 - \tilde{\alpha}_1. \tag{11.13}$$

Then, colluders generate the colluded copy using the two-stage collusion.

During the colluder identification process, because colluder i_1 received only the fingerprinted base layer from the content owner, the detector uses only fingerprints extracted from the base layer to decide whether i_1 is a colluder. The extracted fingerprint from frame $j \in F_b$ in the base layer is

$$\mathbf{Y}_j = \frac{\tilde{\beta}_3 \cdot \mathbf{W}_j^{(i_1)}}{\tilde{K}^{all}} + \sum_{i \in SC^b, i \neq i_1} \frac{\tilde{\beta}_1 \cdot \mathbf{W}_j^{(i)}}{\tilde{K}^b} + \sum_{i \in SC^{b,e1}} \frac{\tilde{\beta}_2 \cdot \mathbf{W}_j^{(i)}}{\tilde{K}^{b,e1}} + \sum_{i \in SC^{all}} \frac{\tilde{\beta}_3 \cdot \mathbf{W}_j^{(i)}}{\tilde{K}^{all}} + \mathbf{n}_j. \tag{11.14}$$

With orthogonal fingerprint modulation, given the colluder set SC, the index of the cheating colluder i_1, and the precollusion processing parameter $CP^{(i_1)} = (F_b, F_b \cup F_{e1} \cup F_{e2})$, if $\mathbf{n}_j$ follows Gaussian distribution $\mathcal{N}(0, \sigma_n^2)$, using the same analysis as that of Zhao and Lio [137], we can show that

$$p\left(T_N^{(i_1)} | SC, i_1, CP^{(i_1)}\right) \sim \mathcal{N}(\tilde{\mu}^{(i_1)}, \sigma_n^2) \quad \text{and} \quad \tilde{P}_d^{(i_1)} = Q\left(\frac{h - \tilde{\mu}^{(i_1)}}{\sigma_n}\right),$$

$$\text{where} \quad \tilde{\mu}^{(i_1)} = \frac{\tilde{\beta}_3}{\tilde{K}^{all}} \sqrt{\sum_{j \in F_b} ||\mathbf{W}_j^{(i_1)}||^2} = \frac{\tilde{\beta}_3 \sqrt{N_b}}{\tilde{K}^{all}} \sigma_w. \tag{11.15}$$

For an honest colluder i_2 who does not process his or her copy before collusion, following the same analysis, we can show that this colluder's detection statistics follow Gaussian distribution $p\left(T_N^{(i_2)} | SC, i_1, CP^{(i_1)}\right) \sim \mathcal{N}(\tilde{\mu}^{(i_2)}, \sigma_n^2)$ where $\tilde{\mu}^{(i_2)} = \tilde{\beta}_1 \sqrt{N_b} \sigma_w / \tilde{K}^b$, and the probability of being detected is $P_d^{(i_2)} = Q\left((h - \tilde{\mu}^{(i_2)})/\sigma_n\right)$.

Note that $\bar{\mu}^{(i_1)}$ in (11.12) does not equal $\tilde{\mu}^{(i_1)}$ in (11.15), and the colluders make an error in estimating the mean of the cheating colluder i_1's detection statistics. This is the result of i_1's precollusion processing behavior; this estimation error helps the cheating colluder further lower his or her risk of being detected.

From (11.10) and (11.15), for fixed h and σ_n^2, comparing the cheating colluder's probability of being detected in these two scenarios is equivalent to comparing $\mu^{(i_1)}$ in (11.10) with $\tilde{\mu}^{(i_1)}$ in (11.15). For a fair comparison, if the constraints in Table 5.1 are satisfied, we fix the frame rate of the colluded copy and let $\tilde{F}^c = F^c$.

To compare the values of the two means, we consider the following scalable fingerprinting systems. We observe that for typical video sequences such as "miss america," "carphone," and "foreman," each frame has approximately 3000 to 7000 embeddable coefficients, depending on the characteristics of the sequences. As an example, we assume that the length of the embedded fingerprints in each frame is 5000, and we test on a total of 40 frames. We choose $F_b = \{1, 5, 9, \ldots\}$, $F_{e1} = \{3, 7, 11, \ldots\}$, and $F_{e2} = \{2, 4, 8, \ldots\}$ as an example of the temporal scalability, and the lengths of the fingerprints embedded in the base layer, enhancement layer 1, and enhancement layer 2 are $N_b = 50000$, $N_{e1} = 50000$, and $N_{e2} = 100000$, respectively. We assume that there are a total of $M = 450$ users and $|\mathbf{U}^b| = |\mathbf{U}^{b,e1}| = |\mathbf{U}^{all}| = 150$. We first generate a unique vector following Gaussian distribution $\mathcal{N}(0, 1/9)$ for each user, and then apply Gram–Schmidt orthogonalization.

We assume that there are a total of $K = 150$ colluders, and $(K^b, K^{b,e1}, K^{all})$ are on the line $\overline{CD}$ in (5.31), which is the boundary of one of the constraints in Table 5.1 to achieve fairness of collusion when generating a colluded copy of the highest resolution. Other values of $(K^b, K^{b,e1}, K^{all})$ and (N_b, N_{e1}, N_{e2}) give the same trend.

Assume that there is only one cheating colluder i_1 and $CP^{(i_1)} = (F_b, F_b \cup F_{e1} \cup F_{e2})$. Figure 11.6 compares $\mu^{(i_1)}$ in (11.10) with $\tilde{\mu}^{(i_1)}$ in (11.15) when $K = 150$ and $(K^b, K^{b,e1}, K^{all})$ takes different values on line $\overline{CD}$ in (5.31). In Figure 11.6, a given value of K^b corresponds to a unique point on Line (5.31) and, therefore, a unique triplet $(K^b, K^{b,e1}, K^{all})$. In Figure 11.6(a), $F^c = F_b \cup F_{e1} \cup F_{e2}$ and the colluded copy has the highest resolution; and in Figure 11.6(b), $F^c = F_b$ and the colluded copy contains only frames in the base layer. From Figure 11.6, increasing the resolution of his or her fingerprinted copy before multiuser collusion can help the cheating colluder further reduce his or her probability of being detected when the colluded copy is of high quality, but it cannot lower the cheating colluder's risk when colluders decide to generate a copy of the lowest frame rate. This is because when $F^c = F_b$, no matter how many frames the cheating colluder i_1 claims that he or she has received, only those in the base layer are used to generate the colluded copy, and those frames are the ones that i_1 received from the content owner. In this scenario, other colluders correctly estimate the mean of i_1's detection statistics during collusion, and increasing the frame rate cannot help the cheating colluder further reduce his or her risk. To generalize, increasing the temporal resolution is effective in reducing a cheating colluder's probability of being captured only if $F^c \supset F^{(i_1)}$.

11.2.1.2 Reducing resolution before multiuser collusion

In this type of precollusion processing, a cheating colluder receives a copy of higher resolution and tells other colluders that he or she has only a copy of lower quality. Shown in Figure 11.7 is an example, in which the cheating colluder i_1 subscribes to all three layers while claiming that he or she has only the fingerprinted base layer. In this example, i_1 simply drops frames in the two enhancement layers during precollusion processing.

When reducing the frame rate of his or her fingerprinted copy, the cheating colluder does not need to forge any frames and, therefore, he or she does not need to worry about the quality constraints. In this scenario, the analysis of the cheating colluder's risk of being detected is similar to that in Section 11.2.1.1 and is thus omitted.

Figure 11.8 compares the means of the cheating colluder's detection statistics when he or she drops frames in the enhancement layers with that when he or she does

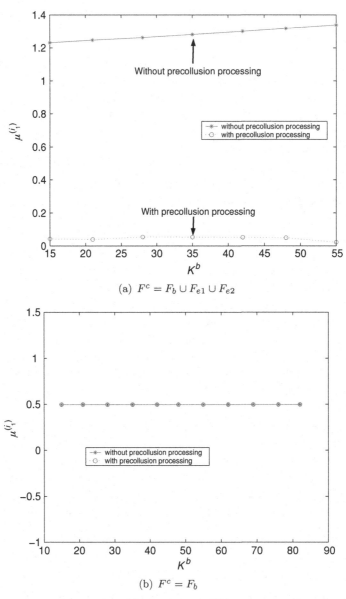

(a) $F^c = F_b \cup F_{e1} \cup F_{e2}$

(b) $F^c = F_b$

Fig. 11.6 Comparison of $\mu^{(i_1)}$ in (11.10) and $\widetilde{\mu}^{(i_1)}$ in (11.15) when $(K^b, K^{b,e1}, K^{all})$ takes different values on line $\overline{CD}$ (5.31)

not apply precollusion processing. The setup of the scalable fingerprinting system in Figure 11.8 is the same as that in Figure 11.6. Similarly, each K^b in Figure 11.8 represents one point on line $\overline{CD}$ in (5.31) and a unique $\left(K^b, K^{b,e1}, K^{all}\right)$ triplet. The precollusion processing parameter is $CP^{(i_1)} = (F_b \cup F_{e1} \cup F_{e2}, F_b)$. $F^c = F_b \cup F_{e1} \cup F_{e2}$ and $F^c = F_b$ in Figure 11.8(a) and (b), respectively. From Figure 11.8, similar to the case in Figure 11.6, when the colluded copy has high resolution, the cheating colluder can

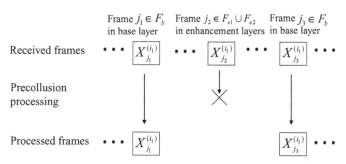

Fig. 11.7 Example of reducing the frame rate before multiuser collusion. $F^{(i_1)} = F_b \cup F_{e1} \cup F_{e2}$ and $\widetilde{F}^{(i_1)} = F_b$

significantly reduce his or her own probability of being detected by reducing the frame rate before multiuser collusion, however, when the colluded copy has low resolution, it cannot further lower the cheating colluder's risk. In general, reducing the temporal resolution before collusion can further reduce the cheating colluder's risk only when $F^c \supset \widetilde{F}^{(i_1)}$.

11.2.2   Performance comparison of different strategies

In the scalable fingerprinting system, each cheating colluder has two choices when modifying the resolution of his or her fingerprinted copy before collusion. For example, for a cheating colluder $i_1 \in SC^{all}$ who receives all three layers from the content owner, during precollusion processing, i_1 can drop the received enhancement layer 2 before collusion and tell other attackers that he or she has a medium-quality fingerprinted copy. The cheating colluder i_1 can also drop both enhancement layers and claim that he or she has the base layer only. This section compares the effectiveness of different precollusion processing strategies in reducing the cheating colluder's risk, assuming that the quality constraints are satisfied and other colluders do not discover the precollusion processing behavior.

From the analysis in the previous section, neither increasing nor reducing the temporal resolution can further reduce the cheating colluder's probability of being detected when the colluded copy contains only frames in the base layer. Therefore, in this section, we consider scenarios in which the colluded copy includes at least one enhancement layer and F^c equals either $F_b \cup F_{e1}$ or $F_b \cup F_{e1} \cup F_{e2}$.

Our simulation setup is similar to that in Section 11.2.1.1. We assume that each frame has 5000 embeddable coefficients and we test on a total of forty frames. We consider a temporally scalable video coding system with $F_b = \{1, 5, 9, \ldots\}$, $F_{e1} = \{3, 7, 11, \ldots\}$, and $F_{e2} = \{2, 4, 8, \ldots\}$; the lengths of the fingerprints embedded in the base layer, enhancement layer 1, and enhancement layer 2 are $N_b = 50000$, $N_{e1} = 50000$, and $N_{e2} = 100000$, respectively. We further assume that there are a total of $M = 450$ users in the system, and $|\mathbf{U}^b| = |\mathbf{U}^{b,e1}| = |\mathbf{U}^{all}| = 150$. For each user, a unique vector is first generated from Gaussian distribution $\mathcal{N}(0, \sigma_w^2)$ with $\sigma_w^2 = 1/9$, and Gram–Schmidt orthogonalization is applied.

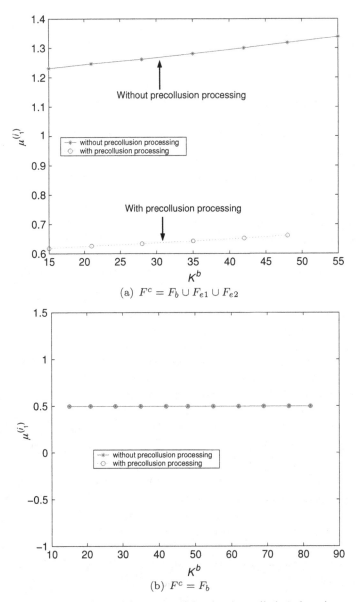

Fig. 11.8 Comparison of the means of the cheating colluder's detection statistics when he or she reduces the frame rate during precollusion processing with that when he or she does not process his or her copy before multiuser collusion

During collusion, we assume that there are a total of $K = 150$ colluders and $(K^b, K^{b,e1}, K^{all})$ takes different values on line $\overline{CD}$ in (5.31). We further assume that the additive noise $\mathbf{n}_j$ in (11.4) follows Gaussian distribution $\mathcal{N}(0, \sigma_n^2)$ with $\sigma_n^2 = 2\sigma_w^2$. In our simulations, we assume that there is only one cheating colluder i_1 and other colluders do not discover his or her precollusion processing.

11.2.2.1 For cheating colluders in subgroup SC^b

For a cheating colluder i_1 who receives the base layer only, he or she can increase the frame rate of his or her fingerprinted copy with two different parameters: $CP_1^{(i_1)} = (F_b, F_b \cup F_{e1})$ and $CP_2^{(i_1)} = (F_b, F_b \cup F_{e1} \cup F_{e2})$. In this section, we compare the effectiveness of these two strategies in reducing i_1's probability of being caught, $P_d^{(i_1)}$.

We fix the probability of accusing a given innocent user P_{fa} as 0.01, and compare $P_d^{(i_1)}$ of different precollusion processing parameters. Figure 11.9 shows our simulation results when $(K^b, K^{b,e1}, K^{all})$ takes different values on line $\overline{CD}$ in (5.31), and each K^b corresponds to a unique point on that line. $F^c = F_b \cup F_{e1}$ and $F^c = F_b \cup F_{e1} \cup F_{e2}$ in Figure 11.9(a) and (b), respectively. From the cheating colluder's point of view, when $F^c = F_b \cup F_{e1}$, the two processing parameters have the same performance. If $F^c = F_b \cup F_{e1} \cup F_{e2}$, $CP_2^{(i_1)} = (F_b, F_b \cup F_{e1} \cup F_{e2})$ gives the cheating colluder a smaller probability of being detected than $CP_1^{(i_1)} = (F_b, F_b \cup F_{e1})$. Therefore, under the quality constraints, a cheating colluder in SC^b should pretend to have received all three layers from the content owner to minimize his or her risk.

In Figure 11.10, we consider two colluders, a cheating colluder i_1 who increases the resolution of his or her copy during precollusion processing and an honest colluder i_2 who does not process his or her copy before collusion, and compare their probabilities of being detected by the fingerprint detector. $i_1 \in SC_b$ and $CP^{(i_1)} = (F_b, F_b \cup F_{e1} \cup F_{e2})$. From Figure 11.10, precollusion processing makes the honest colluder i_2 have a much larger probability of being detected than the cheating colluder i_1, and increases i_2's relative risk when compared with that of i_1. It is certainly a cheating behavior.

11.2.2.2 For cheating colluders in subgroup $SC^{b,e1}$

A cheating colluder $i_1 \in SC^{b,e1}$ who receives the base layer and enhancement layer 1 from the content owner can increase the resolution of his or her copy with parameter $CP_1^{(i_1)} = (F_b \cup F_{e1}, F_b \cup F_{e1} \cup F_{e2})$ during precollusion processing. Colluder i_1 can also drop his or her fingerprinted enhancement layer 1 with parameter $CP_2^{(i_1)} = (F_b \cup F_{e1}, F_b)$.

From the simulation results shown in Figure 11.11(a), when the colluded copy has medium temporal resolution and $F^c = F_b \cup F_{e1}$, dropping enhancement layer 1 with parameter $CP_2^{(i_1)}$ reduces i_1's probability of being detected, but increasing the resolution with parameter $CP_1^{(i_1)}$ cannot further lower the cheating colluder's risk. From Figure 11.11(b), when the colluded copy includes all three layers and $F^c = F_b \cup F_{e1} \cup F_{e2}$, both $CP_1^{(i_1)}$ and $CP_2^{(i_1)}$ can reduce i_1's probability of being captured, and $CP_1^{(i_1)}$ gives the cheating colluder a smaller chance to be detected than $CP_2^{(i_1)}$.

Consequently, for a cheating colluder in subgroup $SC^{b,e1}$ to minimize his or her own risk, when colluders plan to generate a colluded copy of medium temporal resolution, the cheating colluder should drop enhancement layer 1 before multiuser collusion; and when the colluders plan to generate a colluded copy containing all three layers, the cheating colluder should increase the resolution of his or her fingerprinted copy with parameter $CP_1^{(i_1)} = (F_b \cup F_{e1}, F_b \cup F_{e1} \cup F_{e2})$.

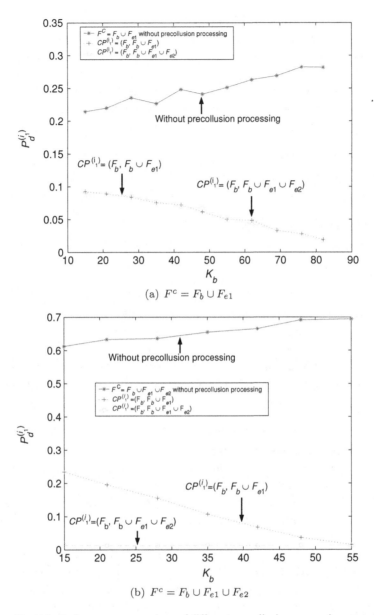

Fig. 11.9 Performance comparison of different precollusion processing strategies for cheating colluders in SC^b

Figure 11.12 investigates the impact of precollusion processing on other colluders' probability of being detected. In Figure 11.12, there are ten cheating colluders who use the same parameter ($F_b \cup F_{e1}$, $F_b \cup F_{e1} \cup F_{e2}$) during precollusion processing, and they process their fingerprinted copies independently. We consider two colluders, a cheating colluder i_1 and an honest colluder i_2 who does not apply precollusion processing. In this scenario, precollusion processing not only reduces the cheating colluders' absolute

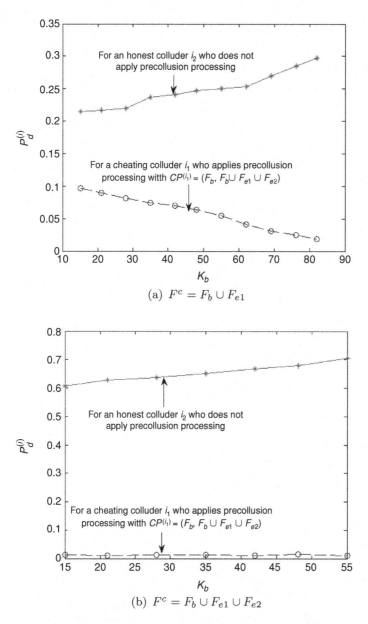

Fig. 11.10 Comparison of different colluders' probabilities of being detected when there exist cheating colluders

risk, but it also decreases other attackers' probability of being detected. However, from Figure 11.12(b), such precollusion processing makes the cheating colluder i_1 take a much smaller chance of being caught than the honest colluder i_2 and increases other colluders' relative risk with respect to the cheating colluders. Therefore, it is still a cheating behavior.

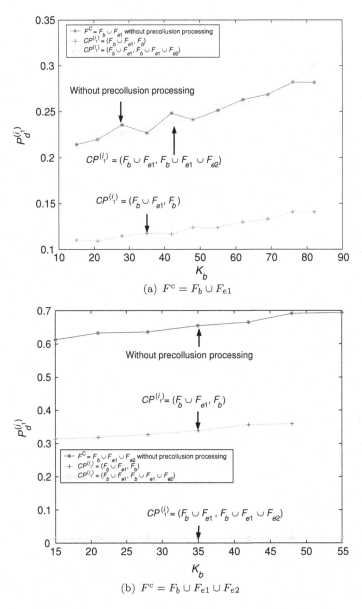

Fig. 11.11 Performance comparison of different precollusion processing strategies for cheating colluders in $SC^{b,e1}$

11.2.2.3   For cheating colluders in SC^{all}

For a cheating colluder i_1 in subgroup SC^{all} who receives all three layers, during precollusion processing, i_1 can reduce the frame rate of his or her fingerprinted copy with two different parameters: $CP_1^{(i_1)} = (F_b \cup F_{e1} \cup F_{e2}, F_b)$ and $CP_2^{(i_1)} = (F_b \cup F_{e1} \cup F_{e2}, F_b \cup F_{e1})$.

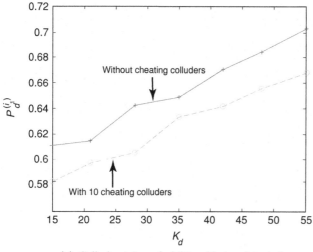

(a) Colluder i_1's probability of being detected

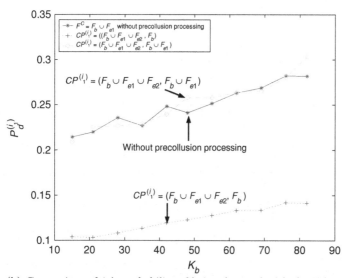

(b) Comparison of i_1's probability of being detected with that of i_2

Fig. 11.12 Impact of precollusion processing on other colluders' probability of being detected

As shown in Figure 11.13(a), when the colluded copy has medium resolution, using $CP_1^{(i_1)}$ further reduces i_1's probability of being detected, whereas $CP_2^{(i_1)}$ does not change the risk. From Figure 11.13(b), if colluders generate a high-resolution colluded copy, both strategies lower the cheating colluder's probability of being captured and $P_d^{(i_1)}$ of $CP_1^{(i_1)}$ is smaller than $P_d^{(i_1)}$ of $CP_2^{(i_1)}$. Consequently, from the cheating colluder's point of view, dropping both enhancement layers before multiuser collusion is preferred for a cheating colluder in subgroup SC^{all} to minimize the risk of being detected.

For an honest colluder i_2 who does not process his or her received copy before collusion, Figure 11.14 shows the impact of cheating colluders' precollusion processing on

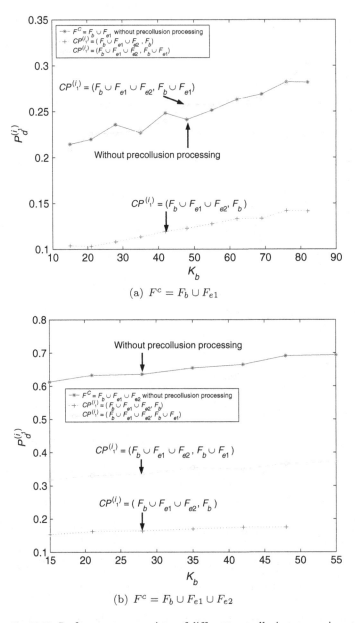

(a) $F^c = F_b \cup F_{e1}$

(b) $F^c = F_b \cup F_{e1} \cup F_{e2}$

Fig. 11.13 Performance comparison of different precollusion processing strategies for cheating colluders in SC^{all}

i_2's probability of being detected when the total number of cheating colluders varies. In Figure 11.14, all cheating colluders select the same precollusion processing parameter $CP = (F_b \cup F_{e1} \cup F_{e2}, F_b)$, but each processes his or her fingerprinted copy independently. From Figure 11.14, dropping enhancement layers before collusion increases others' probability of being detected, and the honest colluder has a greater probability

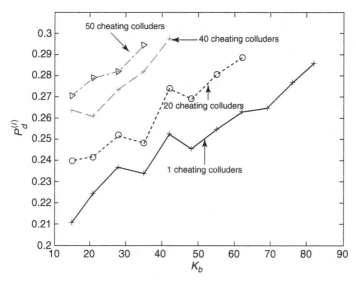

Fig. 11.14 An honest colluder's probability of being detected with different numbers of cheating colluders

of being detected when there are more cheating colluders. In this example, precollusion processing is not only cheating, but also malicious.

11.2.3 Simulation results on real video

We test the effectiveness of changing the resolution of the fingerprinted copy before collusion on real videos, assuming that the quality constraints are satisfied. We choose the first forty frames of the sequence "carphone" as an example. Similar to that in Section 11.2.1.1, we consider a temporally scalable video coding system with $F_b = \{1, 5, 9, \ldots\}$, $F_{e1} = \{3, 7, 11, \ldots\}$ and $F_{e2} = \{2, 4, 8, \ldots\}$. The lengths of the fingerprints embedded in the base layer, enhancement layer 1, and enhancement layer 2 are $N_b = 39988$, $N_{e1} = 39934$, and $N_{e2} = 79686$, respectively. We assume that the total number of users is $M = 450$ and $|\mathbf{U}^b| = |\mathbf{U}^{b,e1}| = |\mathbf{U}^{all}| = 150$. We adopt the human visual model-based spread spectrum embedding of Podilchuk and Zeng [58], and embed the fingerprints in the DCT domain. We first generate independent vectors following Gaussian distribution $\mathcal{N}(0, 1/9)$, and then apply Gram–Schmidt orthogonalization to let the assigned fingerprints be strictly orthogonal and have equal energy. In each fingerprinted copy, similar to that in the work of Su *et al.* [111], fingerprints in adjacent frames are correlated with each other, depending on the similarity between the host frames.

During collusion, we assume that there are a total of $K = 150$ colluders and $(K^b, K^{b,e1}, K^{all})$ takes different values on $\overline{CD}$ in line (5.31). We consider a simple scenario in which there is only one cheating colluder who changes the resolution of his or her received copy before collusion. Furthermore, we assume that no colluders discover the cheating colluder's precollusion processing. In our simulations, we adjust the power

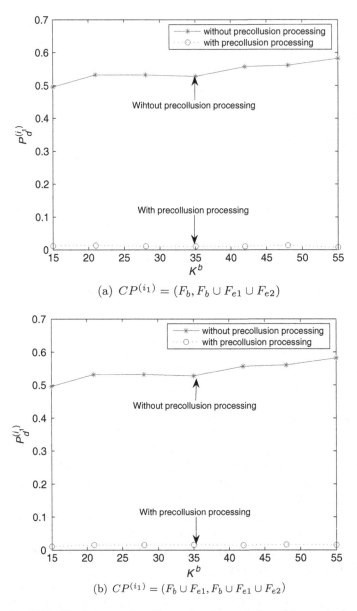

Fig. 11.15 Simulation results of increasing the resolution of the received copies during precollusion processing on the first 40 frames of the sequence "carphone." $F^c = F_b \cup F_{e1} \cup F_{e2}$

of the additive noise $\mathbf{n}_j$ such that $||\mathbf{n}_j||^2/||JND_j \cdot \mathbf{W}_j^{(i)}||^2 = 2$ for every frame in the colluded copy.

Figures 11.15 and 11.16 show the simulation results. From Figures 11.15 and 11.16, under the quality constraints, changing the resolution of the fingerprinted copy can help a cheating colluder further reduce the risk of being caught, especially when the colluded copy has high resolution. The simulation results on real videos agree with our theoretical analysis in Section 11.2.1, and are comparable with the results in Section 11.2.2.

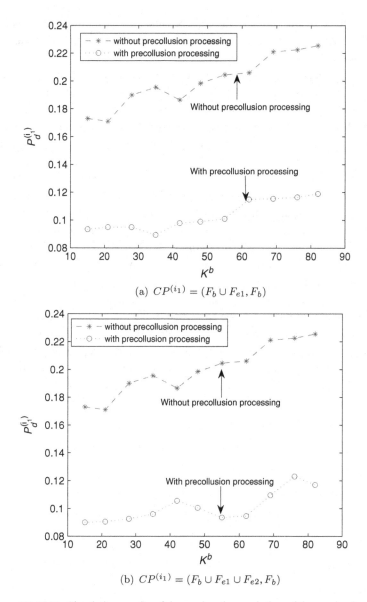

(a) $CP^{(i_1)} = (F_b \cup F_{e1}, F_b)$

(b) $CP^{(i_1)} = (F_b \cup F_{e1} \cup F_{e2}, F_b)$

Fig. 11.16 Simulation results of decreasing the resolution of the received copies during precollusion processing on the first 40 frames of the sequence "carphone." $F^c = F_b \cup F_{e1}$

11.3   Chapter summary

Cheating is likely to happen in media-sharing social networks owing to the cheating nature of the users. Only after understanding cheating can users act against it. Therefore, in this chapter, we analyze the cheating behavior in colluder social networks and formulate the dynamics among attackers during collusion to minimize their own risk of being detected and protect their own interest. We investigate a few precollusion processing strategies that a cheating colluder can use to further reduce his or her chance of being captured by the digital rights enforcer, and analyze their effectiveness. We also analyze the constraints on precollusion processing to maintain the perceptual quality of the fingerprinted copies.

We first investigate the strategies for a cheating colluder to attenuate the energy of the embedded fingerprints before collusion. The cheating colluder can apply temporal filtering to his or her copy and average adjacent frames of similar content before multiuser collusion. We analyze its effectiveness in reducing the cheating colluder's risk as well as the perceptual quality of the fingerprinted copy after temporal filtering. Both our analytical and simulation results show that this temporal filtering reduces the cheating colluder's risk of being captured, at the cost of quality degradation. We then investigate the tradeoff between the risk and the perceptual quality that a cheating colluder needs to address, and derive the optimal filtering coefficients to minimize the probability of being caught while maintaining good quality of the fingerprinted copy.

We then consider the problem of traitors within traitors when attackers receive fingerprinted copies of different resolutions owing to network and device heterogeneity. In such a scenario, a cheating colluder can not only apply temporal filtering to his or her received frames, but can also change the resolution and quality of his or her fingerprinted copy before multiuser collusion. We show that under the quality constraints, changing the resolution of the fingerprinted copy can help a cheating colluder further reduce the probability of being caught, especially when the colluded copy has high quality. For traitors within traitors in scalable fingerprinting systems, we also investigate the selection of the optimal strategy for a cheating colluder to minimize his or her risk under the quality constraints.

12 Attack resistance in peer-to-peer video streaming

Chapter 9 studies cooperation stimulation for P2P video streaming over Internet and wireless networks. One assumption there is that all users in the P2P networks are rational, and their goal is to maximize their own payoffs. As discussed in Chapter 9 and shown in Figure 12.1, they may lie to others about their personal information if they believe cheating can help increase their utilities. There might also exist *malicious* users who aim to exhaust others' resources and attack the system. For example, in P2P systems, they can tamper the media files with the intention of making the content useless (the so-called pollution attack). They can also launch the denial of service (DoS) attack to exhaust other users' resources and make the system unavailable [85]. What is more, once an attacker is detected, he or she can leave the network temporarily, come back later with a new ID, and continue causing damage to the system.

To further proliferate P2P systems and provide reliable service, misbehavior detection and attack resistance are fundamental requirements to stimulate user cooperation even under attacks. A challenging issue in malicious user detection in P2P video streaming is to differentiate between "intentional" misbehavior (for example, intentional modification of the video content) and "innocent" ones (such as transmission error and packet loss in error-prone and congested networks).

In this chapter, we first model the P2P video streaming network over the Internet as a multiplayer game, which includes both rational (selfish) and malicious users. We then study different methods that the attackers may use and analyze the maximum damages that they can cause to the system. Based on such analysis, we explore possible schemes to identify attackers and analyze the attack resistance of the proposed cooperation stimulation strategies. We then extend our study to P2P video streaming over wireless networks. Wireless video streaming social networks often have fewer users, owing to the limited transmission range of mobile devices, and attackers can damage the system more easily. Here, trust management is used to quickly and reliably identify attackers in wireless P2P streaming and to stimulate user cooperation even under attacks.

Because of network congestion and transmission errors, a chunk may be dropped by the intermediate nodes or received incorrectly at the receiver's side. Here, we assume that a transmitted chunk is considered dropped if it does not arrive within one round, and we use P_{ij} to denote the probability that a chunk is successfully transmitted from peer i to peer j in one round. We assume that the chunk request messages can be received immediately and perfectly, because they contain only a few bits

Fig. 12.1 User dynamics in real-world social networks

and can be retransmitted within a short interval, if necessary, to ensure successful transmission.

12.1 Attack-resistant cooperation strategies in P2P video streaming over the Internet

In this section, we consider P2P video streaming over the Internet and investigate how to identify malicious users and design attack-resistant cooperation strategies. Here, we start with a simple scenario in which attackers do not hand-wash; we will address the hand-wash attack in the next section.

12.1.1 Misbehavior in P2P video streaming

We first consider possible strategies that attackers can use to attack the P2P video streaming system; we focus on insider attackers – that is, attackers who also have legitimate identities. In P2P video streaming social networks, attackers can use different strategies to to attack the system, for example:

(1) *Incomplete chunk attack*: The malicious attacker agrees to send the entire requested chunk to a peer, but sends only portions of it or no data at all. This makes the requesting peer waste his or her request quota in this round, and he or she must request the same chunk again in the next round.

(2) *Pollution attack*: The other kind of attack in P2P live streaming is pollution. In P2P streaming system, a malicious user modifies the data chunks, renders the content unusable, and then makes this polluted content available for sharing with other peers. Unable to distinguish polluted chunks from clean (unpolluted) ones, unsuspicious users may download the polluted chunks into their own buffers, and share them with others. In this way, polluted chunks spread through the entire network.

(3) *Hand-wash attack*: Many P2P systems are anonymous in nature, and a user is only identified by the registered ID. If a malicious user is detected, he or she can leave the system temporarily and come back with a new valid ID. P2P streaming systems cannot differentiate between a new user and a returned attacker with a new ID, and the hand-wash attack enables attackers to constantly cause damages to the system.

12.1.2 Game model

For P2P live streaming over the Internet, we model the interactions among peers as the following game:

- *Server:* The streaming server has all the video chunks. Its upload bandwidth is W_s, and the streaming server sends the requested chunks to users in a round-robin fashion.
- *Players and player type:* There are a total of N users/peers in the P2P live streaming social network. Each player $i \in N$ has a type $\theta_i \in \{\text{selfish, malicious}\}$. Let N_s denote the set of all selfish players and $N_m = N \backslash N_s$ is the set including all insider attackers. A selfish user aims to maximize his or her own payoff, and may cheat other peers if cheating can help increase his or her payoff. A malicious user wishes to exhaust other peers' resources and attack the system.
- *Chunk requesting:* As in Section 9.1, in each round, each player has *one* chunk-request quota, with which he or she either *requests a chunk from a peer*, *requests a chunk from the video streaming source*, or *does not request any chunks* in this round.
- *Request answering:* For each player, after receiving a request asking for the upload of a chunk in its buffer, it can either *accept* or *refuse* the request.
- *Cost:* For any player $i \in N$, uploading a chunk to another player incurs a cost of $c_i = M/W_i\tau$, where W_i is player i's upload bandwidth and $W_i \geq W_{min}$, as in Section 9.1.2.
- *Gain:* If a selfish user $i \in N_s$ requests a data chunk from another peer j, and if an unpolluted copy is successfully delivered, his or her gain is g_i where $P_{ji}g_i > c_i$. Here, as in Section 9.1.2, a chunk is considered lost if it does not arrive within one round, and P_{ij} is the probability that a chunk is successfully transmitted from peer i to peer j in one round. Also, as in Chapter 9, user i decides g_i depending on how much he or she wants to watch the video.

In this game, before defining each user's utility function, we first introduce the symbols that we use in this section. For each player $i \in N$,

- $Cr^{(i)}(j, t)$ is the total number of chunks that i has requested from j by time t. Here, j can be either a peer ($j \in N$) or j is the streaming server. $Cr^{(i)}(t) = \sum_{j \in \{N, \text{source}\}} Cr^{(i)}(j, t)$ denotes the total number of chunks that i has requested by time t.
- By time t, peer i has successfully received $Cs^{(i)}(j, t)$ chunks from peer j in time (a chunk is received in time if and only if it is received within the same round that it was requested). $Cs^{(i)}(t) = \sum_{j \in \{N, \text{source}\}} Cs^{(i)}(j, t)$ is peer i's total number of successfully received chunks by time t.
- By time t, $C_p^{(i)}(j, t)$ is the total number of polluted chunks that peer i received from peer j. The total number of successively received unpolluted data chunks that peer i received from peer j is $Cs^{(i)}(j, t) - C_p^{(i)}(j, t)$, and each successfully received unpolluted chunk gives peer j a gain of g_i.
- $Cu^{(i)}(j, t)$ denotes the number of chunks that i has uploaded to player j by time t. $Cu^{(i)}(t) = \sum_{j \in N, j \neq i} Cu^{(i)}(j, t)$ is the total number of chunks that i has uploaded by time t, and the cost of uploading each chunk is c_i for peer i.

Let t_f be the lifetime of the P2P live streaming social network, and let $T^{(i)}(t)$ denote the total time that peer i is in the network by time t. For a selfish player $i \in N_s$, the utility $U_s^{(i)}(t_f)$ is defined as

$$U^{(i)}(t_f) = \frac{\left[Cs^{(i)}(t_f) - \sum_{j \in N} C_p^{(i)}(j, t_f) \right] g_i - Cu^{(i)}(t_f) \frac{M}{W_i \tau}}{Cr^{(i)}(t_f)} \qquad (12.1)$$

Here, the numerator denotes the net profit (i.e., the total gain minus the total cost) that player i receives, and the denominator denotes the total number of chunks that i has requested. This utility function represents the average net profit that i can obtain per requested chunk, which i aims to maximize.

For a malicious player $j \in N_m$, the objective is to maximize its utility $U_m^{(j)}$

$$= \frac{\sum_{i \in N_s} Cu^{(i)}(j, t_f) \frac{M}{W_i \tau} + \sum_{i \in N_s} \left[Cr^{(i)}(j, t_f) - Cs^{(i)}(j, t_f) \right] P_{ji} g_i - Cu^{(j)}(t_f) \frac{M}{W_j \tau}}{T^{(j)}(t_f)}.$$

$$(12.2)$$

The numerator in (12.2) is the net damage caused by j: the first term calculates the total upload bandwidth that other users use to send the requested chunks to the malicious user j; the second term calculates other selfish peers' potential loss in gain as a result of the incomplete chunk attack by peer j; and the last term is peer j's cost to upload chunks to other peers. We normalize the net damage by the lifetime of peer j; this utility function represents the average net damage that j causes to other users per time unit.

12.1.3   Attack-resistant cooperation stimulation strategies

Based on the system description in Section 12.1, we can see that the multiple player game is much more complicated than the two-person game in Section 9.1. First, a peer may request chunks from different peers at different times to maximize the utility. A direct consequence of such a nonrepeated model is that favors cannot be granted simultaneously. This makes cooperation stimulation in P2P live streaming networks an extremely challenging task. In addition, the inevitable packet delay in the Internet can cause severe troubles. For the two-player cheat-proof cooperation strategy, if the link between users is too busy and some packets cannot arrive within one round, the game will be terminated immediately, causing drastic performance degradation. What is more, it is challenging to distinguish "innocent" misbehavior owing to bit errors and packet loss from "intentional" ones by malicious attackers. Thus, direct application of the two-player cooperation strategies in Chapter 9 to the multiple-player scenarios may not work without proper adjustment and accurate schemes to identify malicious attackers.

12.1.3.1   Credit mechanism

We introduce the credit mechanism to address the issue that favors cannot be simultaneously granted in the multiuser game, and to differentiate between intentional uploading of polluted chunks by attackers and unintentional forwarding of polluted chunks by selfish users.

For any two peers i and j,

$$Cc^{(i)}(j, t) = Cu^{(i)}(j, t) - C_p^{(j)}(i, t) \tag{12.3}$$

calculates the total number of *clean* chunks that peer i has uploaded to peer j by time t. We then define

$$D^{(i)}(j, t) = Cc^{(i)}(j, t) - Cc^{(j)}(i, t)$$
$$= \left(Cu^{(i)}(j, t) - C_p^{(j)}(i, t)\right) - \left(Cu^{(j)}(i, t) - C_p^{(i)}(j, t)\right), \tag{12.4}$$

which is the difference between the number of *useful* chunks that peer i has sent to peer j and the number of *useful* chunks that peer j uploaded to peer i.

Similar to the two-player cooperation stimulation strategy in Section 9.1.4, we consider the following strategy: each selfish peer $i \in N_s$ limits the number of chunks that he or she sends to any other peer j such that by any time t, the total number of useful (clean) chunks that i has forwarded to j should be no more than $Cc^{(j)}(i, t)) + D_{max}^{(i)}(j, t)$; that is,

$$D^{(i)}(j, t) \le D_{max}^{(i)}(j, t), \quad \forall t \ge 0. \tag{12.5}$$

Here, $D_{max}^{(i)}(j, t)$ is the "credit line" that user i sets for user j at time t. The credit line is set for two purposes: (1) to prevent egoism when favors cannot be granted simultaneously and to stimulate cooperation between i and j, and (2) to address possible unintentional forwarding of polluted chunks by a selfish user while limiting the possible damages that j can cause to i. By letting $D_{max}^{(i)}(j, t) \ge 0$, i agrees to send some extra, but at most $D_{max}^{(i)}(j, t)$, chunks to j without getting instant payback. Meanwhile, unlike acting fully cooperatively, the extra number of chunks that i forwards to j is bounded to limit the possible damages when j plays noncooperatively or maliciously.

Player i's goal of setting the credit line is to avoid helping player j much more than player j helps i in the long-term view, and vice versa, as neither i nor j has incentive to send more chunks than the other does. Meanwhile, owing to the dynamically changing network conditions, the request rates between i and j may vary from time to time. In this case, the credit line must be large enough, because a small credit line will refuse some requests even when the long-term average request rates between i and j are equal. The ultimate goal of setting the credit line is to make sure that players i and j send asymptotically equal numbers of unpolluted chunks to each other, and

$$\lim_{t \to \infty} Cc^{(i)}(j, t) = \lim_{t \to \infty} Cc^{(i)}(j, t). \tag{12.6}$$

Combining the definition of $D_{max}^{(i)}(j, t)$ with (12.6), $D_{max}^{(i)}(j, t)$ must satisfy

$$\lim_{t \to \infty} \frac{D_{max}^{(i)}(j, t)}{Cr^{(i)}(t)} = 0, \tag{12.7}$$

which also implies that arbitrarily increasing credit lines cannot always increase the number of accepted requests. Equation (12.7) provides an asymptotic upper bound for

$D_{max}^{(i)}(j, t)$. Based on this analysis, $D_{max}^{(i)}(j, t)$ should be large enough in the first few cooperating rounds to stimulate cooperation between user i and j. On the other hand, the ratio $D_{max}^{(i)}(j, t)$ over the number of remaining rounds after time t should be close to 0 to prevent a decrease of user i's utility. Therefore, when choosing $D_{max}^{(i)}(j, t)$, user i should first estimate the number of remaining rounds for the live streaming program, and choose a relatively small number D_{temp}. Then user i should compare D_{temp} with the reciprocal of P_{ij}, so that $D_{max}^{(i)}(j, t)$ should be larger than $1/P_{ij}$ to stimulate cooperation with j. A simple solution to this is to set the credit line to a reasonably large positive constant, which will be shown in Section 12.1.6.

12.1.3.2 Incomplete chunk attack detection

In this section, we focus on the detection of an incomplete chunk attack and design a malicious user detection algorithm that can differentiate between an intentional incomplete chunk attack and packet loss or delay as a result of network congestions.

Recall that P_{ij} is the probability of successful transmitting one chunk from user i to user j within round period τ. Hence, when player i decides to send a chunk to player j, with probability $1 - P_{ij}$, this chunk transmission cannot be completed within one round because of packet dropping or delay because of high traffic volume. We use a Bernoulli random process to model the unsuccessful transmission of a chunk owing to congested networks. Recall that $Cu^{(j)}(i, t)$ denotes the number of chunks that j has agreed to send to user i by time t, and $Cs^{(i)}(j, t)$ is the number of chunks that peer i has successfully received from j. Given the Bernoulli random process model, if user j does not intentionally deploy the incomplete chunk attack, from the central limit theorem [138], for any positive real number x, we can have

$$\lim_{Cu^{(j)}(i,t)\to\infty} Prob\left(\frac{Cs^{(i)}(j, t) - Cu^{(j)}(i, t)P_{ji}}{\sqrt{Cu^{(j)}(i, t)P_{ji}(1 - P_{ji})}} \leq -x\right) = Q(x), \qquad (12.8)$$

where $Q(x) = \frac{1}{\sqrt{2\pi}}\int_x^\infty e^{-t^2/2}dt$ is the Gaussian tail function. If user j does not intentionally sends incomplete chunks, (12.8) indicates that when the P2P live streaming game keeps going and $Cu^{(j)}(i, t)$ is large enough, then $\frac{Cs^{(i)}(j,t)-P_{ji}C_u^{(j)}(i,t)}{\sqrt{Cu^{(j)}(i,t)P_{ji}(1-P_{ji})}}$ can be approximated by a Gaussian random variable with zero mean and unit variance; that is,

$$\frac{Cs^{(i)}(j, t) - P_{ji}C_u^{(j)}(i, t)}{\sqrt{Cu^{(j)}(i, t)P_{ji}(1 - P_{ji})}} \sim \mathcal{N}(0, 1). \qquad (12.9)$$

Therefore, based on (12.9), given a predetermined threshold $h > 0$, every selfish peer i can identify peer j as a malicious user, as follows:

$$j \in N_m^{(i)}(t) \quad \text{if} \quad \frac{Cs^{(i)}(j, t) - Cu^{(j)}(i, t)P_{ji}}{\sqrt{Cu^{(j)}(i, t)P_{ji}(1 - P_{ji})}} \leq -h,$$

$$\text{and } j \in N_s^{(i)}(t) \quad \text{otherwise.} \qquad (12.10)$$

In (12.10), $N_m^{(i)}(t)$ is the set of peers who are marked as malicious by peer i at time t, and $N_s^{(i)}(t)$ is the set of peers who are marked as selfish by peer i at time t. Based on

(12.10), if a malicious user is always sending incomplete chunks to other users, then the probability of correctly identifying the malicious user is $P_d = 1 - Q(h)$, and the probability of falsely accusing a nonmalicious user as malicious is $P_{fa} = Q(h)$.

12.1.3.3 Multiuser cooperation stimulation strategies

Combining the credit line mechanism in (12.7) and the malicious attacker detection algorithm in (12.10), we propose an attack-resistant cooperation stimulation strategy for P2P video streaming in this section.

Recall that P_{ij} is the probability that peer j successfully receives a chunk from peer i in one round. Let P_{si} denote the probability that i successfully receives a chunk from the streaming server in one round, and P_s is the percentage of requests that the streaming server can answer in one round. These probabilities, P_{ij}, P_{si}, and P_s, can be probed or estimated using the methods proposed by Liu *et al.* [94].

We first consider the scenario in which there is a selfish peer i where

$$P_{si} \cdot P_s > P_{ji}, \quad \forall j \in N, \ j \neq i; \tag{12.11}$$

his or her optimal strategy is to always download the video from the streaming server and to reject all chunk requests from other peers. This is because in each round, peer i has a one-chunk request quota and can request a chunk from either the streaming server or one peer, j. If i sends a request to the streaming server, the probability that peer i successfully receives the requested chunk is $P_{si} \cdot P_s$. If i requests a chunk from peer j, the probability that i successfully receives the requested chunk is $P_{ji} \cdot P_{js}$, where $P_{js} \leq 1$ is the probability that j agrees to send the chunk. Because $P_{si} \cdot P_s > P_{ji}$, it is obvious that sending the request to the streaming server gives i a higher probability of receiving the requested chunk. It is the optimal chunk request strategy for each round and, therefore, the optimal strategy for the whole game. Furthermore, because peer i always requests chunks from the server, he or she does not have the incentive to send any chunks to other peers in the network, which decreases peer i's utility, as shown in (12.1). Based on this analysis, peer i will always operate noncooperatively.

The preceding analysis suggests that if a peer has a very good connection with the original streaming server, which is much better than the connections with all other peers, then he or she will always refuse to cooperate, and cooperation cannot be enforced on these peers. In reality, however, there are usually very few peers that can meet this condition, as P2P live streaming social networks are usually very big. Thus, the streaming server is often very busy with a small P_s, and makes the condition $P_{si} \cdot P_s > P_{ji}$ for all $j \in N$ with $j \neq i$ very difficult to satisfy.

The other extreme scenario occurs when peer i is has the worst connection with other peers; that is, for every $j \in N$ and $j \neq i$, there always exists another peer $k \in N$ and $k \neq i, j$ such that $P_{ij} < P_{kj}$. In this scenario, will all other peers in the network refuse to cooperate with i? The answer is no, because of the dynamics in P2P social networks and the assumption of a busy server. In P2P live streaming, different users have different playback times. If peer i's playback time is earlier than that of all other peers in the network, then it is very likely that his or her buffer has chunks that no other peers have, which is the incentive for other peers to cooperate with i under the constraint that $D^{(j)}(i, t) \leq D_{max}^{(j)}(i)$.

To summarize, we can arrive at the following *multiuser cooperation stimulation strategy* for P2P live streaming social networks:

In the peer-to-peer live streaming game, for any selfish peer $i \in N_s$ who does not meet the condition in (12.11), he or she initially marks all other users as selfish. Then, in each round, i uses the following strategy:

- If i receives a chunk request from j, i will accept this request if j has not been marked as malicious by i and if (12.5) holds. Otherwise, i will reject the request.
- When i requests a chunk, he or she will send the request to peer j, who has the chunk and satisfies

$$j = \arg \max_{j \in N_s^{(i)}(t), j \neq i} P_{ji}. \tag{12.12}$$

- Let $1 - Q(h)$ be the maximum allowable false positive probability from i's point of view. When $Cu^{(j)}(i, t)$ is large enough for any user $j \in N$, i will apply the detection rule in (12.10) to detect malicious behavior from j after each chunk request initiated by i.

12.1.4   Strategy analysis without malicious attackers

This section analyzes the optimality of the preceding cooperation strategy for peers who do not satisfy the necessary conditions in (12.11) when there are no malicious users. We first consider an infinite-lifetime situation with $Cr^{(i)}(t) \to \infty$ as $t \to \infty$; the finite-lifetime situation will be discussed later. First, we assume $D_{max}^{(i)}(j, t)$ satisfies (12.7), and i and j send approximately the same number of chunks to each other.

Lemma 12.1. In the P2P live streaming game in which some chunks may be dropped or delayed owing to high traffic volume in the Internet, for a selfish user j, if all other users follow the multiuser attack-resistant cooperation strategy, then playing noncooperatively and sending only part of the requested chunks will not increase j's payoff.

Proof. If user j has agreed to upload a chunk to another user $i \in N$, transmitting only part of the requested chunk can help j reduce his or her cost. However, even though j agrees to upload the chunk, it does not count as a useful chunk unless it is successfully received. In addition, following the multiuser attack-resistant cooperation strategy, player i always ensures that

$$\lim_{t \to \infty} Cs^{(i)}(j, t) \geq \lim_{t \to \infty} Cs^{(i)}(j, t). \tag{12.13}$$

Thus, by sending only part of the requested chunk, player j loses the chance to request a chunk from player i. To get this one-chunk-request chance back, player j has to send another complete chunk successfully to player i. Therefore, intentionally sending partial information of the requested chunks cannot bring any gain to player j.

Lemma 12.2. For a selfish peer $i \in N_s$ in the P2P live streaming game without malicious attackers, once i has received a chunk request from another node $j \in N$, if (12.5) holds and if j follows the multiuser attack-resistant cooperation strategy, then accepting the request is always an optimal decision from player i's point of view.

Proof. From player i's point of view, if (12.7) is satisfied, agreeing to send the requested chunk will not introduce any performance loss, as the average cost to help j goes to zero when $t \to \infty$. Meanwhile, refusing the request may cause $D^{(j)}(i, t) > D_{max}^{(j)}(i, t)$ and thus forbids user i to request chunks from player j in the future. Therefore, accepting the request is an optimal decision.

Lemma 12.3. In the peer-to-peer live streaming game without malicious attackers, a selfish peer $i \in N_s$ has no incentive to cheat on his or her buffer map information.

Proof. From player i's point of view, cheating on his or her buffer information will prevent other peers from requesting chunks from him or her, and thus will decrease the total number of chunks he or she uploads, that is, $Cu^{(i)}(t)$. However, because other users always enforce (12.7) and $Cu^{(j)}(i, t) + D_{max}^{(j)}(i, t) < Cu^{(i)}(j, t)$, a smaller $Cu^{(i)}(t)$ will also decrease the chance of getting chunks from other peers and lower player i's overall payoff, which is similar to the two-player game in Section 9.1.2. Therefore, selfish peers have no incentive to cheat on buffer information.

Theorem 12.1. In the P2P live streaming game without malicious attackers, for all selfish players who do not satisfy the condition in (12.11), the multiuser attack-resistant cooperation strategy forms a subgame perfect Nash equilibrium, and it is also strongly Pareto optimal if $0 < \lim_{t \to \infty} \frac{Cr^{(i)}(t)}{Cr^{(j)}(t)} < \infty$ for any $i, j \in N$.

(1) **Nash equilibrium:** To show that this strategy profile forms a subgame perfect equilibrium, note that this multiuser game can be decomposed into many two-player subgames. Therefore, we need to consider only the two-player subgame between player i and j. Suppose that player i does not follow the above strategy; that is, either i refuses to send chunks to player j when (12.5) is satisfied, or i intentionally sends only part of the requested chunk to player j, or i agrees to send the requested chunks even when (12.5) is not satisfied. First, from Lemmas 12.1 and 12.2, when (12.5) is satisfied, neither refusing to send the requested chunks nor intentionally sending incomplete chunks will give player i any performance gain. Second, when comparing $D^{(i)}(j, t) > D_{max}^{(i)}(j, t)$ with $D^{(i)}(j, t) \leq D_{max}^{(i)}(j, t)$, as long as $D^{(i)}(j, t) > 0$, following the multiuser cooperation stimulation strategy in the previous section, j will always cooperate and send all chunks that i requests. Thus, making $D^{(i)}(j, t)$ go above the credit line cannot help i receive more chunks from j, but rather costs i more upload bandwidth and lowers i's utility. Based on this analysis, we can conclude that this multiuser–attack-resistant and cheat-proof cooperation strategy forms a Nash equilibrium.

(2) **Subgame perfectness:** In every subgame of the equilibrium path, the strategies are: If player j is marked malicious by peer i, player j will play noncooperatively forever, which is a Nash equilibrium. Otherwise, player j follows the multiuser

attack-resistant and cheat-proof strategy, which is also a Nash equilibrium. Therefore, the cooperation stimulation strategy is subgame perfect.

(3) **Strong Pareto optimality:** From the selfish user's utility definition in (12.1), to increase his or her own payoff, a player i can either increase $Cs^{(i)}(t)$ or decrease $Cu^{(i)}(t)$. However, from the preceding analysis, further decreasing $Cu^{(i)}(t)$ will reduce useful chunks successfully received by other peers and lower their payoffs. Thus, to increase his or her payoff, player i should increase $\lim_{t \to \infty} Cs^{(i)}(t)/Cr^{(i)}(t)$, which means that some other players will have to send more chunks to player i. As $\{Cr^{(i)}(t)\}$ are of the same order because $0 < \lim_{t \to \infty} \frac{Cr^{(i)}(t)}{Cr^{(j)}(t)} < \infty$, increasing $\lim_{t \to \infty} Cs^{(i)}(t)/Cr^{(i)}(t)$ (and thus improving player i's payoff) will definitely decrease other players' payoff. Therefore, the preceding strategy profile is strongly Pareto optimal.

Until now, we have focused mainly on the situation in which the game will be played for an infinite duration. In general, a peer will stay in the network for only a finite period of time – for example, until the end of the video streaming. Then, if $D_{max}^{(i)}(j, t)$ is too large, each player i may have helped other users much more than his or her peers have helped i. Meanwhile, if $D_{max}^{(i)}(j, t)$ is too small, player i may not have enough peers to help him or her. How to select a good $D_{max}^{(i)}(j, t)$ is a challenging issue. In Section 12.1.6 we study the tradeoff between the value of $D_{max}^{(i)}(j, t)$ and the peers' utilities using simulations. It is shown there that in a given simulation scenario, a relatively small $D_{max}^{(i)}(j, t)$ achieves near-optimal performance when compared with the scenario in which $D_{max}^{(i)}(j, t)$ is set to infinity. The optimality of the cooperation strategies cannot be guaranteed in finite-duration scenarios. However, we will show in the simulation results that the performance of our attack-resistant cooperation stimulation strategy is very close to optimal.

12.1.5   Strategy analysis with malicious attackers

In this section, we focus on the following two widely used attack models, the pollution attack and the incomplete chunk attack, and analyze the performance of the cooperation stimulation strategy when there are malicious users. To simplify our analysis, we assume that $W_i = W$, $g_i = g$, and $g \frac{M}{W\tau} < \infty$ for all $i \in N$.

Pollution attack: We first study the performance of the cooperation strategy under the pollution attack. By always accepting selfish users' requests and sending polluted chunks to them, malicious attackers waste selfish users' quotas and prevent them from receiving useful chunks in that round. Following the attack-resistant cooperation strategy, every selfish user $i \in N_s$ forces $D^{(i)}(j, t) \leq D_{max}^{(i)}(j, t)$, and the damage caused to selfish user i by one malicious attacker j is upper-bounded by $D_{max}^{(i)}(j, t)g$. With a finite g and $\lim_{t \to \infty} \frac{D_{max}^{(i)}(j,t)}{Cr^{(i)}(t)} = 0$ from (12.7), we have

$$\lim_{t \to \infty} \frac{D_{max}^{(i)}(j, t)g}{Cr^{(i)}(t)} = 0 \qquad (12.14)$$

and, therefore, the overall damage caused by pollution attacks becomes negligible.

Incomplete chunk attack: By sending incomplete chunks to others, malicious users inject useless traffic into the network and waste other peers' chunk-request quotas. With the attacker detection algorithm in (12.10), for a malicious attacker, always sending incomplete chunks may not be a good strategy, because this attacker can be detected easily. Instead, to avoid being detected, attackers should send incomplete and complete chunks alternately. According to the multiuser attack-resistant cooperation strategy in Section 12.1, peer j identifies i as malicious if $Cs^{(j)}(i, t) - Cu^{(i)}(j, t)P_{ij} \leq -h\sqrt{Cu^{(j)}(i, t)P_{ij}(1 - P_{ij})}$. Assume that by time t, user i has agreed to upload a total of n chunks to user j. Therefore, to avoid being identified as malicious by j, i must successfully forward at least $nP_{ij} - h\sqrt{nP_{ij}(1 - P_{ij})}$ complete chunks, and the maximum number of incomplete chunks that i can send to j is upper-bounded by $n(1 - P_{ij}) + h\sqrt{nP_{ij}(1 - P_{ij})}$. Among these $n(1 - P_{ij}) + h\sqrt{nP_{ij}(1 - P_{ij})}$ chunks, $n(1 - P_{ij})$ of them are dropped or delayed by the network owing to the high Internet traffic volume, and the actual number of intentional incomplete chunks sent to j by i is upper-bounded by $h\sqrt{nP_{ij}(1 - P_{ij})}$. Therefore, for user j, the extra damage caused by attacker i's intentional malicious attack is upper-bounded by $h\sqrt{nP_{ij}(1 - P_{ij})}g$. Furthermore, to avoid being identified as malicious, attacker i must successfully forward at least $nP_{ij} - h\sqrt{nP_{ij}(1 - P_{ij})}$ complete chunks to user j, which costs attacker i a utility of $\left[nP_{ij} - h\sqrt{nP_{ij}(1 - P_{ij})}\right]\frac{M}{W\tau}$. Thus, following (12.2), the utility that attacker i receives from intentionally sending incomplete chunks is at most $h\sqrt{nP_{ij}(1 - P_{ij})}(\frac{M}{W\tau} + g) - n(1 - P_{ij})\frac{M}{W\tau}$. Because for any real positive h,

$$\lim_{t \to \infty} \frac{h\sqrt{nP_{ij}(1 - P_{ij})}(\frac{M}{W\tau} + g)}{n(1 - P_{ij})\frac{M}{W\tau}} = 0, \tag{12.15}$$

selectively sending incomplete chunks can bring no gain to the attackers if they want to remain being undetected. In other words, if the game will be played for an infinite duration, the incomplete chunk attack cannot cause damages to selfish nodes.

In summary, when the multiuser attack-resistant strategy is used by all selfish users, the pollution attack can cause only limited damage to the system. Further, the relative damage caused by the incomplete chunk attack asymptotically approaches zero when the game is played for an infinite duration of time. Therefore, except for some false alarms of identifying selfish users as malicious, selfish players' overall payoff will not be affected under attacks. From this analysis, we can also see that no matter what objectives the attackers have and what attacking strategies that they use, as long as selfish peers apply the multiuser attack-resistant cooperation strategy, the selfish users' payoff and the overall system performance can be guaranteed.

Optimal attacking strategy: Based on the preceding analysis on the pollution attack and the incomplete chunk attack, we can conclude that, for the infinite-duration game, an attacker j's overall payoff is upper-bounded by

$$U_m^{(j)} \leq \lim_{t \to \infty} \sum_{i \in N_s} \frac{D_{max}^{(i)}(j, t)}{t} g, \tag{12.16}$$

provided that all selfish users follow the multiuser attack-resistant cooperation strategy. This upper bound can be achieved by the following *optimal attacking strategy in an infinite-game model*: in the P2P live streaming game, upon receiving a request, an attacker $j \in N_m$ should always reject the requests; the attackers should always send requests to selfish users to waste the selfish users' chunk-request quotas, until the selfish users refuse to forward any chunks to the attackers.

When the game will be played for only a finite period of time, the preceding attacking strategy is no longer optimal. In addition to the pollution attack, the attackers can also send incomplete chunks without being detected. The malicious-attacker detection algorithm in Section 12.1.3.2 requires that the game has been played for a long time and peer i and j have interacted for a large number of times to provide an accurate estimation. The algorithm will not be initiated unless $Cu^{(j)}(i, t)$ is large enough to avoid a high false alarm rate. Thus, if the game is played for only a short period of time, users do not have sufficient time to correctly estimate each user's type, and attackers can send incomplete chunks without being detected, which lowers selfish users' utilities and degrades the system performance. In this chapter we focus on the scenario in which the game is played for a reasonably long time and users have enough time to interact with each other and correctly estimate the statistics of chunk transmission. In such a scenario, the malicious attacker detection algorithm in Section 12.1.3.2 can still be used to accurately identify attackers and the relative damage caused by incomplete chunk attack is still insignificant.

12.1.6   Simulation results

In our simulations, we choose the "foreman" CIF video sequence with thirty frames per second, and duplicate it to generate a sixty-minute video. The video is initially stored at an original server with upload bandwidth 3 Mbps. There are 200 DSL peers with 768 kbps uplink bandwidth and 300 cable peers with 300 kbps uplink bandwidth. Each user has a buffer that can store 30 seconds of video. To examine the impact of the system parameters on the performance of the cooperation strategies, we run the simulations under two settings. First, we let the round duration τ be 0.4 second, resulting in a total of 9000 rounds; and the video is encoded into a three-layer bit stream with 50 kbps per layer. The compressed bit stream is divided into one-second layer chunks, and each chunk has 50 kilobits. In our second simulation setup, we let $\tau = 0.2$ second, and the total number of rounds is 1.8×10^4. The video is encoded into a four-layer bit stream with 37.5 kbps per layer. Each chunk is of one-second length and includes 37.5 kilobits. We use the chunk-request algorithm in Section 9.4.1 and let the score weights be $w_1 = 1/2$, $w_2 = 1/6$, and $w_3 = 1/3$. The malicious peers follow the attack strategy in Section 12.1.5 to send incomplete or polluted chunks. The selfish (rational) peers follow the attack-resistant cooperation strategies in Section 12.1.3.2.

We first study how different credit lines affect cooperation stimulation. Figure 12.2 shows the relationship between the credit line and utility when the percentage of attackers is 0, 25, 37, and 50 percent, respectively. The attackers are chosen randomly from the 500 peers. From these results, we can see that, in both simulation setups, when the credit

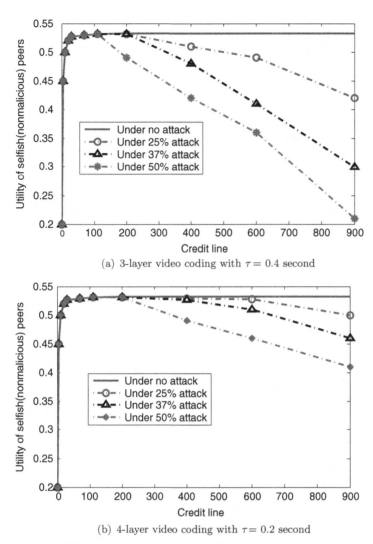

(a) 3-layer video coding with $\tau = 0.4$ second

(b) 4-layer video coding with $\tau = 0.2$ second

Fig. 12.2 Selfish peers' performance with and without attack

line is increased over 50, selfish users' payoffs are saturated. As the credit line continues increasing, selfish users' utilities start to decrease very quickly. It is clear from (12.16) that the maximum damage that attackers can cause is linearly proportional to the credit line. For example, in Figure 12.2(a), when the credit line is larger than 120 and when 50 percent of the users are attackers, the damages caused by attackers are no longer negligible. Also, Figure 12.2 suggests that setting the credit line to be 50 is an optimal choice for both simulation settings, as it stimulates user cooperation to the maximum degree. Nevertheless, arbitrarily increasing the credit line is dangerous for selfish users, because they do not know how many malicious users are in the network.

Next, we examine the robustness of our cooperation strategies against attackers and free riders in terms of PSNR. Because from Figure 12.2, both simulation settings give

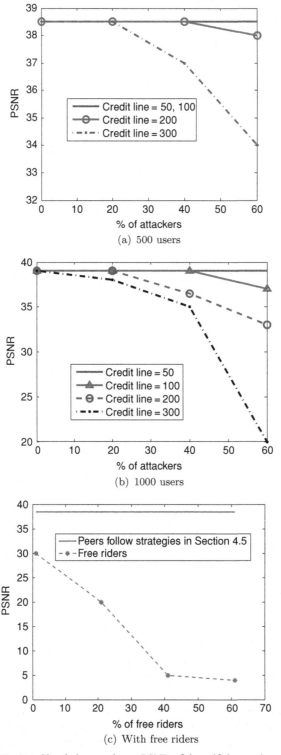

Fig. 12.3 Simulation results on PSNR of the selfish users' received videos

similar trends, here we use simulation setting 1 as an example to demonstrate the robustness. Also, to show how the total number of users affects the optimal credit line, we test our cooperation schemes on 500 users and 1000 users, respectively, and keep the ratio between cable and DSL peers fixed as 3:2. We let the credit lines be equal to 50, 100, 200, and 300, respectively. The malicious peers are selected randomly and follow the optimal attack strategy in Section 12.1.5. Figure 12.3(a) and (b) show the PSNR of a selfish user's video versus the percentage of attackers with different credit lines and different number of users. It is clear that when the credit line is chosen correctly to be around 50, our cooperation strategies can effectively resist attacks. Even when the credit line is increased to 100 and when 60 percent of the users are attackers, the PSNR of selfish users' video does not degrade too much. From this discussion, in general, a credit line between 50 an 100 will simulate cooperation among selfish users even under attacks, will resist cheating behavior, and will give good performance.

Figure 12.3(c) shows the the robustness of the cooperation strategies against the presence of free riders who upload no chunks to others. In Figure 12.3(c), there are a total of 500 users, including free riders, and rational users use the proposed attack-resistant cooperation stimulation strategy. We use a credit line of 50. It is clear from Figure 12.3(c) that the videos received by free riders have poor quality, and our cooperation stimulation strategy makes free riding nonprofitable and thus unattractive. In addition, using the proposed cooperation stimulation strategies, rational users are not affected by the existence of free riders in the network, and the PNSR of their received videos remains constant.

12.2   Attack-resistant cooperation strategies in wireless P2P video streaming

12.2.1   Game model

Following the study of cheat-proof cooperation strategies for P2P video streaming over wireless networks in Chapter 9, we investigate its resistance against attacks and design attack-resistant cooperation strategies here. We use the following game to model the interactions among peers:

- **Server:** The video is originally stored at the streaming server with upload bandwidth W_s, and the server will send chunks in a round-robin fashion to its peers. All players are connected via the same access point to the Internet. This backbone connection has download bandwidth W_d.
- **Players and player type:** There is a finite number of users in the P2P wireless video streaming social network, denoted by N. Each player $i \in N$ has a type $\theta_i \in \{\text{selfish, malicious}\}$. As in the game in P2P video streaming over the Internet, attackers can apply the pollution and the incomplete chunk attacks, and they may also hand-wash from time to time.
- **Chunk request:** In each round, with limited bandwidth B in the channel dedicated for user cooperation, users first bargain for their chunk-request quotas based on the

time-sensitive bargaining solution in Section 9.3.1. As an extension of the two-person time-sensitive bargaining in Section 9.3.1, the N-user bargaining procedure, is as follows: User 1 offers an action N-tuple $\left(a_1^{(1)}, a_2^{(1)}, \ldots, a_N^{(1)}\right)$ first, where $a_i^{(k)}$ is the number of chunks that user i can transmit in user k's offer. Then user 2 decides whether to accept this offer or to reject and offer back another action N-tuple $\left(a_1^{(2)}, a_2^{(2)}, \ldots, a_N^{(2)}\right)$. If user 2 agrees to the offer, user 3 can decide whether to accept or reject and offer back, and the rest of the users make their choices sequentially. This process continues until all players agree on the offer. If users reach agreement at the jth action pair, then g_i is decreased to $\delta_i^{j-1} g_i$ for user i. Therefore, the stationary equilibrium N-tuple $\left\{\left(x_1^{(1)}, \ldots, x_N^{(1)}\right), \left(x_1^{(2)}, \ldots, x_N^{(2)}\right), \ldots, \left(x_1^{(N)}, \ldots, x_N^{(N)}\right)\right\}$ will satisfy

$$\delta_1^{N-1} x_1^{(1)} g_1 - \sum_{i \neq 1} x_i^{(1)} c_1 = \delta_1^{N-2} x_1^{(2)} g_1 - \sum_{i \neq 1} x_i^{(2)} c_1$$

$$= \ldots = x_1^{(N)} g_1 - \sum_{i \neq 1} x_i^{(N)} c_1,$$

$$\delta_2^{N-1} x_2^{(2)} g_2 - \sum_{i \neq 2} x_i^{(2)} c_2 = \delta_2^{N-2} x_2^{(3)} g_2 - \sum_{i \neq 2} x_i^{(3)} c_2$$

$$= \ldots = x_2^{(1)} g_2 - \sum_{i \neq 2} x_i^{(1)} c_2,$$

$$\vdots$$

$$\delta_N^{N-1} x_N^{(N)} g_N - \sum_{i \neq N} x_i^{(N)} c_N = \delta_N^{N-2} x_N^{(1)} g_N - \sum_{i \neq N} x_i^{(1)} c_N$$

$$= \ldots = x_N^{(N-1)} g_N - \sum_{i \neq N} x_i^{(N-1)} c_N,$$

$$\text{and} \sum_{j=1}^{N} x_j^{(i)} \frac{K_j}{P_j} = \tau, \ \forall i \in \{1, 2, \ldots, N\}, \tag{12.17}$$

and all users should accept the first offer.

Given the chunk-request quota, a user can send multiple chunk requests to another user, send the requests to multiple peers, or not request any chunks in this round. The user-cooperation channel is different from the channel between users and the access point. Thus, users can also ask the server for chunks at the same time.

- **Request answering:** After receiving a request, each player, can either *accept* or *reject* the requests.
- **Cost:** For any player $i \in N$, uploading a chunk to player j incurs cost $c_i M P_i / \left[B \log \left(1 + \frac{P_i A_{ij}^2}{d_{ij} \sigma_n^2}\right)\right]$, where c_i is the user-defined cost per unit energy, P_i is the transmission power that player i uses for cooperation, and $P_i \geq P_{min}$.

- *Gain:* If a selfish user $i \in N_s$ requests a data chunk from another peer j, and if a clean copy is successfully delivered to him or her, his or her gain is g_i where $g_i > \max_j c_i M P_i / \left[B \log \left(1 + \frac{P_i A_{ij}^2}{d_{ij} \sigma_n^2} \right) \right]$.

For this game model, the next step is to define each user's utility function. For each player $i \in N$, we first define the following symbols:

- $Cr^{(i)}(j, t)$ is the total number of chunks that i has requested from j by time t. Here, j can be either a peer $(j \in N)$ or the streaming server. $Cr^{(i)}(t) = \sum_{j \in \{N, \text{source}\}} Cr^{(i)}(j, t)$ denotes the total number of chunks that i has requested by time t.
- By time t, peer i has successfully received $Cs^{(i)}(j, t)$ chunks from peer j in time (a chunk is received in time if and only if it is received within the same round that it was requested). $Cs^{(i)}(t) = \sum_{j \in \{N, \text{source}\}} Cs^{(i)}(j, t)$ is peer i's total number of successfully received chunks by time t.
- By time t, $C_p^{(i)}(j, t)$ is the total number of polluted chunks that peer i received from peer j. The total number of successively received unpolluted data chunks that peer i received from peer j is $Cs^{(i)}(j, t) - C_p^{(i)}(j, t)$, and each successfully received unpolluted chunk gives peer j a gain of g_i.
- $Cu^{(i)}(j, t)$ denotes the number of chunks that i has uploaded to player j by time t. $Cu^{(i)}(t) = \sum_{j \in N, j \neq i} Cu^{(i)}(t)$ is the total number of chunks that user i has uploaded to other users by time t.

Let t_f be the lifetime of the P2P video streaming network, and let $T^{(i)}(t)$ denote the total time that peer i has stayed in the network by time t. Then, the utility function for a selfish player $i \in N_s$ is

$$U^{(i)}(t_f) = \frac{\left[Cs^{(i)}(t_f) - \sum_{j \in N} C_p^{(i)}(j, t_f) \right] g_i - \sum_{j \in N} Cu^{(i)}(j, t_f) \frac{c_i M P_i}{B \log \left(1 + \frac{P_i A_{ij}^2}{d_{ij} \sigma^2} \right)}}{Cr^{(i)}(t_f)}.$$

(12.18)

Here, the numerator denotes the net reward (that is, the total gain minus the total cost) that selfish user i obtains, and the denominator denotes the total number of chunks that i has requested. This utility function represents the average net profit that i can obtain per requested chunk, which i aims to maximize. The utility function for a malicious attacker $j \in N_m$ is

$$U_m^{(j)} = \sum_{i \in N_s} Cu^{(i)}(j, t_f) \frac{c_i M P_i}{T^{(j)}(t_f) B \log \left(1 + \frac{P_i A_{ij}^2}{d_{ij} \sigma^2} \right)} + \sum_{i \in N_s} \frac{\left[Cr^{(i)}(j, t_f) - Cs^{(i)}(j, t_f) \right] g_i}{T^{(j)}(t_f)}$$
$$- \sum_{i \in N} Cu^{(j)}(i, t_f) \frac{c_j M P_j}{T^{(j)}(t_f) B \log \left(1 + \frac{P_j A_{ji}^2}{d_{ij} \sigma^2} \right)}.$$

(12.19)

As in (12.2), this utility function represents the average net damage that j causes to other users per time unit.

12.2.2   Attack-resistant cooperation stimulation strategies

12.2.2.1   Hand-wash attack and trust model

To accurately identify an attacker, the statistical malicious user detection algorithm in Section 12.1.3.2 requires that a selfish user interacts with others for a large number of rounds. Thus, in an environment in which attackers can apply the hand-wash attack, the proposed method in Section 12.1.3.2 will fail: an attacker can attack aggressively before selfish users collect sufficient statistics to detect him or her, leave the system once he or she is detected, and then come back with a new ID to continue attacking the system. Thus, selfish users may suffer badly from the hand-wash attacks, and the risk of interacting with unknown users may prevent them from cooperating with one another. To combat the hand-wash attack and to stimulate user cooperation under attacks, selfish users must identify malicious users as soon as possible.

A straightforward solution is to reduce the credit line $D_{max}^{(i)}(j, t)$ defined in (12.5) or the threshold h in (12.10). However, an arbitrary decrease of the credit line or the detection threshold prevents users from cooperating with one another and results in the failure of the whole network. For instance, if a selfish user j unintentionally forwards a polluted chunk to another selfish user i, decreasing $D_{max}^{(i)}(j, t)$ may prevent these two users from cooperating with each other.

To address the hand-wash attack and to speed up the malicious attacker detection process, we introduce the concept of *trust* among selfish users in our framework and let selfish users share their past experience with one another. When determining other users' types, collection of network opinions can help a selfish user detect malicious attackers faster and lower the damage caused by the hand-wash attack. In the following, we discuss our trust model and study how it can help resist the hand-wash attack.

A selfish user i establishes direct trust with another user j upon observations on previous interactions with j. We use the beta-function–based method, in which user i trusts user j at time t with value

$$Tr^{(i)}(j, t) = \frac{Cs^{(i)}(j, t) - C_p^{(i)}(j, t) + 1}{Cr^{(i)}(j, t) + 2}. \tag{12.20}$$

If user j is not malicious and does not receive or upload many polluted chunks, from (12.20), $Tr^{(i)}(j, t)$ should be close to P_{ij}. If user j applies the pollution attack, $C_p^{(i)}(j, t)$ will increase; and if user j applies the incomplete-chunk attack, $Cs^{(i)}(j, t)$ will decrease. Thus, both types of attacks decrease the numerator in (12.20), resulting in a low trust value for malicious users. The trust is directional, which means that the fact that user i trusts user j does not imply that user j trusts user i.

Because trusted selfish users would like to identify malicious users together, the damage caused by attackers to the trusted users are considered collectively. For example, if user i trusts another user j at round t, user i considers the damage that malicious user

k has caused to user j as his or her own damage. This scenario is equivalent to reducing the credit line $D_{max}^{(i)}(k, t)$ in (12.5) to $D_{max}^{(i)}(k, t) - Tr^{(i)}(j, t) \times D_{max}^{(j)}(k, t)$.

There is an effective bad-mouthing attack against the trust system, in which malicious users provide dishonest recommendations to frame good parties and/or boost trust values of malicious users [86]. To resist such a bad-mouthing attack, selfish users should trust only users who have sent them a sufficiently large number of clean chunks. That is, at time t, a selfish user i will trust user j only if user j has sent i more than $Ch^{(i)}(t)$ useful chunks and $Cs^{(i)}(j, t) > Ch^{(i)}(t)$. With the threshold $Ch^{(i)}(t)$, even if a malicious user tries to apply the bad-mouthing attack, he or she must upload a large number of clean chunks before he or she can be trusted. This "cooperative" behavior does not damage, but rather improves, the system performance. In addition, for P2P video streaming over wireless networks, owing to the broadcast nature of wireless communications, everyone can listen to all the chunk requests and chunk answering in the network. Therefore, a malicious user cannot arbitrarily frame a selfish user or boost another attacker with whom he or she has not interacted before.

In summary, in each round, the credit line $D_{max}^{(i)}(j, t)$ in (12.5) is updated as

$$D_{max}^{(i)}(j, t+1) = \max \left\{ 1, D_{max}^{(i)}(j, t) - \sum_{k \in N_{Tr}^{(i)}(t)} Tr^{(i)}(k, t) \times D^{(k)}(j, t) \right\}$$

$$\text{where} \quad N_{Tr}^{(i)}(t) = \left\{ k | k \in N_s^{(i)}(t) \quad \text{and} \quad Cs^{(i)}(k, t) > Ch^{(i)}(t) \right\}. \quad (12.21)$$

Here, $N_{Tr}^{(i)}(t)$ is the set including the indices of all users whom i identifies as "selfish" and each of whom has successfully uploaded at least $Ch^{(i)}(t)$ clean chunks to i by time t. As will be demonstrated in Section 12.2.6, employing the trust model in (12.20) and replacing the modified credit line as in (12.21) will help improve the system's robustness against the hand-wash attack and significantly improve selfish users' utilities.

For a selfish user i, define

$$Cs'^{(i)}(j, t) = \sum_{k \in N_{Tr}^{(i)}(t)} Cs^{(k)}(j, t),$$

$$\text{and} \quad Cu'^{(i)}(j, t) = \sum_{k \in N_{Tr}^{(i)}(t)} Cu^{(k)}(j, t). \quad (12.22)$$

In each round, if $Cu'^{(i)}(j, t)$ is large enough, the malicious user detection algorithm is as follows:

$$j \in N_m^{(i)}(t) \text{ iff } \frac{Cs'^{(i)}(j, t) - Cu'^{(j)}(i, t)p_{ji}}{\sqrt{Cu'^{(j)}(i, t)p_{ji}(1 - p_{ji})}} \leq -h,$$

$$\text{and} \quad j \in N_s^{(i)}(t) \text{ iff } \frac{Cs'^{(i)}(j, t) - Cu'^{(j)}(i, t)p_{ji}}{\sqrt{Cu'^{(j)}(i, t)p_{ji}(1 - p_{ji})}} > -h,$$

$$\text{where} \quad p_{ji} = \frac{1}{|N_{Tr}^{(i)}(t)|} \sum_{k \in N_{Tr}^{(i)}(t)} \overline{P}_{jk}, \quad (12.23)$$

in which $\overline{P}_{jk}$ is the averaged possibility of successful transmission between user j and user k.

12.2.2.2 Multiuser attack-resistant cooperation strategy

In summary, the *multiuser cheat-proof and attack-resistant cooperation stimulation strategy* for P2P video streaming over wireless networks is as follows:

In the P2P wireless video streaming game, a selfish peer $i \in N_s$ initially marks every other user $j \in N$, $j \neq i$ as selfish. Then, in each round t, i uses the following strategy:

- First, i bargains the chunk-request quota with other users in the network.
- Then, i updates the credit line $D_{max}^{(i)}(j, t)$ as in (12.21) and identifies malicious users using (12.23).
- If i receives chunk requests from j, i will accept this request if j has not been marked as malicious by i, and (12.5) holds. Otherwise, i will reject the request.
- When i requests a chunk, he or she will send the request to peer j, who has the chunk in the buffer and satisfies

$$j = \arg \max_{j \in N_s^{(i)}(t), j \neq i} P'_{ji}, \tag{12.24}$$

where $P'_{ji} = P_{ji} \times Cc^{(i)}(j, t)/Cs^{(i)}(j, t)$ is the probability that user i successfully receives a clean chunk from user j.

12.2.3 Strategy analysis

Using the same analysis as in Section 12.1.4, when there are no attackers, the preceding multiuser cooperation strategy can be proved to be a subgame perfect and Pareto optimal Nash equilibrium of the multiuser wireless video streaming game. Similar to the analysis in Section 12.1.5, this cooperation strategy can resist pollution and incomplete chunk attacks; the detailed proof is in reference [119]. In this section, we focus on the analysis of the hand-wash attack and the optimal attack strategy.

As discussed in the previous section, using pollution and incomplete chunk attacks, the damage that each attacker can cause to a selfish user i is bounded by $D_{max}^{(i)}$, which is negligible if the P2P wireless network has infinite lifetime. However, with the hand-wash attack, because selfish users cannot recognize malicious attackers after they hand-wash, malicious users can increase their payoff dramatically. Every (selfish or malicious) user in P2P wireless video streaming, at the beginning of each round, besides the chunk request and answer strategies discussed in Section 12.1, can also choose to hand-wash.

For the P2P wireless video streaming game in which every selfish user follows the cooperation strategy as in Section 12.2.2.2, if a malicious attacker i is not detected by any other users and if $D^{(j)}(i, t) < D_{max}^{(j)}(i, t)$ for all other users $j \in N$, hand-wash will not give the malicious attacker i any further gain; because attacker i is not detected by any other users and (12.5) holds for all $j \in N$, all selfish users will still cooperate with the malicious user i. Using the original identity, the payoff that attacker i receives is the same as that if he or she applies the hand-wash attack. Therefore, hand-wash will not bring the malicious user any extra gain.

On the other hand, if the malicious user i is detected by another user j, or if there exists another user $j \in N$ where $D^{(j)}(i, t) \geq D_{max}^{(j)}(i, t)$, user j refuses to cooperate with attacker i any more. If attackers i reenters the game with a new ID, the selfish user j will cooperate with i again until i is detected again or when (12.5) is not satisfied. Thus, in this scenario, hand-wash causes extra damages to the selfish user j and helps increase attacker i's payoff.

Based on the preceding analysis, *the optimal attacking strategy for malicious attackers* is: Upon receiving a request, attackers should always reject the requests, as sending incomplete or polluted chunks give them no gain; attackers should always send requests to selfish users until they do not agree to help; and attackers should hand-wash once they are identified as malicious by one selfish user in the network. For a malicious attacker to determine whether he or she has been detected, he or she should observe other users' responses to his or her chunk requests: if a selfish user continuously refuses to send the requested chunk(s), it indicates that that selfish user has identified him or her as malicious.

12.2.4 Overrequest for broadcast nature

According to the cooperation strategy in Section 12.2.2.2, in P2P wireless streaming, users first bargain for their chunk-request quotas and ensure that the total bits to be transmitted in one round do not exceed the channel capacity. Thus also ensures that every user is capable of answering all the requests that he or she receives. Based on this analysis, selfish users have incentives to answer all the requests in every round.

For P2P wireless video streaming, all users share the same wireless cooperation channel, which is a broadcast channel and allows all users to listen to one another's signals. Thus, all requests that ask for the same chunk can be combined into one, and only one transmission is necessary. This helps reduce the cost of cooperation, and reduce the total number of bits transmitted in one round. To fully use the channel capacity, we propose the following *overrequest* algorithm:

- Given the quota x_i, a selfish user i first finds x_i chunks that he or she wants the most and mark these chunk requests as 1 (the requests within the bargained quota). The user then finds another $(K - 1)x_i$ chunks that he or she wants, and marks them 0 (requests using the extra quota). Here $K > 1$ is a constant agreed to by all users. Then, user i sends out all Kx_i chunk requests.
- In the request-answering stage, all users who receive chunk requests first choose $q = 1$ chunk to transmit, and exchange this information to confirm that the total number of bits to transmit does not exceed the channel capacity. If the cooperation channel can afford to transmit more bits, users increase q until the channel capacity is fully used. Otherwise, all selfish users answer the chunk requests marked with 1, which ensures that users can answer all received chunk requests without exceeding the channel capacity.

In practice, users may not want to answer all chunk requests that they receive as suggested in the overrequest algorithm, and they need to decide which chunk requests to answer and with whom to cooperate. Because the video streaming social network will last until the end of the video program and has a finite lifetime, selfish users tend to consider the contributions from other peers when choosing which requests to answer. It not only encourages selfish users to be always cooperative in the finite-time model, but also reduces the damage of the hand wash-attack.

We propose the following request-answering algorithm. In round t, among all users who send chunk requests to selfish user i, user i first excludes all users that he or she identifies as malicious, as well as those for whom (12.5) is not satisfied. For the remaining requesting users, user i collects the indices of all chunks that user i has been asked to upload and puts these chunk indices in a set $Cq^{(i)}(t)$. If user i decides to upload q chunks using the overrequest algorithm, for each requested chunk I_j, i first calculates

$$P^{(i)}(I_j, t) = \frac{\sum_{m \in R(I_j, t)}(Cs^{(i)}(m, t) + \epsilon)^{\gamma_i}}{\sum_{I_k \in Cq^{(i)}(t)} \sum_{m \in R(I_k, t)}(Cs^{(i)}(m, t) + \epsilon)^{\gamma_i}} \qquad (12.25)$$

where $R(I_j, t)$ is the set of users that request chunk I_j from user i at round t and ϵ is a small number that gives newcomers who have not sent any chunks to peer i a chance to start cooperation. $P^{(i)}(I_j, t)$ is the probability that user i would choose to transmit chunk I_j. Then user i will randomly choose q chunks to transmit according to $P^{(i)}(I_j, t)$ for all chunks $I_j \in Cq^{(i)}(t)$. γ_i is a parameter that controls the sensitivity of user i to other peers' contributions. If $\gamma_i = 0$, every peer who sends a request to peer i has the same probability of being answered. On the contrary, if γ_i goes to infinity, the request from the user who has sent the most chunks to peer i will be answered.

12.2.5 Attack-resistant P2P wireless video-sharing cooperation strategy with overrequest

From the preceding discussion, the *attack-resistant P2P wireless video streaming cooperation strategy* is as follows:

Any selfish node $i \in N_s$ initially marks every other node $j \in N$, $j \neq i$ as selfish. Then, in round t, i uses the following strategy:

- User i uses (12.10) to identify malicious users and uses (12.21) to update $D^{(i)}_{max}(i, t)$.
- User i bargains with other users and gets the chunk-request quota, which is K times the time-sensitive bargaining solution in (12.25).
- In the chunk-requesting stage, i applies the chunk-request algorithm (12.24), and sends chunk requests to users in $N_s^{(i)}(t)$.
- User i decides q, the number of chunks to transmit in this round, by exchanging information with other users in the network.
- In the request-answering stage, i first identifies the selfish users who satisfy (12.5). Then, i chooses the q chunks to transmit based on the probability distribution in (12.25), and agrees to send the requested chunks to all selfish users who ask for these chunks and who satisfy (12.5).

12.2.6   Simulation settings

We use NS2 and C as the simulation platform. In our simulations, we assume that the users communicate with the access point using IEEE 802.11 within the diameter of 15 meters, and users build their own wireless network that uses a different band dedicated to cooperation. NS2 is used to simulate the wired network from the live streaming server to the access point, and the communication between the access point, and the wireless network users is simulated by C. We use H.264 to encode the video into bit streams that are labeled and stored sequentially in a file for NS2 to read. User cooperation is simulated using the C simulator. NS2 and the C program use the log files to exchange the real-time simulation results.

In our simulations, the link from the wireless access point to the Internet backbone is a DSL link with 1.5 Mbps download bandwidth. There are thirty users in the wireless network using the live streaming service, and there are another five users who are using the Internet resources at the same time. We assume that the traffic generated by the five Internet users is a Poisson process. The thirty live streaming users are distributed randomly in the circle of 15-meter diameter, and they can reach each other via the dedicated cooperation channel. They adopt the enter/leave algorithm from the self-organizing TDMA [139] to access the dedicated cooperation channel. When a user enters the algorithm, he or she must first interrupt an existing user's data slot with a firing message. Then the user waits until the beginning of the next round to join the network and exchange buffer map information with other users. After users exchange the chunk requests and decide how many chunks each user is going to transmit, they take turn to upload the chunks that they agreed to transmit. When a user leaves the network or has nothing to request from others in a certain round, he or she can just keep quiet without doing anything.

We fix the ratio among the laptop, PDA, and PDA2 users as 3:1:1. The video is initially stored at the original streaming server with an upload bandwidth of 3 Mbps, and there are another 800 users on the Internet watching the same video stream. The request round is 0.4 second and the buffer length is 10 seconds with $L_f = 20$ and $L = 20$, which are the buffer length before and after the playback time, respectively. We test the "foreman" and "Akiyo" video sequences with 30 frames per second. We encode the video using H.264 into a three-layer bitstream with 75 kbps per layer, and divide each layer into chunks of 0.1 second. Thus, the layer chunk size is $M' = 7.5$ kilobits. In the wireless network, the chunks are channel-coded using BCH code with rate 15/31, and the chunk size in the wireless video streaming network is $M = 15.45$ kilobits. The thirty live streaming users in the wireless network follow the wireless video streaming cooperation strategy in Section 12.2.5 if they are selfish users, and they follow the optimal attack strategy in Section 12.2.3 if they are malicious attackers. In our simulations, we use $g_i = c_{max} = 1$, $c_{PDA2} = 0.8c_{max}$, and $c_{PDA2}:c_{PDA}:c_{laptop} = 1:0.9:0.4$. $P_{min} = 100$ mW, the noise power is 10 mW, and the available bandwidth is $B = 600$ kHz. The discount measure d in (9.18) is set to 0.7, and γ_i in (12.25) equals 2. PDA2 and PDA users are satisfied with the video quality with the base layer only.

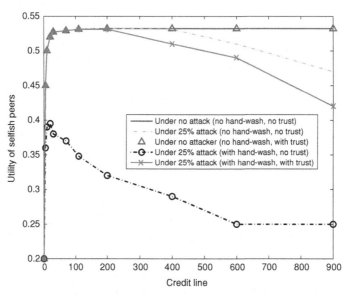

Fig. 12.4 Utility of selfish (nonmalicious) users under attack versus the initial credit line

12.2.7   Performance evaluation

The performance of the cooperation strategies is evaluated by the selfish users' utilities and PSNR of the received videos when compared with the original uncompressed videos.

If attackers apply the hand-wash attack and if selfish users do not use the trust model in Section 12.2.2.1, the selfish users' utilities will be very small no matter which credit line they choose. This case is shown as the dashed circle line in Figure 12.4. With the trust model and the attack-resistant cooperation stimulation strategy in Section 12.2.2.2, as shown in the X'd solid line in Figure 12.4, the trust model can help significantly improve the selfish users' utilities. Here, for a selfish user i, we set $Ch(i)$ in (12.21), the threshold used to determine who are trusted peers, as two times the initial credit line. Also, from Figure 12.4, if the initial credit line is chosen carefully between 50 and 200, the highest utility can be achieved even under the hand-wash attack. In Figure 12.4, we also plot the performance of the proposed cooperation strategies when there are no attackers; our simulation results show that introducing the trust model does not degrade selfish users' utilities when there are no malicious attackers.

Figure 12.5 shows the average selfish users' utility of the overrequest algorithm. Following the observations from Figure 12.4, we choose the initial credit line as 50 and set $Ch(i)$ as 100. When selfish users do not apply the overrequest algorithm in Section 12.2.4 – that is, $K = 1$ in Figure 12.5 – the average selfish user's utility drops by 20 percent when the percentage of attackers increases from zero to 50 percent. However, with $K = 3$, the average selfish user's utility remains the same when the percentage

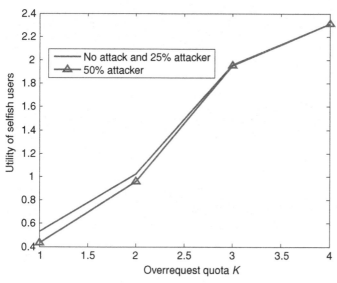

Fig. 12.5 Utility of averaged selfish (nonmalicious) users with or without attack versus the amount of overrequest quota

of attackers increases from zero to 50 percent. This shows that that the overrequest algorithm can effectively increase the selfish users' utilities, and the contribution-based chunk-answering algorithm can also help resist attacks with up to 50 percent malicious attackers.

Figure 12.6 shows the average PSNR of the selfish laptop users who do not worry about power outages and whose cost per unit energy is the smallest. In this figure, attackers apply the hand-wash attack once they are detected, and selfish users use the cooperation strategy in Section 12.2.5. Figure 12.6(a) shows the robustness of different credit lines versus the percentage of attackers. When there are more attackers, a higher credit line gives selfish users a smaller PSNR. This is because the credit line mechanism bounds the maximal damage caused by only one attacker, and the total damage will increase if there are more malicious users in the system. Thus, this phenomenon again suggests that the credit line should be set as the smallest number that can stimulate cooperation among selfish users, which is 50 in this case. Figure 12.6(b) shows the average selfish user's PSNR with different trust thresholds Ch in (12.21) versus the number of rounds. It is clear that after 400 rounds, the selfish user's PSNR is saturated, and $Ch = 0.5D_{max}^{(i)}(j, 0)$ or $Ch = D_{max}^{(i)}(j, 0)$ gives smaller PSNRs than $Ch = 2D_{max}^{(i)}(j, 0)$. These results imply that a smaller trust threshold Ch will cause more damage to the system, as it may cause selfish users to trust malicious users as well. On the other hand, from Figure 12.6(b), a larger Ch needs more rounds for the selfish users' PSNR to saturate, and selfish users need to wait for more rounds before they trust each other. Each user should choose Ch based on how long he or she will stay in the network.

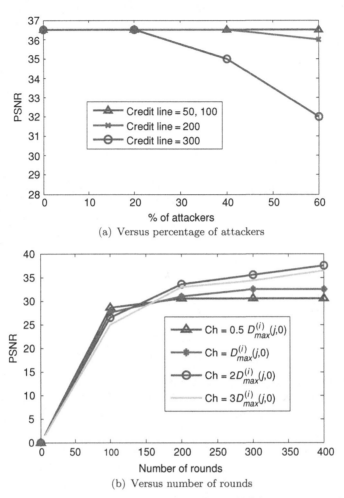

(a) Versus percentage of attackers

(b) Versus number of rounds

Fig. 12.6 Simulation results of the PSNR of the selfish laptop users

Furthermore, we compare our cooperation strategy with the payment-based incentive scheme of Tan and Jarvis [93] and the resource chain trust model for P2P security [140]. The credit line is set to 100, and users choose $K = 3$ in the over request algorithm. We first compare the attack resistance of the three algorithms in Figure 12.7(a) and plot the average PSNR of the videos received by a selfish laptop user. It is clear that our proposed cooperation strategy achieves the best attack resistance among the three, and we do not observe obvious performance drops even when 60 percent of the users are attackers. We also compare the average PDA user's utility of the three algorithms when there no attackers in Figure 12.7(b). We can see that our proposed scheme gives the PDA users a higher utility, as it takes into consideration users' desired resolution. In our proposed scheme, a PDA user will not request higher-layer chunks but instead use all of his or her chunk-request quota to ask for base-layer chunks, which gives him or her a higher utility.

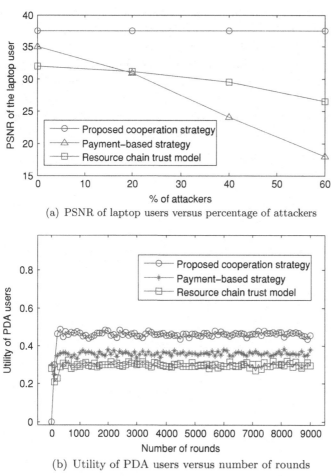

(a) PSNR of laptop users versus percentage of attackers

(b) Utility of PDA users versus number of rounds

Fig. 12.7 Performance comparison of the attack-resistant cooperation strategies, payment-based cooperation strategy, and resource chain trust model

12.3   Chapter summary and bibliographical notes

Misbehavior identification heavily influences the system performance in P2P media sharing social networks. With trustworthy malicious user detection and attack-resistant cooperation strategies, users are encourage to be more cooperative in P2P media sharing social networks, thus leading to much better streaming performance.

In this chapter, we consider the three most common malicious attacks in P2P media sharing systems: the incomplete chunk attack, pollution attack, and hand-wash attack. For each type of attack, we design a detection mechanism to identify attackers and to protect nonmalicious selfish users. We also model the behavior of selfish users and malicious users as a noncooperative game and analyze the attack resistance of the equilibrium, which is the optimal strategy for all users. Simulation results show that the attack-resistant cooperation strategy and the malicious behavior detector can resist up to 50 percent of attackers.

Trust modeling can be much more complicated than what we discussed in this chapter. The beta-function-based method was discussed by Josang and colleagues [141] and the influence of trust among users on social networks was studied by Sun and associates [86]. Interested readers can see reference [7] for the pollution attack. The DoS attack studied by Naoumov and Ross [85] can exhaust users' resources and make the P2P system unavailable. Readers are referred to reference [85] for a detailed discussion of the DoS attack.

Part V

Media-sharing social network structures

13 Misbehavior detection in colluder social networks with different structures

Different social networks may have different structures. The discussions in the previous chapters focused mainly on distributed scenarios. For example, there are no central authorities in the colluder social networks in Chapters 5 and 8, and the P2P systems in Chapters 9 and 12 are fully distributed, meaning that every peer takes the same role. In reality, some social networks have a centralized structure in which there are one or more entities whom all users trust and who can facilitate interaction among users. For example, the first generation P2P file-sharing networks (for example, the Napster music file-sharing system) used a set of central servers to provide content indexing and search services [142]. Although these servers cannot enforce user cooperation, they can facilitate user interaction. Other media-sharing social networks have a distributed structure and a flat topology in which users take the same role – for example, Gnutella and Chord [142]. Distributed schemes should be designed for such social networks. In this chapter, we use colluder social networks in multimedia fingerprinting as an example to investigate the impact of network structure on social networks.

In colluder social networks, as discussed in Chapters 5 and 8, colluders aim to achieve fair collusion, which requires all colluders to report their received fingerprinted copies honestly. As discussed in Chapter 11, the assumption of fair play may not always hold, and some colluders may process their fingerprinted copies before collusion to further lower their risk of being detected. Precollusion processing reduces the cheating colluders' relative risk with respect to their fellow colluders, and in some scenarios, it may even increase other colluders' probability of being detected. Therefore, it is of ample importance to design monitoring mechanisms to detect and identify misbehaving users, and to design cheat-proof cooperation stimulation strategies in media-sharing social networks.

To analyze user dynamics in social networks with cheating users, the first step is to study the cheating strategies that users can use, and then design monitoring mechanisms to detect and identify misbehaving users [143]. Such a monitoring mechanism facilitates the design of cheat-proof strategies, which makes noncooperation nonprofitable; thus, cheating users have no incentives to cheat.

In this chapter, we explore possible strategies to detect and identify cheating colluders, investigate the impact of network structures on misbehavior detection and identification, and analyze the cheat-prevention performance. Addressing different structures in different social networks, we first consider a centralized social network with a trusted ringleader and investigate misbehavior detection with the trusted ringleader's help. We then consider the peer-structured social networks in which all colluders take the same role, and examine autonomous cheating colluder identification.

13.1   Behavior dynamics in colluder social networks

In this chapter, as an example we user equal-risk fairness, in which all colluders take the same risk of being detected. From the discussion in Chapter 11, to further reduce their risk, a cheating colluder may average or swap neighboring frames of similar content or change the temporal resolution of his or her fingerprinted copy. Here, we use the temporal filtering in Section 11.1.1 as an example, and investigate how to detect temporal filtering precollusion processing.

13.1.1   Misbehavior detection and identification

Recall that $\mathbf{S}_j$ is the jth frame in the original host signal, and $\mathbf{X}_j^{(i)} = \mathbf{S}_j + \mathbf{W}_j^{(i)}$ is the fingerprinted frame j that user i receives from the content owner. Here, we drop the term *JND* to simplify the notations. In this chapter, we consider orthogonal fingerprint modulation, in which fingerprints assigned to different users are orthogonal to each other with equal energies. A cheating colluder uses (11.1) to process his or her copy, and uses (11.8) to select the parameter λ_j to minimize the risk of being detected under the quality constraint on the processed copy $\widetilde{\mathbf{X}}_j^{(i)}$.

Define SC as the set containing the indices of all colluders. $SC_s \subseteq SC$ includes all cheating colluders, and $SC_h = SC \backslash SC_s$ is the set with the indices of all honest colluders who do not apply precollusion processing. Let $\widetilde{\mathbf{X}}_j^{(i)}$ be the jth frame from colluder i. For honest colluders k and l, we have

$$\widetilde{\mathbf{X}}_j^{(k)} = \mathbf{X}_j^{(k)} = \mathbf{S}_j + \mathbf{W}_j^{(k)} \quad \text{and} \quad \widetilde{\mathbf{X}}_j^{(l)} = \mathbf{X}_j^{(l)} = \mathbf{S}_j + \mathbf{W}_j^{(l)}. \tag{13.1}$$

For a cheating colluder $i \in SC_s$,

$$\widetilde{\mathbf{X}}_j^{(i)} = \frac{1-\lambda_j}{2}\mathbf{X}_{j-1}^{(i)} + \lambda_j\mathbf{X}_j^{(i)} + \frac{1-\lambda_j}{2}\mathbf{X}_{j+1}^{(i)}$$

$$= \mathbf{S}_j + \Delta\mathbf{S}_j(\lambda_j) + \widetilde{\mathbf{W}}_j^{(i)},$$

$$\text{where} \quad \Delta\mathbf{S}_j(\lambda_j) = (1-\lambda_j)\left(\frac{\mathbf{S}_{j-1}}{2} + \frac{\mathbf{S}_{j+1}}{2} - \mathbf{S}_j\right),$$

$$\text{and} \quad \widetilde{\mathbf{W}}_j^{(i)} = \frac{1-\lambda_j}{2}\mathbf{W}_{j-1}^{(i)} + \lambda_j\mathbf{W}_j^{(i)} + \frac{1-\lambda_j}{2}\mathbf{W}_{j+1}^{(i)}. \tag{13.2}$$

From (13.2), temporal filtering not only averages fingerprints embedded in adjacent frames and attenuates their energies, but also filters neighboring frames in the host signal and introduces extra distortion $\Delta\mathbf{S}_j(\lambda_j)$.

For the jth fingerprinted frames from colluder k and l, define $D_j(k,l) \overset{\triangle}{=} \|\widetilde{\mathbf{X}}_j^{(k)} - \widetilde{\mathbf{X}}_j^{(l)}\|^2$. Because $\{\mathbf{W}_j^{(k)}\}$, $\{\mathbf{W}_j^{(l)}\}$, and $\{\mathbf{W}_j^{(i)}\}$ are orthogonal to each other, from (13.1) and

(13.2), we have

$$D_j(k, l) \approx \|\mathbf{W}_j^{(k)}\|^2 + \|\mathbf{W}_j^{(l)}\|^2,$$

$$\text{and} \quad D_j(k, i) \approx \|\mathbf{W}_j^{(k)}\|^2 + \|\widetilde{\mathbf{W}}_j^{(i)}\|^2 + \|\Delta\mathbf{S}_j(\lambda_j)\|^2,$$

$$\text{where} \quad \|\Delta\mathbf{S}_j(\lambda_j)\|^2 = (1 - \lambda_j)^2 \times \left\|\frac{\mathbf{S}_{j-1}}{2} + \frac{\mathbf{S}_{j+1}}{2} - \mathbf{S}_j\right\|^2. \tag{13.3}$$

For honest colluders k and l, $D_j(k, l)$ can be approximated by the summation of the energies of the two embedded fingerprints $\mathbf{W}_j^{(k)}$ and $\mathbf{W}_j^{(l)}$. For the honest colluder k and the cheating colluder i, in addition to the summation of $\|\mathbf{W}_j^{(k)}\|^2$ and $\|\widetilde{\mathbf{W}}_j^{(i)}\|^2$, $D_j(k, i)$ also includes the additional distortion $\|\Delta\mathbf{S}_j(\lambda_j)\|^2$ introduced by temporal filtering. Therefore, $D_j(k, i)$ has a much larger value than $D_j(k, l)$. For a given video sequence, from (13.3), the difference between $D_j(k, i)$ and $D_j(k, l)$ is more obvious when λ_j takes a smaller value. In addition, $|D_j(k, i) - D_j(k, l)|$ takes a larger value when the scene of the host video sequence changes quickly and the difference between adjacent frames is larger. This observation suggests that $\{D_j(k, l)\}$ can help honest colluders detect precollusion processing and identify cheating colluders.

Before a colluder decides with whom to collude, he or she is unwilling to give others the received fingerprinted copy that contains his or her identification information. Therefore, cheating colluder detection and identification must prevent colluders from accessing the fingerprinted coefficients in others' copies. To meet this antiframing requirement in cheating colluder detection and identification, all fingerprinted copies must be encrypted appropriately during this misbehavior detection process.

Addressing different network structures, in this chapter, we first consider a centralized social structure in which there is a ringleader whom all colluders trust. We investigate how the trusted ringleader can help detect precollusion processing and identify cheating colluders. We then consider the distributed peer social structure of the colluder social networks, and study the autonomous cheating colluder detection and identification, in which colluders help each other detect misbehavior and identify cheating colluders. In this chapter, we consider the scenario in which there are only a few cheating colluders, and most colluders honestly report private information of their fingerprinted copies to others. In such scenarios, we investigate how honest colluders can collaborate with one another to accurately identify misbehaving users and analyze the performance of the misbehaving-user identification algorithm.

13.1.2   Performance criteria

The cheating colluder detection and identification process aims to accurately identify all cheating colluders without falsely accusing any others. To measure the performance of the cheating colluder detection and identification algorithm, we consider two types of detection errors and use the following criteria:

- P_{md}: the rate that an honest colluder misses a cheating colluder during detection; and
- P_{fa}: the rate that an honest colluder falsely accuses another honest colluder as a cheating colluder.

To evaluate the antiframing performance, assume that the fingerprinted frame j that colluder i receives is $\mathbf{X}_j^{(i)} = \left[X_j^{(i)}(1), X_j^{(i)}(2), \ldots, X_j^{(i)}(N_j) \right]$, where $X_j^{(i)}(l)$ is the lth component in $\mathbf{X}_j^{(i)}$, and $\mathbf{X}_j^{(i)}$ is of length N_j. During the cheating colluder detection and identification process, without proper protection, it is possible that another colluder k can access some of the fingerprinted coefficients in $\mathbf{X}_j^{(i)}$. Assume that $\mathrm{Ind}_j^{(k,i)} \subseteq \{1, 2, \ldots, N_j\}$ includes the indices of all the fingerprinted coefficients in $\mathbf{X}_j^{(i)}$ that colluder k can access, and define $\mathrm{Ind}_j^{(k)} \triangleq \bigcup_{i \in SC, i \neq k} \mathrm{Ind}_j^{(k,i)}$. If $\mathrm{Ind}_j^{(k)} = \{1, 2, \ldots, N_j\}$, then colluder k can generate a new copy of high quality that does not contain any information on his or her own fingerprint and that can be used to frame other colluders in SC.

To evaluate the resistance of our algorithms to framing attacks, we define

$$\gamma_j \triangleq \frac{E\left[|\mathrm{Ind}_j^{(k)}| \right]}{N_j}, \quad 0 \leq \gamma_j \leq 1, \tag{13.4}$$

where $E[X]$ returns the statistical mean of X, and $|A|$ is the size of the set A. A smaller γ_j indicates that the cheating colluder detection and identification process is more robust against framing attacks.

13.2   Centralized colluder social networks with trusted ringleaders

In this section, we consider a centralized colluder social network in which there is a trusted ringleader, and we study how to detect and identify cheating colluders there. All colluders believe that the trusted ringleader will not give their fingerprinted copies to others; the ringleader will not frame any colluders; and the ringleader will not modify the cheating colluder detection and identification results.

To identify cheating colluders, each colluder i first generates a secret key $K^{(i)}$ shared with the ringleader only, encrypts his or her fingerprinted copy with $K^{(i)}$ to prevent others' eavesdropping on the communication, and transmits the encrypted version to the ringleader. Because $K^{(i)}$ is known to colluder i and the ringleader only, no one but colluder i and the ringleader can decrypt the transmitted bit stream, and other colluders cannot access the fingerprinted coefficients. After receiving and decrypting the transmitted bit streams from all colluders, the ringleader examines these fingerprinted copies and helps detect and identify cheating colluders. Finally, colluders exclude those identified cheating colluders from multiuser collusion.

In this chapter, we consider the scenario in which colluders receive fingerprinted copies of the same quality (i.e., same SNR). When they receive fingerprinted copies of different quality owing to network heterogeneity and dynamically changing channel conditions, a challenging issue is to differentiate the scenario in which the colluder intentionally changed his or her received fingerprinted copy from another one in which this copy was transmitted through severely congested and erroneous networks. (In our future work, we plan to investigate cheating colluder detection and identification when colluders receive fingerprinted copies of different quality.)

13.2.1 Detection of temporal-interpolation–based precollusion processing

Following the discussion in Section 13.1.1, in this chapter, we use $\{D_j(k,l)\}$ to detect precollusion processing and identify cheating colluders. Figure 13.1 shows an example of the histogram of $\{D_j(k,l)\}_{k,l \in SC}$ for the second frame in sequence "carphone." Other frames and other sequences give the same trend. In our simulations, we adopt the human visual model-based spread spectrum embedding [58] and embed fingerprints in the DCT domain. Fingerprints are generated from Gaussian distribution $\mathcal{N}(0, \sigma_w^2)$ with $\sigma_w^2 = 1/9$. During collusion, we assume that there are 150 colluders and SC is the set containing their indices.

In Figure 13.1(a), there are no cheating colluders and all colluders provide one another with correct information on their fingerprinted copies. In Figure 13.1(b), there is one cheating colluder i_1 who applies temporal filtering before collusion. During pre-collusion processing, i_1 selects the parameter $\lambda_j = 0.6031$ to generate a new frame $\widetilde{\mathbf{X}}_j^{(i_1)}$ with PSNR of 40 dB when compared with the originally received one, $\mathbf{X}_j^{(i_1)}$. In Figure 13.1(c), there are two cheating colluders, i_1 and i_2, who process their fingerprinted copies independently and impose different fidelity constraints. As in Figure 13.1(b), i_1 generates a new frame $\widetilde{\mathbf{X}}_j^{(i_1)}$ of 40 dB. During precollusion processing, i_2 selects the parameter $\lambda_j = 0.7759$ such that the new copy $\widetilde{\mathbf{X}}_j^{(i_2)}$ has a PSNR of 45 dB. From Figure 13.1, when all colluders give one another correct information about their fingerprinted signals, $\{D_j(k,l)\}_{k,l \in SC}$ are from the same distribution with a single mean. If some cheating colluders process their fingerprinted copies before collusion, $\{D_j(k,l)\}_{k,l \in SC}$ are from different distributions with distinct means.

Let us define

$$\mathfrak{D}_j(SC_h, SC_h) \triangleq \{D_j(k,l) : k,l \in SC_h, k \neq l\}$$

$$\mathfrak{D}_j(SC_s, SC_h) \triangleq \{D_j(k,l) : k \in SC_s, l \in SC_h\}. \tag{13.5}$$

From (13.3), the distance between $\mathfrak{D}_j(SC_h, SC_h)$ and $\mathfrak{D}_j(SC_h, SC_s)$ depends on the selected parameter λ_j, as well as on the host video sequence. In Figure 13.1(c), the sample means of $\mathfrak{D}_j(SC_h, SC_h)$, $\mathfrak{D}_j(i_1, SC_h)$, and $\mathfrak{D}_j(i_2, SC_h)$ are 4.7, 10.8, and 6.7, respectively. Thus, the difference between $\mathfrak{D}_j(SC_h, SC_h)$ and $\mathfrak{D}_j(SC_h, SC_s)$ is larger when the cheating colluders select λ_j of smaller values. We also consider video sequences of different characteristics: "carphone," which has moderate motion and "flower," whose scene changes very fast. Figure 13.2 shows the histogram of $\{D_j(k,l)\}$. Here, λ_j is fixed as 0.7. In Figure 13.2(a), $\mathfrak{D}_j(SC_h, SC_h)$ has a sample mean of 4.7 and $\mathfrak{D}_j(i_1, SC_h)$ has a sample mean of 8.2. In Figure 13.2(b), the sample means of $\mathfrak{D}_j(SC_h, SC_h)$ and $\mathfrak{D}_j(i_1, SC_h)$ are 15.8 and 108.2, respectively. Comparing Figure 13.2(b) with Figure 13.2(a), $\mathfrak{D}_j(SC_h, SC_h)$ and $\mathfrak{D}_j(SC_h, SC_s)$ are separated farther away from each other when the scene changes quickly (for example, in sequence "flower"), as the norm $||\Delta \mathbf{S}_j(\lambda_j)||$ is larger.

This analysis suggests that the histogram of $\{D_j(k,l)\}$ can be used to determine the existence of cheating colluders. The ringleader calculates $D(k,l)$ for every pair of colluders (k,l) and broadcasts $\{D_j(k,l)\}$ to all colluders. If $\{D(k,l)\}$ are from the same

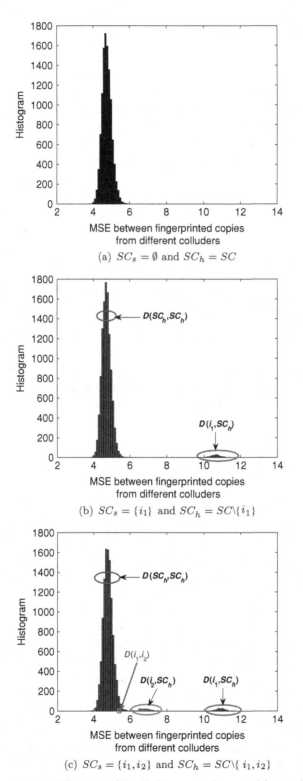

Fig. 13.1 Histogram of $\{D_j(k,l)\}_{k,l \in SC}$ on the second frame of sequence "carphone"

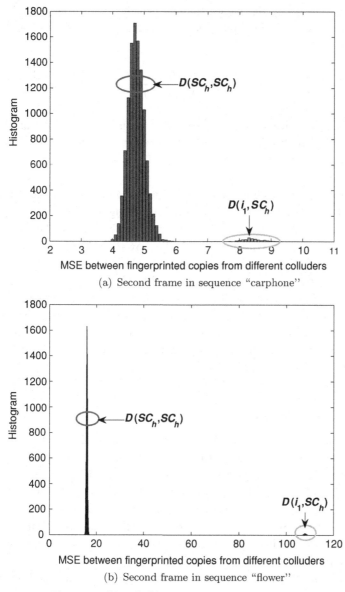

(a) Second frame in sequence "carphone"

(b) Second frame in sequence "flower"

Fig. 13.2 Histogram of $\{D_j(k,l)\}_{k,l \in SC}$

distribution with a single mean, then all colluders keep their fair-collusion agreement and there are no cheating colluders. If $\{D(k,l)\}$ are from two or more distributions with different means, there exists at least one cheating colluder who applies precollusion processing.

In the preceding examples, $\mathfrak{D}_j(SC_h, SC_s)$ and $\mathfrak{D}_j(SC_h, SC_h)$ do not overlap, which enables honest colluders to easily detect the existence of cheating colluders. We now consider the scenario in which $\mathfrak{D}_j(SC_h, SC_s)$ and $\mathfrak{D}_j(SC_h, SC_h)$ overlap. Let $D_j^{max}(SC_h, SC_h) = \max\left(\mathfrak{D}_j(SC_h, SC_h)\right)$ and $D_j^{min}(SC_s, SC_h) =$

$\min \left(\mathfrak{D}_j(SC_s, SC_h) \right)$ be the largest and the smallest values in $\mathfrak{D}_j(SC_h, SC_h)$ and $\mathfrak{D}_j(SC_s, SC_h)$, respectively. Given the total number of colluders K, we define the overlap ratio as

$$c \overset{\triangle}{=} \sum_{k,l \in SC} \frac{I[D_j^{min}(SC_s, SC_h) \leq D_j(k,l) \leq D_j^{max}(SC_h, SC_h)]}{K(K-1)/2}, \qquad (13.6)$$

where $I[\cdot]$ is the indicator function and the denominator is the total number of colluder pairs. The two distributions overlap by a larger ratio when c takes a larger value, and the two distributions do not overlap if $c = 0$; that is, $D_j^{max}(SC_h, SC_h) < D_j^{min}(SC_s, SC_h)$.

We use the second frame in sequence "carphone" as an example, assuming that 10 out of 150 colluders are cheating and process their fingerprinted copies independently before collusion. We observe a similar trend for other frames and other parameters. We intentionally move the two distributions $\mathfrak{D}_j(SC_s, SC_h)$ and $\mathfrak{D}_j(SC_h, SC_h)$ and let them overlap. Figure 13.3(a) and (b) show the resulting histograms of $\{D_j(k,l)\}_{k,l \in SC}$ when $\mathfrak{D}_j(SC_s, SC_h)$ and $\mathfrak{D}_j(SC_h, SC_h)$ overlap by 20 percent and 75 percent, respectively. From Figure 13.3, we can still observe the bimodality of $\{D_j(k,l)\}_{k,l \in SC}$ when c takes a small value. When $c \geq 75\%$, $\mathfrak{D}_j(SC_s, SC_h)$ and $\mathfrak{D}_j(SC_h, SC_h)$ merge, which prevents the detection of cheating behavior. Thus, the bimodality of $\{D_j(k,l)\}$ can help detect the existence of precollusion processing when the overlap between $\mathfrak{D}_j(SC_s, SC_h)$ and $\mathfrak{D}_j(SC_h, SC_h)$ does not exceed 75 percent.

13.2.2   Identification of cheating colluders

After receiving the broadcasted $\{D_j(k,l)\}$ from the ringleader, colluders examine the histogram plot of $\{D_j(k,l)\}$ to determine the existence of cheating colluders. Further identification of cheating colluders requires detailed examination of $\{D_j(k,l)\}$ – in particular, $\mathfrak{D}_j(SC_h, SC_s)$. For each $D_j(k,l) \in \mathfrak{D}_j(SC_h, SC_s)$, the two corresponding colluders, k and l, are in different subgroups: one belongs to SC_h and the other is a cheating colluder in SC_s. Thus, analysis of each individual $D_j(k,l)$ in $\mathfrak{D}_j(SC_h, SC_s)$ can help separate SC into two subgroups and, therefore, enables cheating colluder identification.

To identify cheating colluders, a simple solution is to examine the histogram of all $\{D_j(k,l)\}$ and use a threshold to separate $\mathfrak{D}_j(SC_h, SC_s)$ from $\mathfrak{D}_j(SC_h, SC_h)$. However, the values of $\{D_j(k,l)\}$ change from sequence to sequence, and $\mathfrak{D}_j(SC_h, SC_s)$ and $\mathfrak{D}_j(SC_h, SC_h)$ may overlap. Thus, the thresholding-based method may introduce errors and thus affect the accuracy of the identification algorithm. To address this issue, from Figure 13.3, a larger value of $D_j(k,l)$ gives higher confidence that $D_j(k,l)$ is in $\mathfrak{D}_j(SC_h, SC_s)$ and that the two corresponding colluders, k and l, belong to different subgroups. Thus, our algorithm starts with $D_j(k,l)$ that has the largest value (and thus gives the detector the highest confidence) and determines which of the corresponding two colluders is cheating. Then, it moves to the next largest $D_j(k,l)$. It repeats this procedure until every colluder in SC has been identified as either a cheating colluder

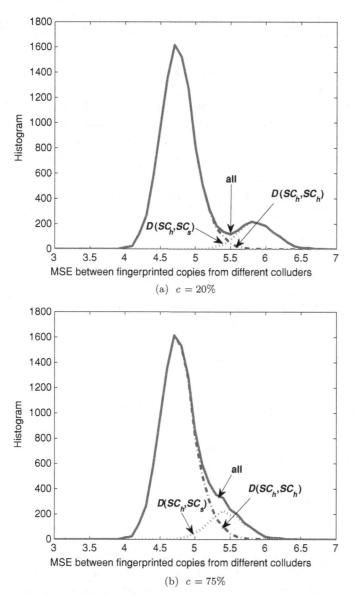

Fig. 13.3 Histogram of $\{D_j(k,l)\}_{k,l \in SC}$ for the second frame in sequence "carphone" with overlapping $\mathfrak{D}_j(SC_s, SC_h)$ and $\mathfrak{D}_j(SC_h, SC_h)$

or an honest colluder. Thus, instead of using all $\{D_j(k,l)\}$, our algorithm uses only those that give higher confidence to accurately identify cheating colluders even when $\mathfrak{D}_j(SC_h, SC_s)$ and $\mathfrak{D}_j(SC_h, SC_h)$ overlap.

In addition, given $\{D_j(k,l)\}$, even if $\mathfrak{D}_j(SC_h, SC_s)$ and $\mathfrak{D}_j(SC_h, SC_h)$ do not overlap, the ringleader can only separate colluders into two subgroups, but cannot tell which contains the honest colluders. Instead, an honest colluder knows that he or she is in SC_h, and given a $D_j(k,l)$ and the two corresponding colluders k and l, he or she can immediately determine the subgroups that they are in. Therefore, in our algorithm, the honest colluders themselves (instead of the ringleader) identify cheating colluders.

Algorithm 13.1 gives the details of how colluder i in SC_h identifies cheating colluders. For a total of KC colluders whose indices are i_1, $i_2, \ldots, i_{KC}$, $\boldsymbol{\Phi}^{(i)} = [\Phi^{(i)}(i_1), \Phi^{(i)}(i_2), \ldots, \Phi^{(i)}(i_{KC})]$. Colluder i sets $\Phi^{(i)}(k) = 1$ when he or she detects that colluder k is a cheating colluder, and $\Phi^{(i)}(k) = 0$ if i believes that colluder k is an honest colluder. $\Psi_t = \{k : \Phi^{(i)}(k) \neq -1\}$ includes the indices of all colluders for whom i has identified which subgroups they belong to in the previous rounds.

Algorithm 13.1 Cheating colluder identification by $u^{(i)}$ in SC_h

---

Set $\Psi_t = \{i\}$, $\boldsymbol{\Phi}^{(i)} = -\mathbf{1}_{1 \times KC}$, $\Phi^{(i)}(i) = 0$, and $m = 0$
while $\Psi_t \neq SC$ **do**
  $m = m + 1$
  select $D_j(k, l)$ with the mth largest value and take the indices of the two corresponding colluders
  if $k \notin \Psi_t$ AND $l \notin \Psi_t$ **then**
    if $D_j(i, k) > D_j(i, l)$ **then**
      $\Phi^{(i)}(k) = 1; \Phi^{(i)}(l) = 0; \Psi_t = \Psi_t \cup \{k, l\}$
    else if $D_j(i, k) < D_j(i, l)$ **then**
      $\Phi^{(i)}(k) = 0; \Phi^{(i)}(l) = 1; \Psi_t = \Psi_t \cup \{k, l\}$
    end if
  else if $k \in \Psi_t$ AND $l \notin \Psi_t$ **then**
    $\Phi^{(i)}(l) = 1 - \Phi^{(i)}(k), \Psi_t = \Psi_t \cup \{l\}$
  else if $l \in \Psi_t$ AND $k \notin \Psi_t$ **then**
    $\Phi^{(i)}(k) = 1 - \Phi^{(i)}(l), \Psi_t = \Psi_t \cup \{k\}$
  end if
end while
Return $\widehat{SC}_s^{(i)} = \{k : \Phi^{(i)}(k) = 1\}$

An honest colluder i first initializes $\boldsymbol{\Phi}^{(i)}$ to an undetermined status -1 and sets $\Phi^{(i)}(i)$ to 0, as he or she is in subgroup SC_h. Then, i examines every $D_j(k, l)$ and starts with the largest one. Given a $D_j(k, l)$, i first checks whether he or she has determined the values of $\Phi^{(i)}(k)$ and $\Phi^{(i)}(l)$ in the previous rounds.

- If both $\Phi^{(i)}(k)$ and $\Phi^{(i)}(l)$ have been decided, i moves to the next largest $D_j(k, l)$.
- If one of them is set to either 0 or 1 while the other is still undetermined with value -1, without loss of generality, assume that $\Phi^{(i)}(k)$ has been determined previously, then i sets $\Phi^{(i)}(l) = 1 - \Phi^{(i)}(k)$.
- If i is unable to determine either $\Phi^{(i)}(k)$ or $\Phi^{(i)}(l)$ in the previous rounds, he or she then compares the values of $D_j(k, i)$ and $D_j(l, i)$. Without loss of generality, assume that $D_j(k, i) > D_j(l, i)$. In this scenario, compared with colluder l, colluder k is more likely to be a cheating colluder. Thus, i sets $\Phi^{(i)}(l) = 0$ and $\Phi^{(i)}(k) = 1$.

Colluder i repeats the above process and stops when $\Psi_t = SC$ and all the components in $\Phi^{(i)}$ have been set to either 0 or 1. Algorithm 13.1 outputs $\widehat{SC}_s^{(i)} = \{k : \Phi^{(i)}(k) = 1\}$, which is the set containing the indices of all colluders whom $u^{(i)}$ detects as cheating colluders.

13.2.3 Cheating colluder detection and identification and performance evaluation

13.2.3.1 Cheating colluder detection and identification

To summarize, if the colluder social network has a centralized structure with a trusted ringleader, the key steps in the cheating colluder detection and identification process are as follows for each frame j.

- **Step 1. Encryption:** Each colluder i first generates a secret key $K^{(i)}$ shared with the ringleader only, encrypts the fingerprinted copy with $K^{(i)}$, and transmits the encrypted copy to the ringleader.
- **Step 2. Calculation of $\{D\}$:** After decrypting the bit streams received from all colluders, the ringleader calculates $D_j(k, l)$ for each pair of colluders (k, l). The ringleader then broadcasts $\{D_j(k, l)\}$ to all colluders, together with his or her digital signature [144].
- **Step 3. Detection of precollusion processing:** Colluders in SC_h first examine the histogram of $\{D_j\}$ to detect precollusion processing. If $\{D_j\}$ are from the same distribution with a single mean, then there are no cheating colluders, and the colluders skip Step 4 and collude with each other. If $\{D_j\}$ are from two or more distributions with different means, there is at least one cheating colluder and honest colluders go to Step 4 to identify cheating colluders.
- **Step 4. Cheating colluder identification:** If Step 3 detects the existence of cheating colluders, each honest colluder in SC_h applies Algorithm 13.1 to estimate the identities of the cheating colluders.

13.2.3.2 Performance evaluation

In this cheating colluder detection and identification process, all the fingerprinted copies are encrypted during transmission. For each copy, only the corresponding colluder and the trusted ringleader can access the fingerprinted coefficients, whereas other colluders do not have the decryption key and cannot decrypt the transmitted bit stream. Therefore, $\gamma_j = 0$ and the cheating colluder detection and identification process is robust against framing attacks.

To evaluate the detection performance, we select three typical video sequences, "miss america," "carphone," and "flower," and test on the first ten frames in each sequence as an example. Other frames and other sequences give the same result. The simulation setup is the same as that in Section 13.2.1. Orthogonal fingerprints are generated from Gaussian distribution $\mathcal{N}(0, \sigma_w^2)$ with $\sigma_w^2 = 1/9$. In each fingerprinted copy, fingerprints that are embedded into neighboring frames are correlated with each other, depending on the similarity between the host frames. Human visual model-based spread spectrum embedding [58] is applied to embed fingerprints into the host signal. We assume that

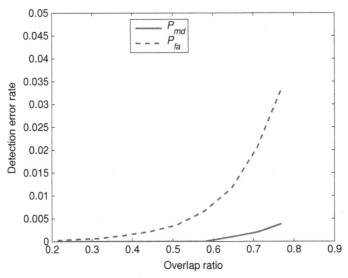

Fig. 13.4 Performance of the cheating colluder identification algorithm with overlapping $\mathfrak{D}(SC_h, SC_s)$ and $\mathfrak{D}(SC_h, SC_h)$

the total number of colluders is 150. There are ten cheating colluders and each processes his or her fingerprinted copy independently before collusion. Among the ten cheating colluders, five of them select the parameter λ_j to generate new frames with PSNR of 40 dB, and the other five cheating colluders generate new frames with PSNR of 45 dB.

For each frame in every sequence, we run 1000 simulation runs to test the performance. In all our simulation runs, Algorithm 13.1 accurately identifies all cheating colluders in SC_s without falsely accusing any honest colluder, and the cheating colluder detection and identification algorithm does not make either type of detection errors. This is because temporal filtering not only averages fingerprints embedded in adjacent frames and reduces the cheating colluder's risk, but also filters adjacent host frames and introduces extra distortion into the host signal. This extra distortion makes the two distributions, $\mathfrak{D}(SC_h, SC_s)$ and $\mathfrak{D}(SC_h, SC_h)$ in Figure 13.1, separate from each other, and it enables our algorithm to correctly identify the cheating colluders without falsely accusing any others.

We then consider the scenario in which $\mathfrak{D}(SC_h, SC_s)$ and $\mathfrak{D}(SC_h, SC_h)$ overlap with each other, and Figure 13.4 shows the simulation results. We use the second frame of "carphone" as an example, and assume that there are ten cheating colluders who process their copies independently. We observe the same trend for other frames and other parameters. In Figure 13.4, we stop the simulations when $c \geq 75$ percent, because in those scenarios, $\mathfrak{D}(SC_h, SC_s)$ and $\mathfrak{D}(SC_h, SC_h)$ merge and the bimodality of $\{D_j(k, l)\}$ cannot be observed. From Figure 13.4, the miss detection rate is below 0.5 percent and the false alarm rate does not exceed 3.5 percent, and our algorithm can accurately identify cheating colluders even if the two distributions overlap.

13.3   Distributed peer-structured colluder social networks

When there is no trusted ringleader, colluders form a peer-structured social network and they help one another detect and identify cheating colluders. In this section, we consider the scenario in which there are only a few cheating colluders, and study autonomous cheating colluder identification. We also address potential attacks on the autonomous cheating colluder identification scheme, and analyze the scheme's attack resistance.

13.3.1   Cheating colluder detection and identification without a trusted ringleader

Without a trusted ringleader, the challenge is to accurately calculate $\{D_j(k, l)\}$ while still protecting the secrecy of the fingerprinted coefficients. In this section, we study how to calculate $D_j(k, l)$ for a given pair of colluders (k, l) without a trusted ringleader. Then we investigate autonomous cheating colluder identification for a group of colluders.

13.3.1.1   Calculation of $D_j(k, l)$

For each pair of colluders (k, l), assume that $\widetilde{\mathbf{X}}_j^{(k)}$ and $\widetilde{\mathbf{X}}_j^{(l)}$ are the fingerprinted copies from colluders k and l, respectively. Colluders k and l cannot calculate $D_j(k, l)$ themselves, because it will leak the fingerprinted coefficients in $\widetilde{\mathbf{X}}_j^{(k)}$ and $\widetilde{\mathbf{X}}_j^{(l)}$ to each other and violate the antiframing requirement. Thus, without a trusted ringleader, they must find a third colluder i to help them calculate $D_j(k, l)$. To prevent colluder i from accessing the fingerprinted coefficients in $\widetilde{\mathbf{X}}_j^{(k)}$ and $\widetilde{\mathbf{X}}_j^{(l)}$, colluders k and l should process their fingerprinted copies beforehand, and let i calculate $D_j(k, l)$ from the processed $\widetilde{\mathbf{X}}_j^{(k)}$ and $\widetilde{\mathbf{X}}_j^{(l)}$.

Define $f(\cdot)$ as the function that k and l use to process their copies, and let $\mathbf{Y}_j^{(k)}$ and $\mathbf{Y}_j^{(l)}$ be the processed copies of $\widetilde{\mathbf{X}}_j^{(k)}$ and $\widetilde{\mathbf{X}}_j^{(l)}$, respectively. To enable colluder i to calculate $D_j(k, l)$ from $\mathbf{Y}_j^{(k)}$ and $\mathbf{Y}_j^{(l)}$, it is required that $f(\cdot)$ does not change the MSE between these two copies and

$$\widetilde{D}_j(k, l) = ||\mathbf{Y}_j^{(k)} - \mathbf{Y}_j^{(l)}||^2 = ||\widetilde{\mathbf{X}}_j^{(k)} - \widetilde{\mathbf{X}}_j^{(l)}||^2 = D_j(k, l). \qquad (13.7)$$

In addition, it is required that given $\mathbf{Y}_j^{(k)}$ and $\mathbf{Y}_j^{(l)}$, i cannot estimate the fingerprinted coefficients in $\widetilde{\mathbf{X}}_j^{(k)}$ and $\widetilde{\mathbf{X}}_j^{(l)}$, respectively.

In this chapter, we use a simple component-wise addition-based method to process $\widetilde{\mathbf{X}}_j^{(k)}$ and $\widetilde{\mathbf{X}}_j^{(l)}$. (Other methods that protect the fingerprinted coefficients and satisfy (13.7), such as the isometry rotation and the permutation-complement–based encryption [145], can also be applied.) Assume that $\widetilde{\mathbf{X}}_j^{(k)}$ and $\widetilde{\mathbf{X}}_j^{(l)}$ are of length N_j. Given a key $K^{k,l}$ shared by colluders k and l only, they use $K^{k,l}$ as the seed of the pseudo random number generator and generate a random sequence $\mathbf{v}_j^{(k,l)}$ of length N_j. The N_j components in $\mathbf{v}_j^{(k,l)}$ are i.i.d. and uniformly distributed in $[-\mathcal{U}, \mathcal{U}]$. Then, colluders k and l add $\mathbf{v}_j^{(k,l)}$ to

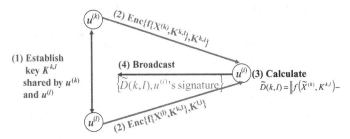

Fig. 13.5 Calculation of $D(k, l)$ without a trusted ringleader

their fingerprinted copies component by component, and calculate

$$\mathbf{Y}_j^{(k)} = f(\widetilde{\mathbf{X}}_j^{(k)}, K^{k,l}) = \widetilde{\mathbf{X}}_j^{(k)} + \mathbf{v}_j^{(k,l)}$$
$$\text{and} \quad \mathbf{Y}_j^{(l)} = f(\widetilde{\mathbf{X}}_j^{(l)}, K^{k,l}) = \widetilde{\mathbf{X}}_j^{(l)} + \mathbf{v}_j^{(k,l)}, \tag{13.8}$$

respectively. Thus, $||\mathbf{Y}_j^{(k)} - \mathbf{Y}_j^{(l)}||^2 = ||\widetilde{\mathbf{X}}_j^{(k)} + \mathbf{v}_j^{(k,l)} - \widetilde{\mathbf{X}}_j^{(l)} - \mathbf{v}_j^{(k,l)}||^2 = ||\widetilde{\mathbf{X}}_j^{(k)} - \widetilde{\mathbf{X}}_j^{(l)}||^2$, and (13.7) is satisfied. To hide information of the embedded fingerprints, colluders should select a large $\mathcal{U}$ and let the random sequence $\mathbf{v}_j^{(k,l)}$ have large amplitude.

Let $\text{Enc}(X, K)$ denote the encryption of message X with key K. To calculate $D_j(k, l)$, as shown in Figure 13.5:

- Colluders k and l first generate a secret key $K^{k,l}$. Then, colluder k generates a secret key $K^{k,i}$ shared with colluder i only, and $K^{l,i}$ is a key shared by colluder l and i.
- Colluder k first processes his or her fingerprinted copy $\widetilde{\mathbf{X}}_j^{(k)}$ using (13.8), then encrypts it with key $K^{k,i}$ to protect the fingerprinted coefficients in $\widetilde{\mathbf{X}}_j^{(k)}$. (Similar to the scenario with a trusted ringleader, encryption here is used only to secure communications between two parties and to prevent eavesdropping, and it will not affect the later steps in cheating colluder identification, as well as multiuser collusion.) Then, colluder k transmits the encrypted copy $\text{Enc}\left(f(\widetilde{\mathbf{X}}_j^{(k)}, K^{k,l}), K^{k,i}\right)$ to i. Colluder l repeats the same process.
- Colluder i calculates $\widetilde{D}_j(k, l) = ||f(\widetilde{\mathbf{X}}_j^{(k)}, K^{k,l}) - f(\widetilde{\mathbf{X}}_j^{(l)}, K^{k,l})||^2$, and broadcasts $\widetilde{D}_j(k, l)$ together with his/her digital signature.

13.3.1.2 Autonomous detection and identification of cheating colluders

To extend the preceding algorithm to a group of colluders, for each frame j in the video sequence:

- Colluders randomly divide themselves into two subgroups SC_1 and SC_2 where $SC_1 \cup SC_2 = SC$ and $SC_1 \cap SC_2 = \emptyset$.† Colluders in SC_1 randomly select an *assistant* $i_1 \in SC_1$ to help colluders in SC_2 calculate $\{D_j(k, l)\}_{k,l \in SC_2}$. Similarly, $i_2 \in SC_2$ is randomly selected to help colluders in SC_1 calculate $\{D_j(k, l)\}_{k,l \in SC_1}$.

---

† We use two subgroups as an example; the algorithm can be easily extended to scenarios with more than two subgroups.

- Assume that K^{SC_1} is a key shared by colluders in SC_1. Each colluder l in SC_1 generates a secret key K^{l,i_2} shared with the selected assistant $i_2 \in SC_2$. Then, colluder l uses (13.8) to process his or her fingerprinted copy $\widetilde{\mathbf{X}}_j^{(l)}$ and generates $f(\widetilde{\mathbf{X}}_j^{(l)}, K^{SC_1})$. Then, colluder l encrypts his or her copy with key K^{l,i_2} and transmits the encrypted version $\text{Enc}\left(f(\widetilde{\mathbf{X}}_j^{(l)}, K^{SC_1}), K^{l,i_2}\right)$ to the selected assistant i_2 in SC_2. Colluders in SC_2 follow the same procedure, process and encrypt their fingerprinted copies, and transmit them to the selected assistant i_1 in SC_1.
- After decrypting the bit streams received from all colluders in SC_1, for each pair of colluders (k, l) in subgroup SC_1, the selected assistant $i_2 \in SC_2$ calculates $\widetilde{D}_j(k, l) = \|f(\widetilde{\mathbf{X}}_j^{(k)}, K^{SC_1}) - f(\widetilde{\mathbf{X}}_j^{(l)}, K^{SC_1})\|^2$, and broadcasts $\{\widetilde{D}_j(k, l)\}_{k,l \in SC_1}$ to colluders in SC_1 together with his or her digital signature. Note that i_2 calculates $\widetilde{D}_j(k, l)$ only when both k and l are in subgroup SC_1. Assistant $i_1 \in SC_1$ repeats the same process to help colluders in SC_2 calculate $\{\widetilde{D}_j(k, l)\}$ for all $k, l \in SC_2$.
- Given $\{\widetilde{D}_j(k, l)\}_{k,l \in SC_1}$, colluders in SC_1 apply the same method as in Section 13.2.3.1 to detect and identify cheating colluders in SC_1. Similarly, colluders in SC_2 examine $\{\widetilde{D}_j(k, l)\}_{k,l \in SC_2}$ and identify selfish colluders in SC_2.

Finally, honest colluders combine the detection results from all frames in the sequence, and exclude identified cheating colluders from collusion.

13.3.2 Performance of the autonomous cheating colluder detection and identification scheme

In this section, we investigate how cheating colluders can *actively* attack our algorithm and manipulate the detection results to avoid being detected. We also provide techniques to ensure accurate identification of cheating colluders even under such attacks. Here, we consider the scenario in which there are only a limited number of cheating colluders. We assume that if honest colluders are selected as assistants to help calculate $\{\widetilde{D}_j\}$, they will give others correct values of $\{\widetilde{D}_j(k, l)\}$.

13.3.2.1 Group of cheating colluders

The performance of the cheating colluder detection and identification algorithm depends on the correctness of $\{\widetilde{D}_j(k, l)\}$. If all the selected assistants give the other colluders correct values of $\{\widetilde{D}_j(k, l)\}$, this autonomous cheating colluder detection and identification scheme has the same performance as that in Section 13.2.3.2, and honest colluders can correctly identify cheating colluders in SC_s without falsely accusing others. However, during the autonomous cheating colluder detection and identification process, it is possible that two or more cheating colluders collaborate with each other to change the detection results. Figure 13.6 shows an example.

In Figure 13.6, the simulation setup is the same as that in Figure 13.1(b). We assume that there are two cheating colluders i_1 and i_2, and they are in different subgroups during the autonomous cheating colluder detection and identification process. Without loss of generality, assume that $i_1 \in SC_1$ and $i_2 \in SC_2$. In SC_2, there are $K_2 = 75$ colluders and we assume that the other 74 colluders in SC_2 do not process their received copies.

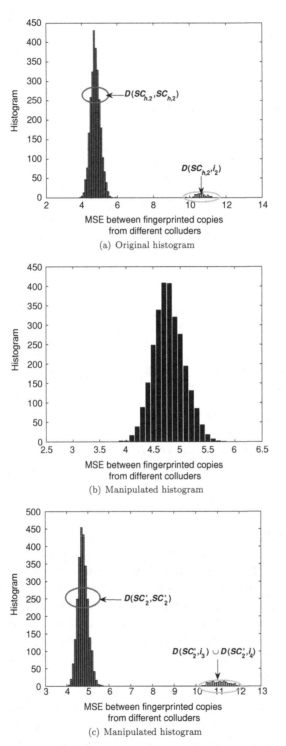

(a) Original histogram

(b) Manipulated histogram

(c) Manipulated histogram

Fig. 13.6 Histogram plots of $\{\widetilde{D}_j(k, l)\}_{k,l\in SC_2}$ for the second frame in sequence "carphone"

Figure 13.6(a) plots the unchanged histogram of $\{D_j(k, l)\}_{k,l \in SC_2}$, from which Algorithm 13.1 can correctly identify colluder i_2 as a cheating colluder. If the cheating colluder i_1 is selected as the assistant to help colluders in SC_2 calculate $\{\widetilde{D}_j(k, l)\}_{k,l \in SC_2}$, i_1 can modify the values of $\{\widetilde{D}_j(k, l)\}_{k,l \in SC_2}$ and let them be from the same distribution – for example, as shown in Figure 13.6(b). Then, Algorithm 13.1 cannot identify i_2 as a cheating colluder and it makes a miss-detection error. The cheating colluder i_1 can also change the values of $\{\widetilde{D}_j(k, l)\}_{k,l \in SC_2}$ and let the histogram be the same as in Figure 13.6(c), where $SC_2' = SC_2 \backslash \{i_3, i_4\}$. Here, Algorithm 13.1 not only misses the real cheating colluder i_2, but it also falsely accuses another two honest colluders, i_3 and i_4. Using the same method, the cheating colluder i_2 can also prevent colluders from detecting i_1's precollusion processing, or make them falsely accuse honest colluders as cheating.

13.3.2.2 Multiple assistants for each subgroup

To reduce the probability that these cheating colluders can successfully change the detection results, in each subgroup, a straightforward solution is to select multiple assistants to calculate $\{\widetilde{D}_j\}$ and use a majority vote when identifying selfish colluders.

For each frame j, as in Section 13.3.1.2, the colluders first randomly divide themselves into two nonoverlapping subgroups, SC_1 and SC_2. To detect and identify cheating colluders in SC_1:

- m colluders are randomly selected from SC_2 to help calculate $\{\widetilde{D}_j(k, l)\}_{k,l \in SC_1}$, and $\mathbf{A}_j(SC_2) = \{i_{2,1}, i_{2,2}, \ldots, i_{2,m}\}$ contains their indices.
- For each selected assistant $i_{2,n} \in \mathbf{A}_j(SC_2)$, colluders in SC_1 follow Step 2 in Section 13.3.1.2, process and encrypt their fingerprinted copies, and transmit them to $i_{2,n}$.
- Each selected assistant $i_{2,n}$ in $\mathbf{A}_j(SC_2)$ follows Step 3 in Section 13.3.1.2 to calculate $\widetilde{D}_j^{i_{2,n}}(k, l) = ||f(\widetilde{\mathbf{X}}_j^{(k)}, K^{SC_1}) - f(\widetilde{\mathbf{X}}_j^{(l)}, K^{SC_1})||^2$ for all $k, l \in SC_1$, and broadcasts the results to colluders in SC_1 together with $i_{2,n}$'s digital signature.
- For every honest colluder $k \in SC_1$, given $\{\widetilde{D}_j^{i_{2,n}}(k, l)\}_{k,l \in SC_1}$ received from the assistant $i_{2,n}$ in $\mathbf{A}_j(SC_2)$, colluder k follows Step 4 in Section 13.3.1.2, examines the histogram of $\{\widetilde{D}_j^{i_{2,n}}(k, l)\}_{k,l \in SC_1}$, and uses Algorithm 13.1 to detect and identify cheating colluders. For every $l \in SC_1$ and for each $i_{2,n} \in \mathbf{A}_j(SC_2)$, colluder k sets $\upsilon_j^{(k)}(n, l) = 1$ if Algorithm 13.1 identifies $l \in SC_1$ as a potential cheating colluder from $\{\widetilde{D}_j^{i_{2,n}}(k, l)\}_{k,l \in SC_1}$, and $\upsilon_j^{(k)}(n, l) = 0$ otherwise. Then colluder k combines the m detection results $\{\upsilon_j^{(k)}(n, l)\}_{n=1,\ldots,m}$ and uses majority vote to determine whether colluder l is a cheating colluder. If $\sum_{n=1}^{m} \upsilon_j^{(k)}(n, l) \geq \lceil m/2 \rceil$, colluder k believes that colluder l processed his or her copy before collusion and sets $\Upsilon_j^{(k)}(l) = 1$. $\Upsilon_j^{(k)}(l) = 0$ otherwise.

The same procedure is used to identify cheating colluders in SC_2.

To further improve the performance of the cheating colluder detection and identification algorithm, colluders in SC_h should jointly consider the detection results from all frames in the video sequence when making the final decision on the identities of the selfish colluders.

For each frame j in the video sequence, in Step 1 of the autonomous cheating colluder detection and identification, define

$$I_j(k, l) \triangleq \begin{cases} 1 & \text{if } k \text{ and } l \text{ are in the same subgroup,} \\ 0 & \text{otherwise.} \end{cases} \tag{13.9}$$

For every pair of colluders (k, l), we further define $F(k, l) \triangleq \{j : I_j(k, l) = 1\}$, which contains the indices of all frames in which colluder k and l are assigned to the same subgroup.

For an honest colluder $k \in SC_h$, to determine whether $\widetilde{\mathbf{X}}^{(l)}$ is the original copy that colluder l received from the content owner, k jointly considers all the detection results $\{\Upsilon_j^{(k)}(l)\}_{j \in F(k,l)}$ that he or she has, and considers colluder l as a cheating colluder if the average of $\{\Upsilon_j^{(k)}(l)\}_{j \in F(k,l)}$ is above a pre-determined threshold α. Colluder k then outputs the estimated cheating colluder set

$$\widehat{SC}_s^{(k)} = \left\{ l : \frac{\sum_{j \in F(k,l)} \Upsilon_j^{(k)}(l)}{|F(k, l)|} > \alpha \right\}. \tag{13.10}$$

A larger α helps lower the false-alarm rate at the cost of a higher miss-detection rate; the selection of the parameter α should address the tradeoff between the false-alarm and the miss-detection rates.

13.3.2.3 Performance analysis

In this section, to address the unique issues in autonomous cheating colluder identification, we investigate how such a group of cheating colluders can manipulate the detection results and how this affects the performance of the autonomous cheating colluder detection and identification scheme. For each frame in the video sequence, define P_{cs} as the probability that the group of cheating colluders can successfully manipulate the detection results and intentionally let others make errors when detecting cheating behavior. In this section, we first analyze P_{cs}, and we then study how it affects the detection error rates.

Terminology definition: Assume that there are a total of KC colluders. For each frame j, during the autonomous cheating colluder detection and identification process, assume that the number of colluders in subgroup SC_1 and SC_2 are KC_1 and KC_2, respectively, with $KC_1 + KC_2 = KC$. $\mathbf{A}_j(SC_1)$ is the set with the indices of the m assistants in SC_1 selected to help colluders in SC_2 calculate $\{\widetilde{D}_j(k, l)\}_{k,l \in SC_2}$, and $\mathbf{A}_j(SC_2)$ contains the indices of the m assistants selected to help colluders in SC_1 calculate $\{\widetilde{D}_j(k, l)\}_{k,l \in SC_1}$.

Let $\mathbf{C}_s$ denote the set with the indices of the cheating colluders who collaborate with each other to avoid being detected by their fellow colluders, and $KC_s = |\mathbf{C}_s|$ is its size.[†] Among the KC_s cheating colluders, $KC_s(SC_1) = |\mathbf{C}_s \cap SC_1|$ of them are in

---

† Note that SC_s contains all cheating colluders who apply precollusion processing to further lower their own risk of being detected by the fingerprint detector, whereas $\mathbf{C}_s$ includes those who work together during the cheating colluder identification process to avoid being detected by their fellow colluders. $\mathbf{C}_s \subseteq SC_s$.

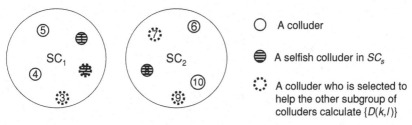

Fig. 13.7 Example to illustrate the terms defined in Section 13.3.2.3

subgroup SC_1, and the other $KC_s(SC_2) = |\mathbf{C}_s \cap SC_2|$ cheating colluders are in SC_2. $KC_s(SC_1) + KC_s(SC_2) = KC_s$ and $0 \leq KC_s(SC_1), KC_s(SC_2) \leq KC_s$. For frame j, we further define $KC_{as}(SC_1) = |\mathbf{C}_s \cap \mathbf{A}_j(SC_1)|$ as the number of cheating colluders in SC_1 who are selected as assistants to help calculate $\{\widetilde{D}_j(k, l)\}_{k,l \in SC_2}$; and $KC_{as}(SC_2) = |\mathbf{C}_s \cap \mathbf{A}_j(SC_2)|$ is the number of cheating colluders in SC_2 who are selected to help calculate $\{\widetilde{D}_j(k, l)\}_{k,l \in SC_1}$.

Figure 13.7 gives an example of the previously defined terms. In this example, there are 10 colluders and $SC = \{1, 2, \ldots, 10\}$. $SC_1 = \{1, 2, 3, 4, 5\}$ and $SC_2 = \{6, 7, 8, 9, 10\}$. $SC_s = \{1, 2, 8\}$ contains the indices of the cheating colluders and $KC_s = 3$. Among these three cheating colluders, colluders 1 and 2 are in SC_1 and colluder 8 is in SC_2. Therefore, $KC_s(SC_1) = 2$ and $KC_s(SC_2) = 1$. In SC_1, colluders 2 and 3 are selected as assistants to help SC_2 calculate $\{D_j(k, l)\}$ and $\mathbf{A}_j(SC_1) = \{2, 3\}$. $\mathbf{A}_j(SC_2) = \{7, 9\}$, and colluder 7 and 9 are selected to help SC_1 calculate $\{D_j\}$. In this example, $\mathbf{C}_s \cap \mathbf{A}_j(SC_1) = \{2\}$ and $KC_{as}(SC_1) = 1$. $\mathbf{C}_s \cap \mathbf{A}_j(SC_2) = \emptyset$ and $KC_{as}(SC_2) = 0$.

Analysis of P_{cs}: In this chapter, we consider the scenario in which $KC_s \ll KC_1$ and $KC_s \ll KC_2$. For subgroup SC_1, among the m selected assistants in $\mathbf{A}_j(SC_2)$, if more than half of them are from $\mathbf{C}_s$ (i.e., $KC_{as}(SC_2) \geq \lceil m/2 \rceil$), even if colluders in SC_1 apply a majority vote as in Section 13.3.2.2, the cheating colluders can still change the values of $\{\widetilde{D}_j(k, l)\}_{k,l \in SC_1}$ and successfully cause others to make detection errors when identifying cheating colluders using frame j. The same holds for subgroup SC_2. Therefore, for each frame in the video sequence, the cheating colluders can change the detection results if and only if either $KC_{as}(SC_1) \geq \lceil m/2 \rceil$ or $KC_{as}(SC_2) \geq \lceil m/2 \rceil$. Define $G_{SC_1}(p) \triangleq \{KC_s(SC_1) = p\}$ as the event that in subgroup SC_1, there are p cheating colluders from $\mathbf{C}_s$; that is, $KC_s(SC_1) = p$. We have

$$P_{cs} = \sum_{p=0}^{KC_s} P\left[\left(KC_{as}(SC_1) \geq \lceil \tfrac{m}{2} \rceil\right) \bigcup \left(KC_{as}(SC_2) \geq \lceil \tfrac{m}{2} \rceil\right) \mid G_{SC_1}(p)\right] P\left[G_{SC_1}(p)\right]$$

$$= \sum_{p=0}^{KC_s} \left\{ 1 - \left(\sum_{p_1=0}^{\min\{\lceil \frac{m}{2} \rceil - 1, p\}} P\left[KC_{as}(SC_1) = p_1 \mid G_{SC_1}(p)\right] \right) \right.$$

$$\left. \times \left(\sum_{p_2=0}^{\min\{\lceil \frac{m}{2} \rceil - 1, KC_s - p\}} P\left[KC_{as}(SC_2) = p_2 \mid G_{SC_1}(p)\right] \right) \right\} P\left[G_{SC_1}(p)\right]. \quad (13.11)$$

In (13.11), for $0 \leq p \leq KC_s$, $0 \leq p_1 \leq \min\{m, p\}$ and $0 \leq p_2 \leq \min\{m, KC_s - p\}$,

$$P\left[KC_{as}(SC_1) = p_1 \mid G_{SC_1}(p)\right] = \binom{p}{p_1}\binom{KC_1 - p}{m - p_1} \bigg/ \binom{KC_1}{m},$$

$$P\left[KC_{as}(SC_2) = p_2 \mid G_{SC_1}(p)\right] = \binom{KC_s - p}{p_2}\binom{KC_2 - (KC_s - p)}{m - p_2} \bigg/ \binom{KC_2}{m},$$

$$\text{and} \quad P\left[G_{SC_1}(p)\right] = \binom{KC_s}{p}\binom{KC - KC_s}{KC_1 - p} \bigg/ \binom{KC}{KC_1}. \tag{13.12}$$

Figure 13.8(a) and (b) plot the simulation results of P_{cs} with a total of $KC = 50$ and $KC = 150$ colluders, respectively. In our simulations, we let SC_1 and SC_2 be of the same size and $KC_1 = KC_2 = KC/2$. The results are based on 4000 simulation runs. From Figure 13.8, selecting multiple assistants in each subgroup significantly reduces P_{cs}. For example, when 10 percent of the colluders are cheating colluders in $\mathbf{C}_s$, choosing $m = 3$ assistants from each subgroup helps lower P_{cs} from 0.2 to 0.05 when compared with the scenario with $m = 1$. In addition, P_{cs} is larger when there are more cheating colluders in $\mathbf{C}_s$.

Simulation results of P_{fa} and P_{md}: The preceding analysis considers one frame in the video sequence. This section studies the performance of our algorithm when the detection results from all frames are considered jointly to identify cheating colluders.

We test on the first 300 frames of sequence "carphone"; our simulation setup is the same as that in Section 13.2.3.2. Human visual model-based spread spectrum embedding [58] is used to embed fingerprints into the host signal, and orthogonal fingerprints are assigned to different users. During precollusion processing, cheating colluders select λ_j such that the newly generated frames have a PSNR of 40 dB when compared with the originally received ones. Each cheating colluder processes his or her copy independently.

For each frame in the video sequence, each subgroup selects $m = 3$ assistants to help the other subgroup calculate $\{\widetilde{D}_j(k, l)\}$, and they apply a majority vote to identify the cheating colluders. We assume that if they are selected as assistants to help calculate $\{\widetilde{D}_j(k, l)\}$, honest colluders tell other colluders correct values of $\{\widetilde{D}_j(k, l)\}$. We further assume that $\mathbf{C}_s = SC_s$, and all cheating colluders who apply precollusion processing collaborate with one another to prevent being detected by other fellow colluders. If a cheating colluder i in subgroup SC_1 is selected to help colluders in SC_2 calculate $\{\widetilde{D}_j(k, l)\}_{k,l \in SC_2}$, we assume that colluder i changes the histogram of $\{\widetilde{D}_j(k, l)\}_{k,l \in SC_2}$ such that none of the cheating colluders in SC_2 can be detected. In addition, colluder i randomly selects an honest colluder $k \in SC_2$, and changes the values of $\{\widetilde{D}_j(k, l)\}_{k,l \in SC_2}$ so that Algorithm 13.1 falsely identifies another colluder k as cheating. This is similar to the situation in Figure 13.6(c). The same holds for cheating colluders in SC_2. The threshold α in (13.10) is set to 0.85.

Based on 4000 simulation runs, Figure 13.9(a) and (b) show the simulation results with $KC = 50$ and $KC = 150$ colluders, respectively. From Figure 13.9, if fewer than 15 percent of the colluders are cheating – that is, $KC_s/KC \leq 15\%$ – the autonomous

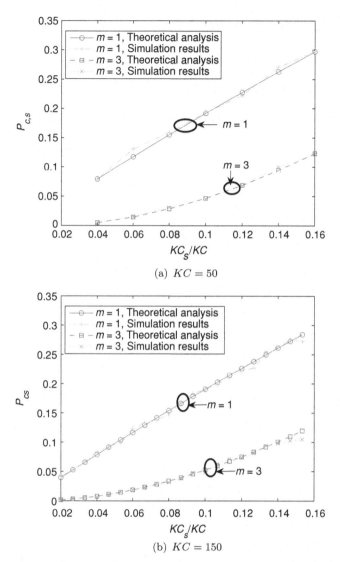

(a) $KC = 50$

(b) $KC = 150$

Fig. 13.8 The probability that a group of cheating colluders modifies the values of $\{\widetilde{D}_j(k, l)\}$ and colluders in SC_h make errors when detecting cheating behavior

cheating colluder identification algorithm can correctly identify all cheating colluders. When $KC_s/KC > 15\%$ and when P_{cs} is larger than $1 - \alpha$, P_{md} increases quickly as the total number of cheating colluders grows.

The false alarm probability (P_{fa}) depends on how cheating colluders change $\{\widetilde{D}_j\}$. In our simulations, when the cheating colluders are selected to help calculate $\{\widetilde{D}_j\}$, they randomly choose one honest colluder and accuse him or her of cheating. In all our 4000 simulation runs, as shown in Figure 13.9, our algorithm does not falsely accuse any honest colluders, even when there are a large number of cheating colluders who cooperate with one another to manipulate the detection results. This is because the

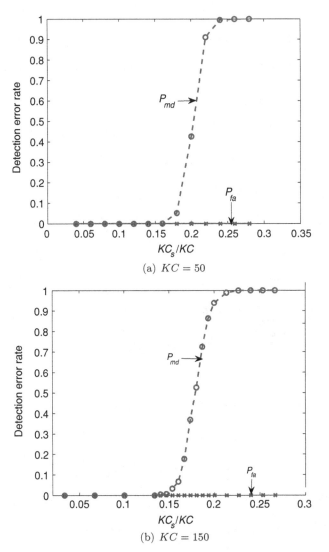

(a) $KC = 50$

(b) $KC = 150$

Fig. 13.9 Simulation results of P_{fa} and P_{md} on the first 300 frames of sequence "carphone"

majority vote and joint consideration of the detection results from all video frames help honest colluders easily correct this false alarm detection error. In another scenario, in which cheating colluders continuously compromise the same colluder in SC_h whenever possible, the false-alarm rate P_{fa} will be similar to the miss-detection rate P_{md}.

13.3.3   Resistance to framing attacks

In addition to actively manipulating the detection results, colluders can also *passively* attack the autonomous cheating colluder identification algorithm. The purpose of this passive attack is not to change the detection results, but to access fingerprinted coefficients in others' copies and frame other colluders. This section analyzes the resistance of the autonomous cheating detection and identification scheme to such framing attacks. We use the term *framing colluders* to denote colluders who try to access fingerprinted coefficients in others' copies and frame other colluders. Framing colluders can be cheating colluders who process their copies before collusion, and it is also possible that framing colluders honestly report their received fingerprinted copies but want to access fingerprinted coefficients in others' copies.

13.3.3.1 Group of framing colluders

During the autonomous cheating colluder detection and identification process in Section 13.3.1.2, every colluder processes and encrypts his or her fingerprinted copy using two different keys. Any other single colluder has at most one key. Therefore, it prohibits a single framing colluder from accessing others' copies, and $\gamma_j = 0$ with one single framing colluder. However, it is possible that a group of framing colluders works collaboratively to access others' fingerprinted copies. For example, in Figure 13.5, colluder l knows $K^{k,l}$ and colluder i has key $K^{k,i}$. If colluder l and i collaborate, they can decrypt $\text{Enc}\left\{ f\left(\widetilde{\mathbf{X}}_j^{(k)}, K^{k,l}\right), K^{k,i}\right\}$ and access the fingerprinted coefficients in $\widetilde{\mathbf{X}}_j^{(k)}$. Let $\mathbf{C}_f$ denote the set containing the indices of framing colluders working together to access others' copies. In this section, we consider the scenario in which there is only a limited number of framing colluders and the size of $\mathbf{C}_f$ is small.

13.3.3.2 Nonoverlapping content to each assistant

To lower γ_j and minimize the framing colluders' chance of successfully accessing others' copies, one possible solution is that for each selected assistant in SC_2, every colluder in SC_1 transmits only part of his or her fingerprinted frame, instead of the entire one. Thus, if only one of the framing colluders in $\mathbf{C}_f$ is selected to help calculate the MSE between different copies, they can decrypt only part of the fingerprinted copies, and decrypting the entire fingerprinted frames requires that multiple framing colluders are selected as assistants.

Assume that the jth fingerprinted frame from colluder i is

$$\widetilde{\mathbf{X}}_j^{(i)} = \left[\widetilde{X}_j^{(i)}(1), \widetilde{X}_j^{(i)}(2), \ldots, \widetilde{X}_j^{(i)}(N_j)\right]. \tag{13.13}$$

As in Section 13.3.2, for each frame j, the colluders first divide themselves into two nonoverlapping subgroups SC_1 and SC_2. Then, m colluders in SC_2 are selected as assistants, and $\mathbf{A}_j(SC_2) = \{i_{2,1}, \ldots, i_{2,m}\} \subset SC_2$ is the set containing their indices. Colluders in SC_1

• First randomly shuffle the vector $[1, 2, \ldots, N_j]$, and let

$$\mathbf{a1} = [a1(1), a1(2), \ldots, a1(N_j)] \tag{13.14}$$

be the returned shuffled vector. $a1(l) \in \{1, 2, \ldots, N_j\}$ for $l = 1, \ldots, N_j$, and $a1(l_1) \neq a1(l_2)$ if $l_1 \neq l_2$.

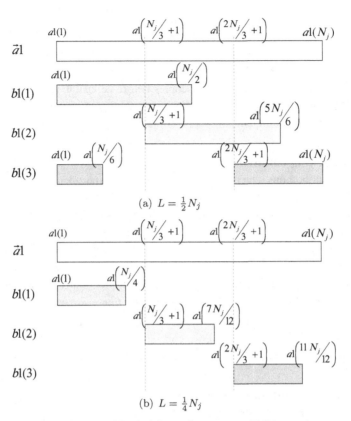

(a) $L = \frac{1}{2} N_j$

(b) $L = \frac{1}{4} N_j$

Fig. 13.10 Examples of $\{\mathbf{b}1(n)\}$ for each $i_{2,n} \in \mathrm{Ass}_j(SC_2)$, $n = 1, \ldots, m$

- For each $i_{2,n} \in \mathbf{A}_j(SC_2)$, let

$$\mathbf{b}1(n) \triangleq \left\{ a1\left(\mathrm{mod}(l, N_j)\right) : \frac{n-1}{m} N_j + 1 \leq l \leq \frac{n-1}{m} N_j + L \right\}, \qquad (13.15)$$

where $1 \leq L \leq N_j$. Figure 13.10 shows examples of $\{\mathbf{b}1(n)\}$ for each $i_{2,n} \in \mathbf{A}_j(SC_2)$ with $m = 3$. For $i_{2,p}, i_{2,q} \in \mathbf{A}_j(SC_2)$ where $p \neq q$, $\mathbf{b}1(p)$ and $\mathbf{b}1(q)$ are of the same length L.

- For each $i_{2,n} \in \mathbf{A}_j(SC_2)$, every colluder $k \in SC_1$ selects

$$\widetilde{\mathbf{X}}_{j,n}^{(k)} \triangleq \left\{ \widetilde{X}_j^{(k)}(l) : l \in \mathbf{b}1(n) \right\}, \qquad (13.16)$$

processes and encrypts $\widetilde{\mathbf{X}}_{j,n}^{(k)}$ in the same way as in Section 13.3.2.2, and then transmits it to assistant $i_{2,n}$.

Note that $L = N_j/m$ corresponds to a random partitioning. Colluders in SC_2 repeat the same process: generate a shuffled vector $\mathbf{a}2$, select $\mathbf{b}2(n)$ for each assistant $i_{1,n} \in \mathbf{A}_j(SC_1)$, and transmit the encrypted version of $\widetilde{\mathbf{X}}_{j,n}^{(k \in SC_2)} \triangleq \left\{ \widetilde{X}_j^{(k)}(l) : l \in \mathbf{b}2(n) \right\}$ to $i_{1,n}$. Finally, colluders follow the same procedure as in Section 13.3.2.2 to detect and identify cheating colluders.

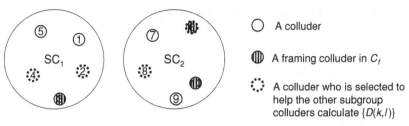

Fig. 13.11 An example to illustrate the terms defined in Section 13.3.3.3

13.3.3.3 Performance analysis

In this section, we first calculate γ_j defined in (13.4) for the autonomous cheating colluder detection and identification algorithm. We then quantify its robustness against framing attacks and evaluate the maximum number of framing colluders that the autonomous cheating colluder detection and identification algorithm can withstand.

Terminology definition: Assume that there are KC_1 and KC_2 colluders in subgroups SC_1 and SC_2, respectively. $KC_f = |\mathbf{C}_f|$ is the number of framing colluders. Among the KC_f framing colluders, $KC_f(SC_1) = |\mathbf{C}_f \cap SC_1|$ of them are in subgroup SC_1 and the other $KC_f(SC_2) = |\mathbf{C}_f \cap SC_2|$ are in SC_2. $KC_f(SC_1) + KC_f(SC_2) = KC_f$ and $0 \leq KC_f(SC_1), KC_f(SC_2) \leq KC_f$. We further define $KC_{af}(SC_1) \triangleq |\mathbf{C}_f \cap \mathbf{A}_j(SC_1)|$ as the number of framing colluders that are selected to help colluders in SC_2 calculate $\{\widetilde{D}_j(k, l)\}_{k,l \in SC_2}$, and $KC_{af}(SC_2) \triangleq |\mathbf{C}_f \cap \mathbf{A}_j(SC_2)|$ is the number of framing colluders that are selected to help colluders in SC_1 calculate $\{\widetilde{D}_j(k, l)\}_{k,l \in SC_1}$.

Figure 13.11 gives an example of the above defined terms. In this example, there are ten colluders and $SC = \{1, 2, \ldots, 10\}$. $SC_1 = \{1, 2, 3, 4, 5\}$ and $SC_2 = \{6, 7, 8, 9, 10\}$. $\mathbf{C}_f = \{3, 6, 10\}$ includes all the framing colluders and $KC_f = 3$. Among these three framing colluders, colluder 3 is in SC_1 and colluders 6 and 10 are in SC_2. $KC_f(SC_1) = 1$ and $KC_f(SC_2) = 2$. In SC_1, colluders 2 and 4 are selected to help SC_2 calculate $\{D_j(k, l)\}$ and $\mathbf{A}_j(SC_1) = \{2, 4\}$. $\mathbf{A}_j(SC_2) = \{6, 8\}$, and colluders 6 and 8 are selected as assistants. $\mathbf{C}_f \cap \mathbf{A}_j(SC_1) = \emptyset$ and $KC_{af}(SC_1) = 0$. $\mathbf{C}_f \cap \mathbf{A}_j(SC_2) = \{6\}$ and $KC_{af}(SC_2) = 1$.

Analysis of γ_j: In this section, we consider the scenario in which $KC_f \ll KC_1$ and $KC_f \ll KC_2$. We consider two scenarios: $L = N_j$, in which every colluder in SC_1 transmits his or her entire fingerprinted frame to all the selected assistants in $\mathbf{A}_j(SC_2)$, and $L < N_j$, in which every colluder in SC_1 gives only part of his or her copy to each assistant in $\mathbf{A}_j(SC_2)$.

(1) $\boldsymbol{L = N_j}$: In this scenario, for each frame j in the video sequence, an assistant in $\mathbf{A}_j(SC_1)$ receives the entire fingerprinted frame from each colluder in SC_2. If both SC_1 and SC_2 contain framing colluders in $\mathbf{C}_f$ (i.e., $KC_f(SC_1) > 0$ and $KC_f(SC_2) > 0$), and if at least one framing colluder is selected to help calculate $\{\widetilde{D}_j(k, l)\}$ (i.e., $K_{af}(SC_1) > 0$ or $KC_{af}(SC_2) > 0$), then the framing colluders are able to obtain both keys and access

others' fingerprinted coefficients. They can generate a new frame of high quality that does not contain any information on their own fingerprints. Recall that $\mathrm{Ind}_j^{(k,i)}$ is the set including all the indices of the fingerprinted coefficients in $\mathbf{X}_j^{(i)}$ that $u^{(k)}$ could access. $\mathrm{Ind}_j^{(k)} = \bigcup_{i \in SC, i \neq k} \mathrm{Ind}_j^{(k,i)}$, and $\mathrm{Ind}_j^{(\mathbf{C}_f)} = \bigcup_{k \in \mathbf{C}_f} \mathrm{Ind}_j^{(k)}$. Therefore,

$$
|\mathrm{Ind}_j^{(\mathbf{C}_f)}| = \begin{cases} N_j & \text{if } \left(0 < KC_f(SC_1), KC_f(SC_2) < K_f\right) \\ & \bigcap \left(\{KC_{af}(SC_1) \geq 1\} \bigcup \{KC_{af}(SC_2) \geq 1\}\right), \\ 0 & \text{otherwise.} \end{cases} \tag{13.17}
$$

For $0 \leq p \leq KC_f$, let $G_f(p) \triangleq \{KC_f(SC_1) = p\}$ denote the event that p of the framing colluders are in SC_1. For $0 \leq p \leq KC_f$, $0 \leq m_1 \leq \min\{m, p\}$ and $0 \leq m_2 \leq \min\{m, KC_f - p\}$, we have

$$
\gamma_j = \sum_{p=1}^{KC_f - 1} P\left[\left(KC_{af}(SC_1) \geq 1\right) \bigcup \left(KC_{af}(SC_2) \geq 1\right) \mid G_f(p)\right] \times P\left[G_f(p)\right]
$$

$$
= \sum_{p=1}^{KC_f - 1} \left\{1 - P\left[KC_{af}(SC_1) = 0 \mid G_f(p)\right] \cdot P\left[KC_{af}(SC_2) = 0 \mid G_f(p)\right]\right\}
$$

$$
\times P\left[G_f(p)\right], \tag{13.18}
$$

where

$$
P\left[KC_{af}(SC_1) = m_1 \mid G_f(p)\right] = \binom{KC_1 - p}{m - m_1}\binom{p}{m_1} \bigg/ \binom{KC_1}{m},
$$

$$
P\left[KC_{af}(SC_2) = m_2 \mid G_f(p)\right] = \binom{KC_2 - (KC_f - p)}{m - m_2}\binom{KC_f - p}{m_2} \bigg/ \binom{KC_2}{m},
$$

$$
\text{and } P\left[G_f(p)\right] = \binom{KC_f}{p}\binom{KC - KC_f}{KC_1 - p} \bigg/ \binom{KC}{KC_1}. \tag{13.19}
$$

(2) $L < N_j$: We use $L \leq N_j/m$ as an example to analyze the performance of the cheating colluder detection and identification algorithm. The analysis for $N_j/m < L < N_j$ is similar and is thus omitted here. From Figure 13.10(b), if $L \leq N_j/m$, $\mathbf{b1}(p) \bigcap \mathbf{b1}(q) = \emptyset$ for any $i_{2,p}, i_{2,q} \in \mathbf{A}_j(SC_2)$ where $p \neq q$. Similarly, $\mathbf{b2}(p) \bigcap \mathbf{b2}(q) = \emptyset$ for any $i_{1,p}, i_{1,q} \in \mathbf{A}_j(SC_1)$ where $p \neq q$.

For each frame j, among all the K_f framing colluders in $\mathbf{C}_f$, assume that $KC_{af}(SC_1) = m_1 \leq m$ of them are selected to help colluders in SC_2 calculate $\{\widetilde{D}_j(k, l)\}_{k,l \in SC_2}$, and $KC_{af}(SC_2) = m_2 \leq m$ of the framing colluders in $\mathbf{C}_f$ are selected to help colluders in SC_1 calculate $\widetilde{D}_j(k, l)\}_{k,l \in SC_1}$. Let $\mathbf{A}(m_1, m_2)$ denote the event that $KC_{af}(SC_1) = m_1, KC_{af}(SC_2) = m_2, 0 < K_f(SC_1), KC_f(SC_2) < KC_f$. By

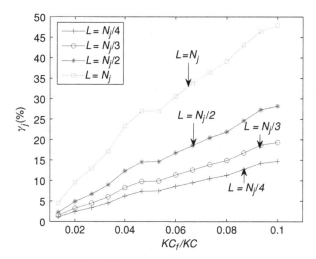

Fig. 13.12 γ_j when L takes different values. There are a total of $KC = 150$ colluders. $KC_1 = KC_2 = KC/2$, $m = 3$. Simulation results are based on 4000 simulation runs

combining all the decrypted fingerprinted coefficients that they have, we can show that

$$E\left[||\text{Ind}_j^{(C_f)}|| \, \mathbf{A}(m_1, m_2)\right] = \min\left\{m_1 L + m_2 L - m_1 m_2 \frac{L^2}{N_j}, N_j\right\}. \quad (13.20)$$

Therefore, we have

$$\gamma_j = \sum_{p=1}^{KC_f-1} \sum_{m_1=0}^{\min\{m,p\}} \sum_{m_2=0}^{\min\{m,KC_f-p\}} E\left[\frac{||\text{Ind}_j^{(C_f)}||}{N_j} \,\Big|\, \mathbf{A}(m_1, m_2), G_f(p)\right]$$

$$\times P\left[KC_{af}(SC_1) = m_1, KC_{af}(SC_2) = m_2, G_f(p)\right]$$

$$= \sum_{p=1}^{KC_f-1} \sum_{m_1=0}^{\min\{m,p\}} \sum_{m_2=0}^{\min\{m,KC_f-p\}} \min\left\{m_1 \frac{L}{N_j} + m_2 \frac{L}{N_j} - m_1 m_2 \left(\frac{L}{N_j}\right)^2, 1\right\}$$

$$\times P\left[KC_{af}(SC_1) = m_1 | G_f(p)\right]$$

$$\times P\left[KC_{af}(SC_2) = m_2 | G_f(p)\right] \times P\left[G_f(p)\right], \quad (13.21)$$

where $P\left[KC_{af}(SC_1) = m_1 | G_f(p)\right]$, $P\left[KC_{af}(SC_2) = m_2 | G_f(p)\right]$ and $P\left[G_f(p)\right]$ are the same as in (13.19).

Figure 13.12 shows the simulation results of γ_j when L takes different values. There are a total of $KC = 150$ colluders. $KC_1 = KC_2 = KC/2$ and $m = 3$. As seen in Figure 13.12, transmitting only part of the fingerprinted frames to each selected assistant can significantly reduce γ_j and help improve the robustness against framing attacks. For example, with $KC_f/KC = 0.1$, γ_j equals 50 percent when $L = N_j$ and is reduced to 15 percent if $L = N_j/4$. In addition, γ_j has a smaller value when there are fewer framing colluders in $\mathbf{C}_f$.

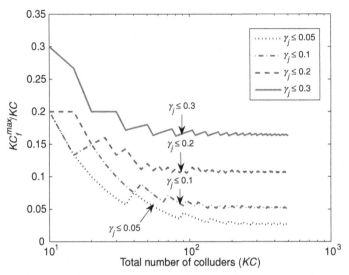

Fig. 13.13 Maximum number of framing colluders in $\mathbf{C}_f$ that the autonomous cheating colluder detection and identification process can resist

From Figure 13.12, γ_j has a smaller value when L decreases and, therefore, a smaller L is preferred to minimize γ_j and resist framing attacks. On the other hand, for each assistant $i_{2,n} \in \mathbf{A}_j(SC_2)$, $\{\widetilde{\mathbf{X}}_{j,n}^{(k)}\}_{k \in SC_1}$ must be long enough that given $\{\widetilde{d}_j^{(i_{2,n})}(k,l) \triangleq \|f(\widetilde{\mathbf{X}}_{j,n}^{(k)}, K^{SC_1}) - f(\widetilde{\mathbf{X}}_{j,n}^{(l)}, K^{SC_1})\|^2\}_{k,l \in SC_1}$ that are received from assistant $i_{2,n}$, Algorithm 13.1 can correctly detect and identify cheating colluders in SC_1. We use the second frame in the "carphone" sequence an example, and assume that 10 out 150 colluders are selfish and process their fingerprinted copies before collusion. They select the parameter λ_j to generate new frames with PSNR 45 dB. When $L = N_j$, $\mathfrak{D}_j(SC_h, SC_s)$ and $\mathfrak{D}_j(SC_h, SC_h)$ do not overlap, and Algorithm 1 can accurately identify all 10 selfish colluders. With $L = N_j/4$ and $L = N_j/8$, the overlap ratios c defined in (13.6) are 14.54 percent and 34.24 percent, respectively, and Algorithm 13.1 starts to make detection errors. Thus, a larger L should be used to ensure the performance of Algorithm 13.1. To address this tradeoff, for the example in Figure 13.12, $L \approx N_j/m$ with $m \approx 3$ is often preferred; that is, $L = N_j/3$ or $L = N_j/4$.

Resistance to framing attacks: In this section, we quantify the robustness of the cheating colluder identification algorithms against framing attacks. For any fingerprinted copy, given the requirement that framing colluders can access no more than θ percent of the fingerprinted coefficients (i.e., $\gamma \leq \theta$), we define $KC_f^{max} \triangleq \arg\max_{KC_f}\{\gamma_j \leq \theta\}$, which is the maximum number of framing colluders that it can resist.

Figure 13.13 plots the ratio KC_f^{max}/KC versus KC when θ takes different values. In Figure 13.13, the two subgroups SC_1 and SC_2 are of the same size, $KC/2$, and there are $m = 3$ assistants selected in each subgroup to help calculate $\{\widetilde{D}_j(k, l)\}$. $L = N_j/3$. From Figure 13.13, with hundreds of colluders, if no more than 5 percent of them are framing colluders, then others can be sure that the framing colluders can access no more than 10 percent of the fingerprinted coefficients in their copies. If KC_f does not exceed 10 percent of the total number of colluders, then the framing colluders can access fewer than 20 percent of the fingerprinted coefficients in others' copies.

13.3.4 Autonomous cheating colluder detection and identification process

To summarize, in peer-structured colluder social networks, the key steps in the autonomous selfish colluder detection and identification process are as follows. For each frame j in the video sequence:

- **Step 1. Grouping:** Colluders randomly divide themselves into two subgroups, SC_1 and SC_2, with $SC_1 \cup SC_2 = SC$ and $SC_1 \cap SC_2 = \emptyset$. m colluders in SC_1, $\mathbf{A}_j(SC_1) = \{i_{1,1}, \ldots, i_{1,m}\}$, are randomly selected to calculate $\{\widetilde{D}_j(k, l)\}_{k,l \in SC_2}$ for colluders in SC_2. Similarly, colluders in SC_2 randomly select m assistants $\mathbf{A}_j(SC_2) = \{i_{2,1}, \ldots, i_{2,m}\}$ to help colluders in SC_1 calculate $\{\widetilde{D}_j(k, l)\}_{k,l \in SC_1}$.
- **Step 2. Encryption:** Assume that K^{SC_1} is a key that is shared by colluders in SC_1. For each selected assistant $i_{2,n} \in \mathbf{A}_j(SC_2)$, every colluder $k \in SC_1$ generates a secret key $K^{k,i_{2,n}}$ shared with the assistant $i_{2,n} \in \mathbf{A}_j(SC_2)$. For each $i_{2,n} \in \mathbf{A}_j(SC_2)$, colluders in SC_1 follow the same procedure as in Section 13.3.3.2 to generate $\mathbf{b}1(n)$. Then, every colluder k in SC_1 selects $\widetilde{\mathbf{X}}_{j,n}^{(k)} = \{\widetilde{X}_j^{(k)}(l) : l \in \mathbf{b}1(n)\}$, processes and encrypts it with keys K^{SC_1} and $K^{k,i_{2,n}}$, respectively, in the same way as in Section 13.3.1.2. Finally, colluder k transmits the encrypted $\widetilde{\mathbf{X}}_{j,n}^{(k)}$ to assistant $i_{2,n}$. Colluders in SC_2 follow the same procedure, process and encrypt their fingerprinted copies, and transmit them to the corresponding assistants in $\mathbf{A}_j(SC_1)$.
- **Step 3. Calculation of $\{D_j\}$:** After decrypting the bit streams received from all colluders in SC_1, each selected assistant $i_{2,n} \in \mathbf{A}_j(SC_2)$ follows the same procedure in Section 13.3.1.1 to calculate

$$\widetilde{d}_j^{i_{2,n}}(k, l) = ||f(\widetilde{\mathbf{X}}_{j,n}^{(k)}, K^{SC_1}) - f(\widetilde{\mathbf{X}}_{j,n}^{(l)}, K^{SC_1})||^2 \tag{13.22}$$

for all $k, l \in SC_1$, and broadcasts the results to colluders in SC_1 together with his or her digital signature. Each selected assistant $i_{1,n}$ in $\mathbf{A}_j(SC_1)$ repeats the same process to help colluders in SC_2 calculate $\left\{\widetilde{d}_j^{i_{1,n}}(k, l)\right\}_{k,l \in SC_2}$.
- **Step 4. Cheating colluder detection and identification:** For every honest colluder $k \in SC_1$, given $\left\{\widetilde{d}_j^{i_{2,n}}(k, l)\right\}_{k,l \in SC_1}$ received from the selected assistant $i_{2,n} \in \mathbf{A}_j(SC_2)$, colluder k follows Step 4 in Section 13.3.1.2 and sets $v_j^{(k)}(n, l) = 1$ if Algorithm 13.1 outputs $l \in SC_1$ as a cheating colluder. $v_j^{(k)}(n, l) = 0$ otherwise. Then, for every colluder $l \neq k$ in SC_1, k combines the m detection results $\{v_j^{(k)}(n, l)\}_{n=1,\ldots,m}$, considers

colluder l as a potential cheating colluder, and sets $\Upsilon_j^{(k)}(l) = 1$ if $\sum_{n=1}^m v_j^{(k)}(n,l) > \lceil m/2 \rceil$. $\Upsilon_j^{(k)}(l) = 0$ otherwise.

Finally, each honest colluder $k \in SC_h$ combines the detection results from all frames in the video sequence and outputs the estimated cheating colluder set $\widehat{SC}_s^{(k)} = \left\{ l : \frac{\sum_{j \in F(k,l)} \Upsilon_j^{(k)}(l)}{|F(k,l)|} > \alpha \right\}$ where α is a predetermined threshold. Then, colluders in SC_h exclude those identified cheating colluders from collusion.

From the analysis in Section 13.3.2.3, the preceding autonomous cheating colluder detection and identification process can accurately identify cheating colluders without falsely accusing others when there are a limited number of cheating colluders. From Section 13.3.3.3, this algorithm also helps resist framing attacks and prevent colluders from accessing the fingerprinted coefficients in others' copies when the number of framing colluders is small.

13.4   Chapter summary and bibliographical notes

This chapter analyzes the impact of network structures on misbehavior detection and identification in colluder social networks. We also evaluate the performance of our algorithms and analyze their resistance against framing attacks.

We first consider the centralized colluder social networks in which there exists a ringleader whom all colluders can trust, and we design an algorithm by which the trusted ringleader helps to detect and identify cheating colluders. The trusted ringleader calculates the difference between fingerprinted copies from different colluders, and the colluders analyze the histogram of this difference to detect precollusion processing and identify cheating colluders. We show that our cheat-proof scheme can accurately cheating colluders without falsely accusing others even if $\mathcal{D}_j(SC_h, SC_s)$ and $\mathcal{D}_j(SC_h, SC_h)$ overlap. The cheating user detection algorithm also protects the fingerprinted coefficients in all copies and prevents colluders from framing one another.

We then consider the peer structure in which there is not a trusted ringleader, and we design an autonomous algorithm by which colluders help each other detect precollusion processing and identify cheating colluders. In this scenario, all the fingerprinted copies must be processed and encrypted appropriately during cheating colluder detection to prevent framing attacks. From our analytical and simulation results, when detecting cheating behavior, the detection algorithm can accurately identify cheating colluders even if a small group of cheating colluders collaborate with one another to change the detection results. We also evaluate its antiframing performance, and quantify the maximum number of framing colluders that it can resist. Our results show that framing colluders can access no more than 10 percent of the fingerprinted coefficients in others' copies, if the number of framing colluders does not exceed 5 percent of the total number of colluders.

Interested readers can refer to recent studies on the cheating behavior in different social network structures. The free-riding behavior in P2P file sharing was studied by Locher and colleagues and by Bougman *et al.* [146,147] who analyzed cheating in P2P gaming. Cheating behavior in visual cryptography for secret sharing was investigated by Horng and co-workers [148].

14 Structuring cooperation for hybrid peer-to-peer streaming

In the previous chapter, using colluder social networks in multimedia fingerprinting as an example, we showed that the network structure can affect misbehavior detection and and the overall system performance. In this chapter, we investigate the impact of network structure on the optimal cooperation strategies in hybrid P2P streaming networks, in which some users with very high interconnection bandwidth act jointly as one user to interact with the rest of the peers.

Although P2P video streaming systems have achieved promising results, they also introduce a large number of unnecessary traverse links, which consequently leads to substantial network inefficiency. However, in reality, every peer can have a large number of geographically neighboring peers with large intragroup upload and download bandwidth, such as peers in the same lab, building, or campus. If these peers have special collaboration among themselves and work jointly as one user toward the rest of the network, the unnecessary traverse links can be reduced. In this chapter, we denote those geographically neighboring peers with large intragroup upload and download bandwidths as *group peers*. To reduce the unnecessary traverse links and improve network efficiency, instead of considering each peer's strategy independently, we investigate possible cooperation among group peers and study their optimal collaboration strategy.

Because of the heterogeneous network structures, different group peers might take different roles during cooperation. In this chapter, we investigate the optimal cooperation strategies under different compositions of group peers, including the scenarios with or without central authorities, and examine whether peers can have different amounts of resources.

The structure of the social network plays a key role during the cooperation among the group peers. For example, because peers are selfish, they tend to act as free riders to improve their own utilities. If a central authority is presented, the equilibrium cooperation strategy can be enforced. Nevertheless, if there is no such an authority, the peers may take out-of-equilibrium strategies owing to the uncertainty of other peers' strategies. Under such a circumstance, a robust Nash equilibrium solution is desired for every peer. To solve such a problem, we formulate the cooperative streaming problem as an evolutionary game and derive the evolutionarily stable strategy (ESS) for every peer, which is the desired stable Nash equilibrium.

Furthermore, the central authority can calculate the optimal cooperation strategy for every group peer. But in a fully distributed network structure, the cooperative streaming

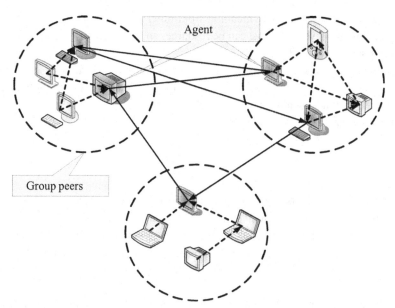

Fig. 14.1 Example of group peers and the P2P streaming network

scheme should be distributed. Therefore, we also provide a distributed algorithm for every peer to achieve the ESS by learning from his or her own past payoff history.

14.1 System model and utility function

14.1.1 System model

Figure 14.1 shows an example of the P2P streaming networks with group peers. There is a set of *group peers* (three in this example) who want to view a real-time video streaming simultaneously.[†] We assume that the upload and download bandwidth among members in a group is large. The peers in the same group will cooperate with one another and choose *k* representative peers, called *agents*, to download video streams from agents in other groups. Then, agents will distribute the video streams to the other peers in the group. To achieve good streaming performance through cooperation, two questions need to be addressed: given a group of peers, how many agents should be chosen, and which peers should be chosen as agents.

14.1.2 Utility functions

In a P2P network, a peer not only acts as a client to download video data from other peers, but also acts as a server to upload video data to other peers. Therefore, although

---

[†] How to group peers itself is an interesting problem. However, in this chapter, we assume that peers have already been grouped and focus mainly on how group peers cooperate with one another to achieve better streaming performance.

a peer can benefit from downloading from other peers, he or she also incurs a cost in uploading data, when the cost can be resources spent to upload data, such as bandwidth, buffer size, and the like.

Given peers $u_1, u_2, \ldots, u_N$ in one group, we assume that they cooperate with each other and choose k agents to download multimedia data from peers outside the group. Suppose that the download rates of the k agents are $r_1, r_2, \ldots, r_k$, then the total download rate of the *group peers* is given by

$$y_k = \sum_{i=1}^{k} r_i. \tag{14.1}$$

Because the agents randomly and independently select peers outside the group to download data, the download rates r_i are random variables. According to [149], the cumulative distribution function (CDF) of a peer's download bandwidth can be modelled as a linear function, which means that the probability density Function (PDF) of a peer's download bandwidth can be modeled as a uniform distribution; that is, r_is are uniformly distributed.

Obviously, if the total download rate y_k is not smaller than the source rate r, then the group peers can achieve real-time streaming, and all group peers can obtain a certain gain G. Otherwise, there will be some delay, and in this case we assume that the gain is zero. Therefore, given the total download rate y_k and the source rate r, if peer u_i chooses to be an agent, then the utility function of u_i is given by

$$U_{A,i}(k) = P(y_k \geq r)G - C_i, \forall k \in [1, N], \tag{14.2}$$

where C_i is the cost of u_i when he or she serves as an agent, and $P(y_k \geq r)$ is the probability of achieving real-time streaming.

Theorem 14.1. If $r_1, r_2, \ldots, r_k$ are i.i.d. uniformly distributed in $[r^L, r^U]$, then $P(y_k \geq r)$ is given by

$$P(y_k \geq r) = \frac{1}{2k!} \sum_{l=0}^{k} (-1)^l \binom{k}{l} \left[(k-l)^k - \mathrm{sgn}(\hat{r} - l)(\hat{r} - l)^k \right], \tag{14.3}$$

where $\hat{r} = \frac{r - r^L}{r^U - r^L}$. When k is sufficiently large, $P(y_k \geq r)$ can be approximated as

$$P(y_k \geq r) \approx Q\left(\frac{\hat{r} - \frac{k}{2}}{\sqrt{\frac{k}{12}}} \right), \tag{14.4}$$

where $Q(x)$ is the Gaussian tail function $\int_x^{\infty} \frac{1}{\sqrt{2\pi}} \exp^{-t^2/2} dt$.

Proof. Let $\hat{r}_l = \frac{r_l - r^L}{r^U - r^L}$ for an agent l. Since $\hat{r}_1, \hat{r}_2, \ldots, \hat{r}_k$ are i.i.d. uniformly distributed in $[0, 1]$, the characteristic function of $\hat{r}_l$ is

$$\phi(t) = \frac{j(1 - e^{jt})}{t}, \tag{14.5}$$

where j is the imaginary unit. Let $\hat{y}_k = \sum_{l=1}^{k} \hat{r}_l$, and its characteristic function is

$$\phi_{\hat{y}_k}(t) = \left(\frac{j(1 - e^{jt})}{t} \right)^k. \tag{14.6}$$

Therefore, the probability density function of $\hat{y}_k$ is

$$f_{\hat{y}_k}(y) = \mathcal{F}_t^{-1} \left[\left(\frac{i(1 - e^{it})}{t} \right)^k \right](y)$$

$$= \frac{1}{2(k-1)!} \sum_{l=0}^{k} (-1)^l \binom{k}{l} \operatorname{sgn}(y - l)(y - l)^{k-1}. \tag{14.7}$$

Because $P(y_k \geq r) = P(\hat{y}_k \geq \hat{r})$, according to (14.7), we have

$$P(y_k \geq r) = P(\hat{y}_k \geq \hat{r}) = \int_{\hat{r}}^{\infty} f_{\hat{y}_k}(y) dy$$

$$= \frac{1}{2k!} \sum_{l=0}^{k} (-1)^l \binom{k}{l} \left[(k - l)^k - \operatorname{sgn}(\hat{r} - l)(\hat{r} - l)^k \right], \tag{14.8}$$

which proves (14.3).

When k is sufficiently large, according to the central limit theorem, the distribution of $\hat{y}_k$ can be approximated as Gaussian distribution $\mathcal{N}(\frac{k}{2}, \frac{k}{12})$. Therefore, we have

$$P(y_k \geq r) = P(\hat{y}_k \geq \hat{r}) \approx Q \left(\frac{\hat{r} - \frac{k}{2}}{\sqrt{\frac{k}{12}}} \right), \tag{14.9}$$

which proves (14.4).

Because the upload bandwidth and the download bandwidth in the group are large, the cost to update data to other peers in the group can be neglected. In such a case, if peer u_i chooses not to be an agent, then there is no cost for u_i and the utility function becomes

$$U_{N,i}(k) = \begin{cases} P(y_k \geq r)G, & \text{if } k \in [1, N-1]; \\ 0, & \text{if } k = 0. \end{cases} \tag{14.10}$$

14.2   Agent selection within a homogeneous group

In the previous section, we discussed the system model and the peer's utility function. To optimize the streaming performance, proper peers should be selected as agents to download data from the peers outside the group. In this section, we discuss how to select agents within a homogeneous group when the cost of all peers serving as an agent is assumed to be the same. We will first discuss the scenario in which there is a central authority and then investigate the optimal agent selection in a fully distributed peer group.

14.2.1   Centralized agent selection

If there is a central controller who can choose which peers should act as agents, then a straightforward criterion for selecting proper agents is to maximize the social welfare, which is the sum of all peers' utilities.

Let $C_i = C$ be the cost of a peer serving as an agent in a homogeneous group. Then the social welfare of an N-peer group with k agents can be calculated by

$$SW(k) = P(y_k \geq r)GN - kC. \tag{14.11}$$

Based on (14.11), the agent selection problem to maximize the social welfare can be formulated as

$$\max_k SW(k) = P(y_k \geq r)GN - kC, \tag{14.12}$$

where $k \in \{1, 2, \ldots, N\}$.

By solving (14.12), we can find the optimal $k^\star$ that maximizes the social welfare. Then, the central controller can choose $k^\star$ peers from the group as agents to download data from the peers outside the group based on some mechanism, such as peers taking turns to serve as agents. However, because peers' behaviors are highly dynamic, they may join or leave the P2P network at any time. In such a case, the centralized approach may not be practical.

14.2.2   Distributed agent selection

To overcome the drawback of the centralized approach, we consider a distributed approach in which each peer acts as an agent with probability x. Then, according to (14.2) and (14.10), the group's social welfare can be computed by

$$U_{total}(x) = \sum_{i=1}^{N} \binom{N}{i} x^i (1-x)^{N-i} \left[P(y_i \geq r)GN - iC \right]. \tag{14.13}$$

Then the problem of finding an optimal x to maximize the social welfare can be formulated as

$$\max_x \sum_{i=1}^{N} \binom{N}{i} x^i (1-x)^{N-i} \left[P(y_i \geq r)GN - iC \right]$$

$$s.t. \quad 0 \leq x \leq 1. \tag{14.14}$$

However, because peers are selfish by nature, they are not as cooperative as the system designer/controller desires. By solving (14.14), we can find the optimal $x^\star$ that maximizes the social welfare, but $x^\star$ can not maximize each peer's own utility. Therefore, the social welfare maximizer $x^\star$ is not attainable when peers are selfish. Moreover, the solution to the optimization problem shown in (14.14) is not stable, as any perturbation will lead to a new solution.

14.2.3   Evolutionary cooperative streaming game

To provide a robust equilibrium strategy for the selfish peers, we adopt the concept of ESS [150,151], which is defined as follows:

Definition 14.1. A strategy $a^\star$ is an ESS if and only if $\forall a \neq a^\star$, $a^\star$ satisfies

- Equilibrium condition: $U_i(a, a^\star) \leq U_i(a^\star, a^\star)$, and
- stability condition: if $U_i(a, a^\star) = U_i(a^\star, a^\star)$, $U_i(a, a) < U_i(a^\star, a)$,

where $U_i(a_1, a_2)$ is player i's utility when he or she uses strategy a_1 and another player uses strategy a_2.

Because all peers are selfish, they will cheat if cheating can improve their payoffs, which means that all peers are uncertain of other peers' actions and utilities. In such a case, to improve their utilities, peers will try different strategies in every play and learn from the strategic interactions using the methodology of understanding by building. During this process, the percentage of peers using a certain pure strategy may change. Such a population evolution can be modeled by replicator dynamics. Specifically, let x_a stand for the probability that a peer uses the pure strategy $a \in \mathcal{A}$, where $\mathcal{A} = \{A, N\}$ is the set of pure strategies, including being an agent (A) and not being an agent (N). By replicator dynamics, the evolution dynamics of x_a are given by the following differential equation:

$$\dot{x}_a = \eta[\bar{U}(a, x_{-a}) - \bar{U}(x_a)]x_a, \tag{14.15}$$

where $\bar{U}(a, x_{-a})$ is the peers' average payoff using pure strategy a, x_{-a} is the set of peers who use pure strategies other than a, $\bar{U}(x_a)$ is the average payoff of all peers, and η is a positive scale factor.

From (14.15), we can see that if adopting pure strategy a can lead to a higher payoff than the average level, the probability of a peer using a will grow and the growth rate $\dot{x}_a/x_a$ is proportional to the difference between the average payoff of using strategy a ($\bar{U}(a, x_{-a})$) and the average payoff of all peers ($\bar{U}(x_a)$).

14.2.4   Analysis of the cooperative streaming game

According to (14.2) and (14.10), if a peer chooses to be an agent, his or her average payoff can be computed by

$$\bar{U}_A(x) = \sum_{i=0}^{N-1} \binom{N-1}{i} x^i (1-x)^{N-1-i} \left[P(y_{i+1} \geq r)G - C \right], \tag{14.16}$$

where x is the probability of a peer being an agent, and $\binom{N-1}{i} x^i (1-x)^{N-1-i}$ is the probability that among the remaining $N-1$ peers in the group, i of them choose to be agents.

Similarly, the average payoff of a peer if he or she chooses not to be an agent is given by

$$\bar{U}_N(x) = \sum_{i=1}^{N-1} \binom{N-1}{i} x^i (1-x)^{N-1-i} P(y_i \geq r)G. \tag{14.17}$$

According to (14.16) and (14.17), the average payoff of a peer is

$$\bar{U}(x) = x\bar{U}_A(x) + (1-x)\bar{U}_N(x). \tag{14.18}$$

Substituting (14.18) back to (14.15), we have

$$\dot{x} = \eta x(1-x)[\bar{U}_A(x) - \bar{U}_N(x)]. \tag{14.19}$$

At equilibrium $x^\star$, no player will deviate from the optimal strategy, which means $\dot{x}^\star = 0$, and we can obtain $x^\star = 0$, 1, or the solutions to $\bar{U}_A(x) = \bar{U}_N(x)$. However, because $\dot{x}^\star = 0$ is only the necessary condition for $x^\star$ to be an ESS, we examine the sufficient conditions for each ESS candidate and draw the following conclusions with the proofs shown in **Theorems 14.2 through 14.4**.

- $x^\star = 0$ is an ESS only when $P(y_1 \geq r)G - C \leq 0$.
- $x^\star = 1$ is an ESS only when $P(y_N \geq r)G - P(y_{N-1} \geq r)G \geq C$.
- Let $x^\star$ be the solution to $\bar{U}_A(x) = \bar{U}_N(x)$, and $x^\star \in (0, 1)$. Then, $x^\star$ is an ESS.

Lemma 14.1. Let $f(x) = \bar{U}_A(x) - \bar{U}_N(x)$; then $f'(x) < 0$ for all $x \in [0, 1]$.

Proof. From (14.16) and (14.17), we have

$$f(x) = \sum_{i=0}^{N-1} \binom{N-1}{i} x^i (1-x)^{N-1-i} w_i - C, \tag{14.20}$$

where $w_i = [P(y_{i+1} \geq r) - P(y_i \geq r)]G$.

To prove Lemma 14.1, we first calculate the first-order derivative of $f(x)$ with respect to x, which is

$$f'(x) = \sum_{i=0}^{N-1} \binom{N-1}{i} x^{i-1}(1-x)^{N-2-i}[i - (N-1)x]w_i,$$

$$= \sum_{i=0}^{i_1} \binom{N-1}{i} x^{i-1}(1-x)^{N-2-i}[i - (N-1)x]w_i$$

$$+ \sum_{i=i_1+1}^{N-1} \binom{N-1}{i} x^{i-1}(1-x)^{N-2-i}[i - (N-1)x]w_i, \tag{14.21}$$

where i_1 is the integer satisfies $i_1 \leq (N-1)x$ and $i_1 + 1 > (N-1)x$.

Note that w_i stands for the additional gain by introducing one more agent into the i-agent system, as shown in Figure 14.2; it is a decreasing function of i. That is, $w_i \geq w_{i_1}$

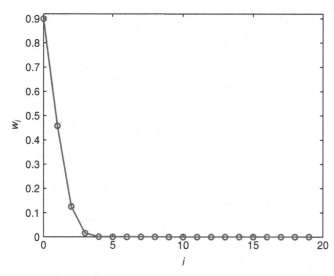

Fig. 14.2 The deceasing property of w_i

for all $i \leq i_1$ and $w_i \leq w_{i_1}$ for all $i > i_1$. Therefore, from (14.21), we have

$$f'(x) < \sum_{i=0}^{i_1} \binom{N-1}{i} x^{i-1}(1-x)^{N-2-i}[i-(N-1)x]w_{i_1}$$

$$+ \sum_{i=i_1+1}^{N-1} \binom{N-1}{i} x^{i-1}(1-x)^{N-2-i}[i-(N-1)x]w_{i_1},$$

$$= w_{i_1} \sum_{i=0}^{N-1} \binom{N-1}{i} x^{i-1}(1-x)^{N-2-i}[i-(N-1)x],$$

$$= w_{i_1} \frac{d\left[\sum_{i=0}^{N-1} \binom{N-1}{i} x^i (1-x)^{N-1-i}\right]}{dx} = 0. \tag{14.22}$$

Therefore, $f'(x) < 0$ for all $x \in (0, 1)$.

Second, we find the first-order derivative of $f(x)$ at $x = 0$, which is

$$f'(0) = \lim_{\varepsilon \to 0} \frac{f(\varepsilon) - f(0)}{\varepsilon}$$

$$= \lim_{\varepsilon \to 0} \frac{\sum_{i=0}^{N-1} \binom{N-1}{i} \varepsilon^i (1-\varepsilon)^{N-1-i} w_i - w_0}{\varepsilon}$$

$$= \lim_{\varepsilon \to 0} \frac{(1-\varepsilon)^{N-1} w_0 - w_0}{\varepsilon} + \lim_{\varepsilon \to 0} \frac{(N-1)\varepsilon(1-\varepsilon)^{N-2} w_1}{\varepsilon}$$

$$= (N-1)(w_1 - w_0) < 0, \tag{14.23}$$

where the last inequality comes from the fact that w_i is a decreasing function in terms of i.

Similarly, the first-order derivative of $f(x)$ at $x = 1$ can be computed by

$$f'(1) = \lim_{\varepsilon \to 0} \frac{f(1) - f(1 - \varepsilon)}{\varepsilon}$$

$$= \lim_{\varepsilon \to 0} \frac{w_{N-1} - \sum_{i=0}^{N-1} \binom{N-1}{i}(1 - \varepsilon)^i \varepsilon^{N-1-i} w_i}{\varepsilon}$$

$$= \lim_{\varepsilon \to 0} \frac{w_{N-1} - (1 - \varepsilon)^{N-1} w_{N-1}}{\varepsilon} + \lim_{\varepsilon \to 0} \frac{-(N-1)(1 - \varepsilon)^{N-2} \varepsilon w_{N-2}}{\varepsilon}$$

$$= (N - 1)(w_{N-1} - w_{N-2}) < 0, \tag{14.24}$$

where the last inequality comes from the fact that w_i is a decreasing function of i.

To summarize, $f'(x) < 0$ for all $x \in [0, 1]$. This completes the proof of the lemma.

Theorem 14.2. The necessary condition for $x^* = 0$ to be an ESS is $P(y_1 \geq r)G - C \leq 0$.

Proof. According to (14.16)–(14.18), the utility that a peer using mixed strategy x and the other peers use mixed strategy $x^* = 0$ can be written as

$$\bar{U}(x, 0) = \bar{U}_N(0) + (\bar{U}_A(0) - \bar{U}_N(0))x,$$

where $\bar{U}_A(0) = P(y_1 \geq r)G - C$ and $\bar{U}_N(0) = 0$.

- If $P(y_1 \geq r)G - C > 0$, that is, $\bar{U}_A(0) > \bar{U}_N(0)$, every peer will deviate to $x = 1$ to obtain $\bar{U}_A(0)$ rather than $\bar{U}_N(0)$.
- If $P(y_1 \geq r)G - C < 0$, that is, $\bar{U}_A(0) < \bar{U}_N(0)$, every peer will stay at $x = 0$ to obtain $\bar{U}_N(0)$ rather than $\bar{U}_A(0)$.
- If $P(y_1 \geq r)G - C = 0$, that is, $\bar{U}_A(0) = \bar{U}_N(0)$, then $\bar{U}(x, 0) = 0 \ \forall x$, and $f(0) = \bar{U}_A(0) - \bar{U}_N(0) = 0$. According to **Lemma 14.1**, we know that $f'(x) < 0 \ \forall x \in [0, 1]$, so $f(x) = \bar{U}_A(x) - \bar{U}_N(x) < 0$. In such a case, $\bar{U}(0, x) = \bar{U}_N(x) > \bar{U}(x, x) = \bar{U}_N(x) + (\bar{U}_A(x) - \bar{U}_N(x))x$, which means $x^* = 0$ is an ESS according to **Definition 14.1**.

Therefore, $x^* = 0$ is an ESS only when $P(y_1 \geq r)G - C \leq 0$.

Theorem 14.3. The necessary condition for $x^* = 1$ to be an ESS is $P(y_N \geq r)G - P(y_{N-1} \geq r)G \geq C$.

Proof. According to (14.16)–(14.18), the utility that a peer uses mixed strategy x and the other peers use strategy $x^* = 1$ can be written as

$$\bar{U}(x, 1) = \bar{U}_N(1) + (\bar{U}_A(1) - \bar{U}_N(1))x,$$

where $\bar{U}_A(1) = P(y_N \geq r)G - C$ and $\bar{U}_N(1) = P(y_{N-1} \geq r)G$.

- If $P(y_N \geq r)G - P(y_{N-1} \geq r)G < C$, i.e., $\bar{U}_N(1) > \bar{U}_A(1)$, every peer will deviate to $x = 0$ to obtain $\bar{U}_N(1)$ rather than $\bar{U}_A(1)$.
- If $P(y_N \geq r)G - P(y_{N-1} \geq r)G > C$, i.e., $\bar{U}_N(1) < \bar{U}_A(1)$, every peer will stay at $x = 1$ to obtain $\bar{U}_A(1)$ rather than $\bar{U}_N(1)$.

- If $P(y_N \geq r)G - P(y_{N-1} \geq r)G = C$, i.e. $\bar{U}_N(1) = \bar{U}_A(1)$, then $\bar{U}(x, 1) = \bar{U}_N(1)$ $\forall x$, and $f(1) = \bar{U}_A(1) - \bar{U}_N(1) = 0$. According to **Lemma 14.1**, we know that $f'(x) < 0$ $\forall x \in [0, 1]$, so $f(x) = \bar{U}_A(x) - \bar{U}_N(x) > 0$. In such a case, $\bar{U}(1, x) = \bar{U}_N(x) + (\bar{U}_A(x) - \bar{U}_N(x))1 > \bar{U}(x, x) = \bar{U}_N(x) + (\bar{U}_A(x) - \bar{U}_N(x))x$, which means $x^\star = 1$ is an ESS according to **Definition 14.1**.

Therefore, $x^\star = 1$ is an ESS only when $P(y_N \geq r)G - P(y_{N-1} \geq r)G \geq C$.

Theorem 14.4. If $x^\star \in (0, 1)$ is a solution to $\bar{U}_A(x) = \bar{U}_N(x)$, then $x^\star$ is an ESS.

Proof. Let $\bar{U}_i(x, x^\star)$ be the utility of player i when player i uses mixed strategy x and other users use mixed strategy $x^\star$. Then, we have

$$\bar{U}_i(x, x^\star) = x\bar{U}_A(x^\star) + (1 - x)\bar{U}_N(x^\star). \tag{14.25}$$

Because $x^\star$ is a solution to $\bar{U}_A(x) = \bar{U}_N(x)$, we have $\bar{U}_A(x^\star) = \bar{U}_N(x^\star)$. Therefore, (14.25) becomes

$$\bar{U}_i(x, x^\star) = \bar{U}_A(x^\star) = \bar{U}_i(x^\star, x^\star), \tag{14.26}$$

which means $x^\star$ satisfies the equilibrium condition shown in **Definition 14.1**.
Moreover, according to (14.18), we have

$$\bar{U}_i(x, x) = \bar{U}_N(x) + (\bar{U}_A(x) - \bar{U}_N(x))x, \tag{14.27}$$

and

$$\bar{U}_i(x^\star, x) = \bar{U}_N(x) + (\bar{U}_A(x) - \bar{U}_N(x))x^\star. \tag{14.28}$$

Therefore, we have

$$\bar{U}_i(x^\star, x) - \bar{U}_i(x, x) = (\bar{U}_A(x) - \bar{U}_N(x))(x^\star - x). \tag{14.29}$$

From **Lemma 14.1**, we know that $f(x) = \bar{U}_A(x) - \bar{U}_N(x)$ is a monotonically decreasing function. Because $\bar{U}_A(x^\star) = \bar{U}_N(x^\star)$, $\bar{U}_A(x) - \bar{U}_N(x) > 0$ if $x < x^\star$, and $\bar{U}_A(x) - \bar{U}_N(x) < 0$ if $x > x^\star$. Therefore, $(\bar{U}_A(x) - \bar{U}_N(x))(x^\star - x) > 0$, $\forall x \neq x^\star$; that is,

$$\bar{U}_i(x^\star, x) > \bar{U}_i(x, x), \forall x \neq x^\star, \tag{14.30}$$

which means $x^\star$ satisfies the stability condition shown in **Definition 14.1**.
According to (14.26) and (14.30), we know that $x^\star$ is an ESS.

14.3 Agent selection within a heterogeneous group

In this section, we discuss how to select agents within a heterogeneous group in which the cost functions of the peers acting as agents are different.

Let x_{i,a_i} stand for the probability of peer u_i using pure strategy $a_i \in \mathcal{A}$. By replicator dynamics, the evolution dynamics of x_{i,a_i} are given by the following differential equation:

$$\dot{x}_{i,a_i} = \eta[\bar{U}_i(a_i, x_{-i}) - \bar{U}_i(x_i)]x_{i,a_i}, \tag{14.31}$$

Table 14.1 Utility table of a two-player game

	A	N
A	$(B_2 - C_1, B_2 - C_2)$	$(B_1 - C_1, B_1)$
N	$(B_1, B_1 - C_2)$	$(0, 0)$

where $\bar{U}_i(a_i, x_{-i})$ is the average payoff of peer u_i using pure strategy a_i, $\bar{U}_i(x_i)$ is the average payoff of peer u_i using mixed strategy x_i, and η is a positive scale factor.

Because it is generally very difficult to represent $\bar{U}_i(a_i, x_{-i})$ and $\bar{U}_i(x_i)$ in a compact form, in the following, we first analyze a two-player game to gain some insight. Then, we generalize the observation in the two-player game to the multiplayer game.

14.3.1   Two-player game

Let x_1 and x_2 be the probabilities of u_1 and u_2 being an agent, respectively. Let $B_1 = P(y_1 \geq r)G$ and $B_2 = P(y_2 \geq r)G$. The payoff matrix of u_1 and u_2 can then, be written as in Table 14.1.

Therefore, the average payoff $\bar{U}_1(A, x_2)$ can be computed by

$$\bar{U}_1(A, x_2) = (B_2 - C_1)x_2 + (B_1 - C_1)(1 - x_2), \tag{14.32}$$

and the average payoff $\bar{U}_1(x_1)$ becomes

$$\bar{U}_1(x_1) = (B_2 - C_1)x_1 x_2 + (B_1 - C_1)x_1(1 - x_2) + B_1(1 - x_1)x_2. \tag{14.33}$$

According to Cressman (14.31), the replicator dynamics equation of u_1 is given by

$$\dot{x}_1 = \eta x_1(1 - x_1)[B_1 - C_1 - (2B_1 - B_2)x_2]. \tag{14.34}$$

Similarly, the replicator dynamics equation of u_2 can be computed by

$$\dot{x}_2 = \eta x_2(1 - x_2)[B_1 - C_2 - (2B_1 - B_2)x_1]. \tag{14.35}$$

At equilibrium, we know that $\dot{x}_1 = 0$ and $\dot{x}_2 = 0$. According to (14.34) and (14.35), we can get five equilibria: $(0, 0)$, $(0, 1)$, $(1, 0)$, $(1, 1)$, and the mixed strategy equilibrium $\left(\frac{B_1 - C_2}{2B_1 - B_2}, \frac{B_1 - C_1}{2B_1 - B_2}\right)$.

According to Cressman [152], if an equilibrium of the replicator dynamics equations is a locally asymptotically stable point in a dynamic system, it is an ESS. Therefore, by viewing (14.34) and (14.35) as a nonlinear dynamic system and analyzing the corresponding Jacobian matrix, we can examine whether the five equilibria are ESSs. By taking partial derivatives of (14.34) and (14.35), the Jacobian matrix can be written as

$$J = \begin{bmatrix} \frac{\partial \dot{x}_1}{\partial x_1} & \frac{\partial \dot{x}_1}{\partial x_2} \\ \frac{\partial \dot{x}_2}{\partial x_1} & \frac{\partial \dot{x}_2}{\partial x_2} \end{bmatrix} = \begin{bmatrix} J_{11} & J_{12} \\ J_{21} & J_{22} \end{bmatrix}, \tag{14.36}$$

where $J_{11} = (1 - 2x_1)(B_1 - C_1 - (2B_1 - B_2)x_2)$, $J_{12} = x_1(1 - x_1)(2B_1 - B_2)$, $J_{21} = x_2(1 - x_2)(2B_1 - B_2)$, and $J_{22} = (1 - 2x_2)(B_1 - C_2 - (2B_1 - B_2)x_1)$.

The asymptotic stability requires that $\det(J) > 0$ and $trace(J) < 0$ [152]. Substituting the five equilibria, $(0, 0)$, $(0, 1)$, $(1, 0)$, $(1, 1)$, and $\left(\frac{B_1-C_2}{2B_1-B_2}, \frac{B_1-C_1}{2B_1-B_2}\right)$, into (14.36), we conclude that

- If $B_2 - B_1 - C_1 > 0$ and $B_2 - B_1 - C_2 > 0$, there is a unique ESS $(1, 1)$, where both u_1 and u_2 converge to be agents.
- If $B_2 - B_1 - C_1 > 0$ and $B_2 - B_1 - C_2 < 0$, there is a unique ESS $(1, 0)$, where u_1 converges to be an agent and u_2 converges to be a free rider.
- If $B_2 - B_1 - C_1 < 0$ and $B_2 - B_1 - C_2 > 0$, there is a unique ESS $(0, 1)$, where u_2 converges to be an agent and u_1 converges to be a free rider.
- Otherwise, there are two ESSs $(0, 1)$ and $(1, 0)$, where the converged strategy profiles depend on the initial strategy profiles.

From this analysis, we can see that when the gain of being an agent $(B_2 - B_1)$ is greater than the cost of being an agent (C_1 or C_2), the peer tends to be an agent. Also, the peer with a higher cost tends to be a free rider and relies on the other peer with a lower cost.

14.3.2   Multiplayer game

From the analysis of the two-player game, we can infer that a peer with a higher cost (C_i) tends to take advantage of another peer with a lower cost. This observation can be extended to a multiplayer game. If there are more than two peers in the game, the strategy of peers with higher $C_i's$ will converge to N with a larger probability. The peers with smaller $C_i's$ tend to be agents, as they suffer relatively larger losses if no one serves as an agent.

Algorithm 14.1: Distributed learning algorithm for ESS

---

1. Given the step size η and the slot index $t = 0$, each peer u_i initializes x_i with $x_i(0)$.
2. During slot t, for $q = 1 : M$,

   - u_i tosses a coin with probability $x_i(t)$ being head. If the outcome is head, u_i serves as an agent and downloads data from the peers outside the group with download rate $r_i(t, q)$. On the other hand, if the outcome is tail, u_i acts as a free rider and downloads the data from the agents.
   - u_i computes his or her utility using (14.39).
   - u_i computes the indicator function using (14.38).

3. Then u_i approximates $\bar{U}_i(A, x_{-i}(t))$ and $\bar{U}_i(x_i(t))$ using (14.40) and (14.41).
4. Finally, u_i updates the probability of being an agent $x_i(t)$ using (14.37).

14.4   Distributed learning algorithm for ESS

From the previous two sections, we can see that the ESS can be found by solving the replicator dynamics equations (14.19) or (14.31). However, solving the replicator dynamics equations requires the exchange of private information and strategies adopted by other peers. In this section, we present a distributed learning algorithm that can gradually converge to ESS without information exchange.

We first discretize the replicator dynamics equation shown in (14.31) as

$$x_i(t+1) = x_i(t) + \eta \left[\bar{U}_i(A, x_{-i}(t)) - \bar{U}_i(x_i(t)) \right] x_i(t), \qquad (14.37)$$

where t is the slot index and $x_i(t)$ is the probability of u_i being an agent during slot t. Here, we assume that each slot can be further divided into M subslots and each peer can choose to be an agent or not at the beginning of each subslot.

From (14.37), we can see that to update $x_i(t+1)$, we need to first compute $\bar{U}_i(A, x_{-i}(t))$ and $\bar{U}_i(x_i(t))$. Let us define an indicator function $I_i(t, k)$ as

$$I_i(t, q) = \begin{cases} 1, & \text{if } u_i \text{ is an agent at subslot } q \text{ in slot } t, \\ 0, & \text{else}, \end{cases} \qquad (14.38)$$

where q is the subslot index.

The immediate utility of u_i at subslot q in slot t can be computed by

$$U_i(t, q) = \begin{cases} G - C_i, & \text{if } u_i \text{ is an agent and } r^t \geq r, \\ -C_i, & \text{if } u_i \text{ is an agent and } r^t < r, \\ G, & \text{if } u_i \text{ is not an agent and } r^t \geq r, \\ 0, & \text{if } u_i \text{ is not an agent and } r^t < r, \end{cases} \qquad (14.39)$$

where r^t is the total download rate of the agents and r is the source rate.

Then, $\bar{U}_i(A, x_{-i}(t))$ can be approximated using

$$\bar{U}_i(A, x_{-i}(t)) = \frac{\sum_{q=1}^{M} U_i(t, q) I_i(t, q)}{\sum_{q=1}^{M} I_i(t, q)}. \qquad (14.40)$$

Similarly, $\bar{U}_i(x_i(t))$ can be approximated as

$$\bar{U}_i(x_i(t)) = \frac{1}{M} \sum_{q=1}^{M} U_i(t, q). \qquad (14.41)$$

Based on (14.37)–(14.41), u_i can gradually learn the ESS. In Algorithm 14.1, we summarize the detailed procedures of the distributed learning algorithm.

14.5   Simulation results

In all simulations, the parameters G, r^L, and r^U are set to be 1, 50, and 800, respectively. For convenience, in the rest of this chapter, we denote the centralized approach maximizing the social welfare shown in (14.12) as **MSW-C**, the distributed approach maximizing

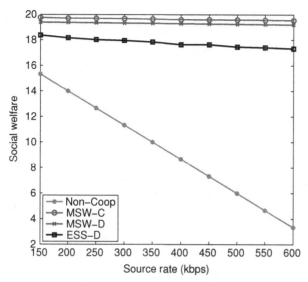

Fig. 14.3 The social welfare comparison among **Non-Coop, MSW-C, MSW-D**, and **ESS-D**

the social welfare shown in (14.14) as **MSW-D**, and the ESS-based approach as **ESS-D**. We compare the hybrid P2P structure with the traditional P2P noncooperation method, which is denoted as **Non-Coop**.

In the first simulation, we show the social welfare (the sum of all peers' utilities) comparison among different approaches, in which we assume that there are 20 homogenous peers and the cost C is 0.1. As shown in Figure 14.3, **MSW-C** achieves the best social welfare performance because its objective function is to maximize the social welfare with pure strategy. By using the mixed strategy to maximize the social welfare, **MSW-D** achieves the second-best social welfare performance. However, as discussed in Section 14.2.2, the solution to **MSW-D** is not stable. With **ESS-D**, a stable Nash equilibrium solution can be obtained at the cost of a slight loss in social welfare. Nevertheless, all three algorithms perform much better than the **Non-Coop** method. In **Non-Coop**, the social welfare performance decreases linearly in terms of the source rate. With cooperation and adaptively selecting the proper number of agents, all three algorithms can preserve a high social welfare performance even with a large source rate.

In the second simulation, we evaluate the convergence property of the **ESS-D**. In Figure 14.4, we show the replicator dynamic of the cooperation streaming game with homogeneous peers, where $C = 0.1$ and $r = 500$. We can see that starting from a high initial value, all peers gradually reduce their probabilities of being an agent, as being a free rider more often can bring a higher payoff. However, because too low a probability of being an agent increases the chance of having no peer be an agent, the probability of being an agent will finally converge to a certain value that is determined by the number of peers.

In Figure 14.5, we show the replicator dynamic of the cooperation streaming game with 20 heterogeneous peers, where $r = 500$ and the cost C_i is randomly chosen from

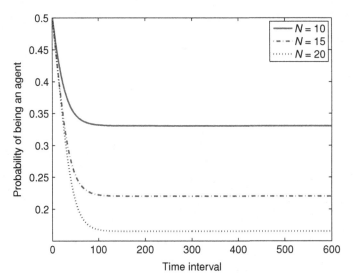

Fig. 14.4 Behavior dynamic of a homogeneous group of peers

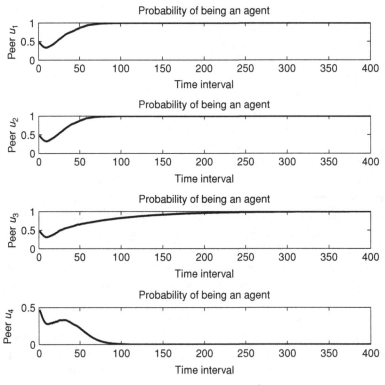

Fig. 14.5 Behavior dynamic of a heterogeneous group of peers

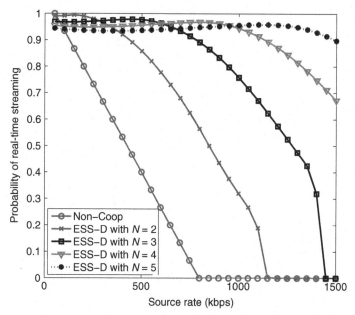

Fig. 14.6 Probability of real-time streaming comparison between **Non-Coop** and **ESS-D**

[0.1, 0.3]. We further assume that C_i is monotonically increasing in i where u_1 has the lowest cost and u_{20} has the highest cost. From Figure 14.5, we can see that the peers with lower costs (u_1, u_2, and u_3 in this simulation) converge to be agents, whereas the peers with higher costs (u_4 to u_{20} in this simulation) converge to be free riders. This observation coincides with our conclusion in Section 14.3.2, which is that *the peers with lower costs tend to be agents because they suffer relatively higher losses if no one serves as an agent*. Because of space limitations, we show only the behavior dynamics of u_1 through u_4. All other peers, u_5 through u_{20}, have similar behavior dynamics to u_4, and they all converge to be free riders.

In the third simulation, we compare the probability of real-time streaming performance between **Non-Coop** and **ESS-D**. The simulation results are shown in Figure 14.6. We can see that with cooperation, the probability of real-time streaming can be significantly improved, especially at the high source rate region. We also find that at the high source rate region, the probability of real-time streaming increases as N increases. To give more insight into the algorithms, in this chapter we assume that there is no buffering effect. However, the analysis and conclusion can be extended to the case in which a buffering effect is considered.

We then show the simulation result of the source rate versus the utility. As shown in Figure 14.7, without cooperation, if the peer requires a utility around 0.8, the source rate cannot be larger than 130 kbps. However, with cooperation, the source rate can be more than 400 kbps even when there are only two peers. Therefore, with cooperation, the peers can enjoy much higher quality video with the same utility.

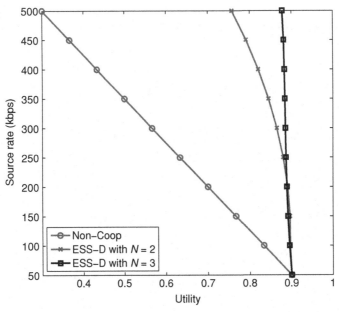

Fig. 14.7 Single-source rate comparison between **Non-Coop** and **ESS-D**

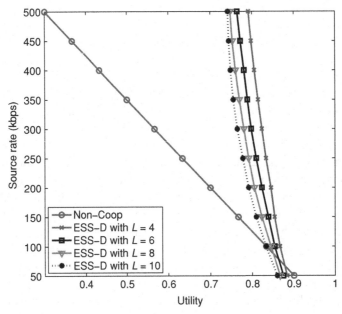

Fig. 14.8 Multisource rate comparison between **Non-Coop** and **ESS-D**

In the last simulation, we consider the case in which the peers in the same group are viewing multiple channels, with L being the number of channels. We assume that the source rate is the same for all channels and there are twenty homogenous peers with the cost $C = 0.1$. Similar to the view-upload decoupling (VUD) scheme [153], the uploading and downloading are decoupled in the **ESS-D** algorithm in this case. We allow cooperation among all the peers, wherein the agent may download source data that he or she is not viewing. As shown in Figure 14.8, without cooperation, if the peer requires a utility around 0.8, the source rate can not be larger than 130 kbps in the **Non-Coop** method. However, with the **ESS-D** algorithm, the source rate can be around 240 kbps even when the peers are viewing eight different channels. This phenomenon fully demonstrates the efficiency of the method discussed in this chapter.

14.6   Chapter summary and bibliographical notes

In this chapter, we study how the hybrid P2P cooperative streaming scheme changes with different network structures. We analyze the optimal cooperation strategies among selfish peers who have large intragroup upload and download bandwidths among themselves. We formulate the problem as an evolutionary game and derive the ESS for every peer. To combat the difficulties in a fully distributed network structure, a distributed learning algorithm for each peer to converge to the ESS by learning from his or her own past payoff history is also studied. From the simulation results, we can see that compared with the traditional noncooperative P2P schemes, the cooperation in the hybrid P2P network achieves much better social welfare, higher probability of real-time streaming, and better video quality (higher source rate). Moreover, with the recent VUD scheme, the cooperative streaming scheme studied in this chapter also allows the peers who are viewing different videos to cooperate with one another and mutually improve the streaming performance.

There has been research on video streaming over the Internet, such as the client-server service model [154,155], in which the video is streamed directly from a server to clients. However, with the client–server service model, the upload bandwidth of the server grows proportionally with the number of clients [156], which makes large-scale video streaming impractical. How to overcome the drawbacks of P2P networks, such as the unnecessary traverse links, has already drawn attentions of the research community. Interested readers can see references [157,158] for the usage of locality-aware P2P schemes to reduce the unnecessary traverse links within and across ISPs and thus reduce the download time. Purandare and Guha [159] studied an alliance-based peering scheme to reduce the playback time lag and improve the quality of service. The P4P architecture, which allows cooperative traffic control between applications and network providers, was introduced by Xie and Yang [160].

References

[1] D. A. Williamson. Social network ad spending: A brighter outlook next year. *eMarketer*, July 2009; summary at http://www.emarketer.com/Report.aspx?code=emarketer_2000592.

[2] D. A. Williamson. Social network ad spending: 2010 outlook. *eMarketer*, December 2009; summary at http://www.emarketer.com/Report.aspx?code=emarketer_2000621.

[3] Report: Social network ad revenue to double in 2010, reaching revenues of $421M. *Mobile-MarketingWatch*, March 25, 2010.

[4] R. Kwok. Phoning in data. *Nature*, **458**:959–961, April 2009.

[5] J. G. Breslin, S. Decker, M. Hauswirth, *et al.* Integrating social networks and sensor networks. *W3C Workshop on the Future of Social Networking*, January 2009.

[6] K. Mayer-Patel. Systems challenges of media collectives supporting media collectives with adaptive MDC. *Proceedings of the 15th International Conference on Multimedia* pp. 625–630, 2007.

[7] J. Liang, R. Kumar, Y. Xi, and K. W. Ross. Pollution in P2P file sharing systems. *IEEE INFOCOM*, **2**:1174–1185, March 2005.

[8] D. Rosenblum. What anyone can know: the privacy risks of social networking sites. *IEEE Security & Privacy*, **5**(3):40–49, May–June 2007.

[9] M. A. Moreno, N. C. Fost, and D. A. Christakis. Research ethics in the MySpace era. *Pediatrics*, **121**:157–161, 2008.

[10] A. Fallon. Libel: Ex-friend's Facebook revenge costs £22,000 in damages at high court. *The Guardian, http://www.guardian.co.uk/uk/2008/Jul/25/law.facebook*, July 25, 2008.

[11] M. Hines. Scammers gaming YouTube ratings for profit. *InfoWorld*, http://www.infoworld.com/d/security-central/scammers-gaming-youtube-ratings-profit-139, May 16, 2007.

[12] E. Mills. Study: eBay sellers gaming the reputation system? *CNET News*, http://news.cnet.com/8301-10784_3-6149491-7.html, Jan. 11, 2007.

[13] X. Hei, C. Liang, J. Liang, Y. Liu, and K. Ross. A measurement study of a large-scale P2P IPTV system. *IEEE Trans. Multimedia*, **9**(8):1672–1687, December 2007.

[14] B. Wellman and S. D. Berkowitz. *Social Structures: A Network Approach*. Cambridge University Press, 1988.

[15] D. Knoke and S. Yang. *Social Network Analysis*, 2nd ed. SAGE, 2008.

[16] S. Wasserman and K. Faust. *Social Network Analysis: Methods and Applications*. Cambridge University Press, 1994.

[17] R. Dunbar and M. Spoor. Social networks, support cliques and kinship. *Human Nature*, **6**:273–290, 1995.

[18] P. Blau, D. Ruan, and M. Ardelt. Interpersonal choice and networks in China. *Social Forces*, **69**:1037–1062, 1991.

[19] E. D. Kolaczyk. *Statistical Analysis of Network Data*. Springer, 2009.

[20] E. D. Kolaczyk. Statistical network analysis: models, issues and new directions: a workshop at the 23rd International Conference on Machine Learning. *Lecture Notes in Computer Science series*. Springer, 2006.

[21] S. Asur, S. Parthasarathy, and D. Ucar. An event-based framework for characterizing the evolutionary behavior of interaction graphs. *ACM SIGKDD International Conference on Knowledge Discovery and Data Mining*, pp. 913–921, 2007.

[22] R. Kumar, J. Novak, and A. Tomkins. Structure and evolution of online social networks. *ACM SIGKDD International Conference on Knowledge Discovery and Data Mining*, pp. 611–617, 2006.

[23] J. Sun, C. Faloutsos, S. Papadimitriou, and P. S. Yu. GraphScope: parameter-free mining of large time-evolving graphs. *ACM SIGKDD International Conference on Knowledge Discovery and Data Mining*, pp. 687–696, 2007.

[24] M. Spiliopoulou, I. Ntoutsi, Y. Theodoridis, and R. Schult. MONIC: modeling and monitoring cluster transitions. *ACM SIGKDD International Conference on Knowledge Discovery and Data Mining*, pp. 706–711, 2006.

[25] Y.-R. Lin, Y. Chi, S. Zhu, H. Sundaram, and B. L. Tseng. Analyzing communities and their evolutions in dynamic social networks. *ACM Trans. Knowledge Discovery from Data*, 3(2): 8:1–8:31, April 2009.

[26] K. Yang, M. Wang, and H. J. Zhang. Active tagging for image indexing. *IEEE International Conference on Multimedia and Expo (ICME)*, pp. 1620–1623, 2009.

[27] X. S. Hua and G. J. Qi. Online multi-label active annotation: toward large-scale content-based search. *ACM International Conference on Multimedia*, pp. 141–150, 2008.

[28] J. Li and J. Z. Wang. Real-time computerized annotation of pictures. *ACM International Conference on Multimedia*, pp. 911–920, 2006.

[29] W. Hsu, T. Mei, and R. Yan. Knowledge discovery over community-sharing media: From signal to intelligence. *IEEE International Conference on Multimedia and Expo (ICME)*, pp. 1448–1451, 2009.

[30] M. Lew, N. Sebe, C. Djeraba, and R. Jain. Content-based multimedia information retrieval: state of the art and challenges. *ACM Trans. Multimedia, Computing, Communications, and Applications*, 2(1):1–19, Feb. 2006.

[31] L. Ahn and L. Dabbish. Labeling images with a computer game. *ACM Proceedings of SIGCHI Conference on Human Factors in Computing Systems*, pp. 319–326, 2004.

[32] Y. Yang and P. T. Wu. Contextseer: content search and recommendation at query time for shared consumer photos. *ACM International Conference on Multimedia*, pp. 199–208, 2008.

[33] X. S. Hua, D. Liu, L. Yang, M. Wang, and H. J. Zhang. Tag ranking. *International World Wide Web Conference*, pp. 351–360, 2009.

[34] A. Sun and S. S. Bhowmick. Image tag clarity: in search of visual-representative tags for social images. *ACM International Conference on Multimedia, the First SIGMM Workshop on Social Media*, pp. 19–22, 2009.

[35] H. Sundaram. Making sense of meaning: leveraging social processes to understand media semantics. *IEEE International Conference on Multimedia and Expo (ICME)*, pp. 1648–1650, 2009.

[36] Y. G. Jiang, J. Wang, S. F. Chang, and C. W. Ngo. Domain adaptive semantic diffusion for large scale context-based video annotation. *IEEE International Conference on Computer Vision (ICCV)*, pp. 1420–1427, 2009.

[37] J. Yang, R. Yan, and A. G. Hauptmann. Cross-domain video concept detection using adaptive SVMs. *ACM International Conference on Multimedia*, pp. 188–197, 2007.

[38] W. Jiang, E. Zavesky, S.-F. Chang, and A. Loui. Cross-domain learning methods for high-level visual concept classification. *IEEE International Conference on Image Processing (ICIP)*, pp. 161–164, 2008.

[39] A. Zunjarwad, H. Sundaram, and L. Xie. Contextual wisdom: Social relations and correlations for multimedia event annotation. *ACM International Conference on Multimedia*, pp. 615–624, 2007.

[40] P. Cudré-Mauroux. *Emergent Semantics: Interoperability in Large-Scale Decentralized Information Systems*. EPFL and CRC Press, 2008.

[41] M. D. Choudhury, H. Sundaram, Y. R. Lin, A. John, and D. D. Seligmann. Connecting content to community in social media via image content, user tags and user communication. *IEEE Internatinal Conference on Multimedia and Expo (ICME)*, pp. 1238–1241, 2009.

[42] Y. R. Lin, H. Sundaram, Y. Chi, J. Tatemura, and B. L. Tseng. Blog community discovery and evolution based on mutual awareness expansion. *IEEE/WIC/ACM International Conference on Web Intelligence*, pp. 48–56, 2007.

[43] Y. R. Lin, H. Sundaram, and A. Kelliher. Summarization of large scale social network activity. *IEEE International Conference on Acoustics, Speech and Signal Processing (ICASSP)*, pp. 3481–3484, 2009.

[44] Y. R. Lin, M. D. Choudhury, H. Sundaram, and A. Kelliher. Temporal patterns in social media streams: Theme discovery and evolution using joint analysis of content and context. *IEEE International Conference on Multimedia Expo (ICME)*, pp. 1456–1459, 2009.

[45] M. D. Choudhury, H. Sundaram, A. John, and D. D. Seligmann. Dynamic prediction of communication flow using social context. *ACM Conference on Hypertext and Hypermedia*, pp. 49–54, 2008.

[46] M. D. Choudhury, H. Sundaram, A. John, and D. D. Seligmann. Multi-scale characterization of social network dynamics in the blogosphere. *ACM Conference on Information and Knowledge Management*, pp. 1515–1516, 2008.

[47] M. D. Choudhury, H. Sundaram, A. John, and D. D. Seligmann. Social synchrony: predicting mimicry of user actions in online social media. *IEEE International Conference on Social Computing (SocialCom)*, pp. 151–158, 2009.

[48] J. Mitchell, W. Pennebaker, C. Fogg, and D. LeGall. *MPEG Video Compression Standard*. Chapman & Hall, 1997.

[49] J. Chen, U. Koc, and K. J. R. Liu. *Design of Digital Video Coding Systems: A Complete Compressed Domain Approach*. Dekker, 2002.

[50] K. Ngan, C. Yap, and K. Tan. *Video Coding for Wireless Communication Systems*. Marcel Dekker, 2001.

[51] MPEG committee. MPEG-21 overview ISO/IEC JTC1/SC29/WG11/N4801, http://www.cselt.it/mpeg/standards/mpeg-21/mpeg-21.htm.

[52] D. Ensor. Drone may have spotted bin Laden in 2000 – CIA to investigate leak of tape to media. *CNN News*, http://www.cnn.com/2004/WORLD/asiapcf/03/17/predator.video/index.html, March 17, 2004.

[53] The International Intellectual Property Alliance (IIPA). http://www.iipa.com/.

[54] Movie "screener" suspect charged – Chicago man violated Hollywood studios copyright. Associated Press, http://www.msnbc.msn.com/id/4037016/, Jan. 23, 2004.

[55] F. Hartung and M. Kutter. Multimedia watermarking techniques. *Proc. IEEE*, **87**(7):1079–1107, July 1999.

[56] I. Cox, M. Miller, J. Bloom, J. Fridrich, and T. Kalker. *Digital Watermarking and Steganography*, 2nd ed. Morgan Kaufmann Publishers, 2007.

[57] M. Wu and B. Liu. *Multimedia Data Hiding*. New York: Springer-Verlag, 2002.

[58] C. Podilchuk and W. Zeng. Image adaptive watermarking using visual models. *IEEE J. Select. Areas Commun.*, **16**(4):525–540, May 1998.

[59] I. Cox, J. Killian, F. Leighton, and T. Shamoon. Secure spread spectrum watermarking for multimedia. *IEEE Trans. Image Processing*, **6**(12):1673–1687, December 1997.

[60] H. V. Poor. *An Introduction to Signal Detection and Estimation*, 2nd ed. Springer Verlag, 1999.

[61] Z. J. Wang, M. Wu, H. V. Zhao, W. Trappe, and K. J. R. Liu. Anti-collusion forensics of multimedia fingerprinting using orthogonal modulation. *IEEE Trans. Image Processing*, **14**(6):804–821, June 2005.

[62] W. Trappe, M. Wu, Z. Wang, and K. J. R. Liu. Anti-collusion fingerprinting for multimedia. *IEEE Tran. Signal Processing*, **51**(4):1069–1087, April 2003.

[63] F. Ergun, J. Killian, and R. Kumar. A note on the limits of collusion-resistant watermarks. *Advances in Cryptology – EuroCrypto '99. Lecture Notes in Computer Science*, **1592**:140–149, 2001.

[64] J. Killian, T. Leighton, L. R. Matheson, T. G. Shamoon, R. Tajan, and F. Zane. Resistance of digital watermarks to collusive attacks. Technical Report TR-585-98, Department of Computer Science, Princeton University, 1998.

[65] J. Su, J. Eggers, and B. Girod. Capacity of digital watermarks subject to an optimal collusion attacks. *European Signal Processing Conference (EUSIPCO)*, pp. 1981–1984, 2000.

[66] H. Stone. Analysis of attacks on image watermarks with randomized coefficients. Technical Report 96-045, NEC Research Institute, 1996.

[67] H. V. Zhao, M. Wu, Z. J. Wang, and K. J. R. Liu. Forensic analysis of nonlinear collusion attacks for multimedia fingerprinting. *IEEE Trans. Image Processing*, **14**(5):646–661, May 2005.

[68] D. Kirovski and M. K. Mihcak. Bounded gaussian fingerprints and the gradient collusion attack. *IEEE International Conference on Acoustics, Speech, and Signal Processing (ICASSP)*, II:1037–1040, March 2005.

[69] F. Zane. Efficient watermark detection and collusion security. *Proc. of Financial Cryptography. Lecture Notes in Computer Science*, **1962**:21–32, February 2000.

[70] S. B. Wicker. *Error Control Systems for Digital Communication and Storage*. Prentice-Hall, 1995.

[71] S. He and M. Wu. Joint coding and embedding techniques for multimedia fingerprinting. *IEEE Trans. Inform. Forensics and Security*, **1**(2):231–247, June 2006.

[72] J. Dittmann, P. Schmitt, E. Saar, J. Schwenk, and J. Ueberberg. Combining digital watermarks and collusion secure fingerprints for digital images. *SPIE J. Electronic Imaging*, **9**(4):456–467, October 2000.

[73] Z. J. Wang, M. Wu, W. Trappe, and K. J. R. Liu. Group-oriented fingerprinting for multimedia forensics. *EURASIP J. Applied Signal Processing, Special Issue on Multimedia Security and Rights Management*, **14**:2142–2162, Nov. 2004.

[74] A. Varna, S. He, A. Swaminathan, M. Wu, H. Lu, and Z. Lu. Collusion-resistant fingerprinting for compressed multimedia signals. *IEEE International Conference on Acoustics, Speech and Signal Processing (ICASSP)*, pp. 165–168, April 2007.

[75] K. J. R. Liu, W. Trappe, Z. J. Wang, M. Wu, and H. Zhao. *Multimedia Fingerprinting Forensics for Traitor Tracing*. EURASIP Book Series on Signal Processing and Communications, Hindawi Publishing Corporation, 2005.

[76] Y. Wang, J. Ostermann, and Y. Zhang. *Video Processing and Communications*, 1st ed. Prentice Hall, 2001.

[77] H. V. Zhao, W. S. Lin, and K. J. R. Liu. A case study in multimedia fingerprinting: Behavior modeling and forensics for multimedia social networks. *IEEE Signal Proc. Mag.*, 26(1):118–139, January 2009.

[78] YouTube surpasses 100 million U.S. viewers for the first time. *ComScore*, March 4, 2009.

[79] Y. Chu, S. G. Rao, S. Seshan, and H. Zhang. A case for end system multicast. *IEEE J. Select. Areas Commun.*, 20(8):1456–1471, October 2002.

[80] N. Magharei, R. Rejaie, and Y. Guo. Mesh or multiple-tree: A comparative study of live P2P streaming approaches. *IEEE INFOCOM*, pp. 1424–1432, 2007.

[81] X. Zhang, J. Liu, B. Li, and Y. S. P. Yum. CoolStreaming/DONet: A data-driven overlay network for Peer-to-Ppeer live media streaming. *IEEE INFOCOM*, vol. 3, pp. 2102–2111, 2005.

[82] S. Annapureddy, S. Guha, C. Gkantsidis, D. Gunawardena, and P. R. Rodriguez. Is high-quality VoD feasible using P2P swarming? *International Conference on World Wide Web*, pp. 903–912, 2007.

[83] C. Dana, D. Li, D. Harrison, and C. N. Chuah. BASS: BitTorrent assisted streaming system for video-on-demand. *IEEE Workshop on Multimedia Signal Processing (MMSP)*, pp. 1–4, 2005.

[84] R. Kumar, Y. Liu, and K. W. Ross. Stochastic fluid theory for P2P streaming systems. *IEEE INFOCOM*, pp. 919–927, 2007.

[85] N. Naoumov and K. Ross. Exploiting P2P systems for DDoS attacks. *Proceedings of the 1st International Conference on Scalable Information Systems*, 2006.

[86] Y. Sun, Z. Han, and K.J. Ray Liu. Defense of trust management vulnerabilities in distributed networks. *IEEE Commun. Mag.*, 42(6):112–119, February 2008.

[87] R. Rejaie and A. Ortega. PALS: peer-to-peer adaptive layered streaming. *International Workshop on Network and Operating Systems Support for Digital Audio and Video (NOSDAV)*, pp. 153–161, 2003.

[88] M. Hefeeda, A. Habib, B. Botev, D. Xu, and B. Bhargava. PROMISE: peer-to-peer media streaming using CollectCast. *ACM International Conference on Multimedia*, pp. 45–54, 2003.

[89] M. Adler, R. Kumar, K. W. Ross, D. Rubenstein, T. Suel, and D. Yao. Optimal peer selection for P2P downloading and streaming. *IEEE InfoCom*, vol. 3, pp. 1538–1549, 2005.

[90] A. Habib and J. Chuang. Service differentiated peer selection: an incentive mechanism for Peer-to-Peer media streaming. *IEEE Trans. Multimedia*, 8(3):610–621, June 2006.

[91] D. Xu, M. Hefeeda, S. Hambrusch, and B. Bhargava. On peer-to-peer media streaming. *IEEE International Conference on Distributed Computing Systems*, pp. 363–371, 2002.

[92] Y. Chu, J. Chuang, and H. Zhang. A case for taxation in peer-to-peer streaming broadcast. *ACM SIGCOMM Workshop on Practice and Theory of Incentives and Game Theory in Networked Systems (PINS)*, pp. 205–212, 2004.

[93] G. Tan and S. A. Jarvis. A payment-based incentive and service differentiation mechanism for peer-to-peer streaming broadcast. *IEEE International Workshop on Quality of Service (IWQoS)*, pp. 41–50, June 2006.

[94] Z. Liu, Y. Shen, S. Panwar, K. W. Ross, and Y. Wang. Using layered video to provide incentives in P2P streaming. *Proceedings of the 2007 workshop on Peer-to-peer Streaming and IP-TV*, pp. 311–316, 2007.

[95] X. Hei, Y. Liu, and K. W. Ross. Inferring network-wide quality in P2P live streaming systems. *IEEE J. Select. Areas Commun.*, **25**(9): 1640–1654, December 2007.

[96] M. J. Osborne and A. Rubinstein. *A Course in Game Theory*. MIT Press, 1994.

[97] G. Owen. *Game Theory*, 3rd ed. Academic Press, 1995.

[98] R. Myerson. *Game Theory: Analysis of Conflicts*. Harvard University Press, 1997.

[99] G. Dantzig. *Linear Programming and Extensions*. Princeton University Press, 1963.

[100] Z. J. Wang, M. Wu, H. Zhao, W. Trappe, and K. J. R. Liu. Resistance of orthogonal Gaussian fingerprints to collusion attacks. *IEEE International Conference on Acoustics, Speech and Signal Processing (ICASSP)*, vol. 4, pp. 724–727, April 2003.

[101] M. Swanson, B. Zhu, and A. Tewfik. Multiresolution scene-based video watermarking using perceptual models. *IEEE J. Select. Areas Commun.*, **16**(4):540–550, May 1998.

[102] M. Holliman and N. Memon. Counterfeiting attacks and blockwise independent watermarking techniques. *IEEE Trans. Image Processing*, **9**(3):432–441, March 2000.

[103] D. Kirovski and F. A. P. Petitcolas. Blind pattern matching attack on watermarking systems. *IEEE Trans. Signal Proc.*, **51**(4):1045–1053, 2003.

[104] G. Doerr, J. L. Dugelay, and L. Grange. Exploiting self-similarities to defeat digital watermarking systems: case study on still images. *Proceedings of the 2004 ACM Multimedia and Security Workshop*, pp. 133–142, 2004.

[105] I. Cox and J. P. Linnartz. Some general methods for tampering with watermarking. *IEEE J. Select. Areas Commun.*, **16**(4):587–593, May 1998.

[106] F. Petitcolas, R. Anderson, and M. Kuhn. Attacks on copyright marking systems. *2nd Workshop on Information Hiding. Lecture Notes in Computer Science*, pp. 218–238, April 1998.

[107] A. J. Goldsmith and P. P. Varaiya. Capacity of fading channels with channel side information. *IEEE Trans. Inform. Theory*, **43**(6):1986–1992, 1997.

[108] W. S. Lin, H. V. Zhao, and K. J. R. Liu. Behavior forensics with side information for multimedia fingerprinting social networks. *IEEE Trans. on Inform. Forensics and Security*, **4**(4):911–927, December 2009.

[109] Y. Chen, W. S. Lin, and K. J. R. Liu. Risk-distortion analysis for video collusion attacks: A mouse-and-cat game. *IEEE Trans. Image Processing*, **19**(7):1798–1807, July 2010.

[110] K. Su, D. Kundur, and D. Hatzinakos. Statistical invisibility for collusion-resistant digital video watermarking. *IEEE Trans. Multimedia*, **7**(1):43–51, February 2005.

[111] K. Su, D. Kundur, and D. Hatzinakos. Spatially localized image-dependent watermarking for statistical invisibility and collusion resistance. *IEEE Trans. Multimedia*, **7**(1):52–66, February 2005.

[112] N. Jayant and P. Noll. *Digital Coding of Waveforms*. Prentice-Hall, 1984.

[113] H. M. Hang and J. J. Chen. Source model for transform video coder and its application-part I: Fundamental theory. *IEEE Trans. Circuits Syst. Video Technol.*, **7**(2):287–298, April 1997.

[114] S. Boyd and L. Vandenberghe. *Convex Optimization*. Cambridge University Press, 2004.

[115] P. Cheung and J. T. Kwok. A regularization framework for multiple-instance learning. *Proceedings of the International Conference on Machine Learning*, pp. 193–200, 2006.

[116] A. Smola, S. Vishwanathan, and T. Hofmann. Kernel methods for missing variables. *Proceedings of the International Workshop on Artificial Intelligence and Statistics*, pp. 325–332, 2005.

[117] A. L. Varna, S. He, A. Swaminathan, and M. Wu. Analysis of nonlinear collusion attacks on fingerprinting systems for compressed multimedia. *Proceedings of the IEEE International Conference on Image Processing (ICIP)*, vol. 2, pp. 133–136, 2007.

[118] Y. Wu. Nonlinear collusion attack on a watermarking scheme for buyer authentication. *IEEE Trans. Multimedia*, **3**(3):626–629, June 2006.

[119] W. S. Lin, H. V. Zhao, and K. J. R. Liu. Incentive cooperation strategies for peer-to-peer live streaming social networks. *IEEE Trans. Multimedia*, **11**(3):396–412, April 2009.

[120] W. S. Lin, H. V. Zhao, and K. J. R. Liu. Cooperation stimulation strategies for peer-to-peer wireless live video-sharing social networks. *IEEE Trans. Image Processing*, **19**(7):1768–1784, Jul. 2010.

[121] W. Yu and K. J. R. Liu. Game theoretic analysis of cooperation and security in autonomous ad hoc networks. *IEEE Trans. Mobile Comput.*, **6**(5):507–521, 2007.

[122] J. Nocedal and S. J. Wright. *Numerical Optimization*, 2nd ed. Springer Publishing, 2000.

[123] D. Marpe, H. Schwarz, and T. Wiegand. Joint scalable video model (JSVM) 2. Joint Video Team, Doc. JVT-O202, April 2005.

[124] A. Habib and J. Chuang. Incentive mechanism for peer-to-peer media streaming. *International Workshop on Quality of Service (IWQoS)*, pp. 171–180, June 2004.

[125] S. C. Kim, M. G. Kim, and B. H. Rhee. Seamless connection for mobile P2P and conventional wireless network. *IEEE International Conference on Advanced Communications Technology*, pp. 1602–1605, February 2007.

[126] M. Cagalj, S. Capkun, and J. P. Hubaux. Key agreement in peer-to-peer wireless networks. *Proc. IEEE*, **94**(2):467–478, February 2006.

[127] S. Ghandeharizadeh, B. Krishnamachari, and S. Song. Placement of continuous media in wireless peer-to-peer networks. *IEEE Trans. Multimedia*, **6**(2):335–342, April 2004.

[128] S. Ghandeharizadeh and T. Helmi. An evaluation of alternative continuous media replication techniques in wireless peer-to-peer networks. *ACM International Workshop on Data Engineering for Wireless and Mobile Access (MobiDe)*, pp. 77–84, 2003.

[129] G. Gualdi, A. Prati, and R. Cucchiara. Video streaming for mobile video surveillance. *IEEE Trans. Multimedia*, **10**(6):1142–1154, October 2008.

[130] D. F. S. Santos and A. Perkusich. Granola: a location and bandwidth aware protocol for mobile video-on-demand systems. *International Conference on Software, Telecommunications and Computer Networks (SoftCom)*, pp. 309–313, September 2008.

[131] S. Sudin, A. Tretiakov, R. H. R. M. Ali, and M. E. Rusli. Attacks on mobile networks: an overview of new security challenge. *International Conference on Electronic Design*, pp. 1–6, December 2008.

[132] International Telecommunication Union. http://www.itu.int/itud/ict/statistics/ict/graphs/mobile.jpg.

[133] H. Lee, Y. Lee, J. Lee, D. Lee, and H. Shin. Design of a mobile video streaming system using adaptive spatial resolution control. *IEEE Trans. Consumer Electronics*, **55**(3):1682–1689, August 2009.

[134] G. Aniba and S. Aissa. A general traffic and queueing delay model for 3G wireless packet networks. *International Conference on Telecommunications (ICT). Lecture Notes in Computer Science*, pp. 942–949, 2004.

[135] H. V. Zhao and K. J. R. Liu. Traitor-within-traitor behavior forensics: strategy and risk minimization. *IEEE Trans. Inform. Forensics and Security*, **1**(4):440–456, December 2006.

[136] S. Yoon and N. Ahuja. Frame interpolation using transmitted block-based motion vectors. *IEEE International Conference on Image Processing (ICIP)*, pp. 856–859, October 2001.

[137] H. V. Zhao and K. J. R. Liu. Behavior forensics for scalable multiuser collusion: fairness versus effectiveness. *IEEE Tran. Inform. Forensics and Security*, 1(3):311–329, September 2006.

[138] O. Kallenberg. *Foundations of Modern Probability*. Springer-Verlag, 1977.

[139] J. Degesys, I. Rose, A. Patel, and R. Nagpal. Desync: self-organizing desynchronization and TDMA on wireless sensor networks. *International Conference on Information Processing in Sensor Networks (IPSN)*, pp. 11–20, 2007.

[140] S. Lee, S. Zhu, and Y. Kim. P2P trust model: The resource chain model. *ACIS International Conference on Software Engineering, Artificial Intelligence, Networking and Parallel/Distributed Computing (SNPD)*, Vol. 2, pp. 357–362, 2007.

[141] A. Josang, R. Ismail, and C. Boyd. A survey of trust and reputation systems for online service provision. *Decision Support Systems*, 42(2):618–644, 2005.

[142] E. Lua, J. Crowcroft, M. Pias, R. Sharma, and S. Lim. A survey and comparison of peer-to-peer overlay network schemes. *IEEE Commun. Surveys and Tutorial*, 7(2):72–93, March 2004.

[143] H. V. Zhao and K. J. R. Liu. Impact of social network structure on misbehavior detection and identification. *IEEE J. Selected Topics in Signal Proc.*, 4(4):687–703, Aug. 2010.

[144] A. Menezes, P. Oorschot, and S. Vanstone. *Handbook of Applied Cryptography*. CRC Press, 1996.

[145] A. Tosun and W. Feng. On error preserving encryption algorithms for wireless video transmission. *ACM Conference on Multimedia*, 9:302–308, 2001.

[146] T. Locher, P. Moor, S. Schmid, and R. Wattenhofer. Free riding in BitTorrent is cheap. *Fifth Workshop on Hot Topics in Networks (HotNets)*, pp. 85–90, November 2006.

[147] N. E. Baughman, M. Liberatore, and B. N. Levine. Cheat-proof playout for centralized and peer-to-peer gaming. *IEEE/ACM Trans. Networking*, 15(1):1–13, February 2007.

[148] G. Horng, T. Chen, and D. Tsai. Cheating in visual cryptography. *Designs, Codes and Cryptography*, 38(2):219–236, February 2006.

[149] C. Huang, J. Li, and K. W. Ross. Can Internet video-on-demand be profitable? *ACM SIGCOMM Conference on Applications, Technologies, Architectures, and Protocols for Computer Commun.*, pp. 133–144, 2007.

[150] J. M. Smith. *Evolution and the Theory of Games*. Cambridge University Press, 1982.

[151] B. Wang, K. J. R. Liu, and T. C. Clancy. Evolutionary cooperative spectrum sensing game: how to collaborate? *IEEE/ACM Trans. Networking*, 58(3):890–900, March 2010.

[152] R. Cressman. *Evolutionary Dynamics and Extensive Form Games*. MIT Press, 2003.

[153] D. Wu, C. Liang, Y. Liu, and K. W. Ross. View-upload decoupling: a redesign of multi-channel P2P video systems. *IEEE INFOCOM*, pp. 2726–2730, 2009.

[154] S. Deering and D. Cheriton. Multicast routing in datagram inter-networks and extended LANs. *ACM Trans. Computer Systems*, 8(2):85–110, May 1990.

[155] L. Kontothanassis, R. Sitaraman, J. Wein, D. Hong, R. Kleinberg, B. Mancuso, D. Shaw, and D. Stodolsky. A transport layer for live streaming in a content delivery network. *Proc. IEEE*, 92(9):1408–1419, 2004.

[156] Y. Liu, Y. Guo, and C. Liang. A survey on peer-to-peer video streaming systems. *J. Peer-to-Peer Networking and Applications*, 1:18–28, 2008.

[157] T. Karagiannis, P. Rodriguez, and K. Papagiannaki. Should Internet service providers fear peer-assisted content distribution? *ACM SIGCOMM Conference of Internet Measurement*, 2005.

[158] H. V. Madhyastha, T. Isdal, M. Piatek, *et al.* iPlane: An information plane for distributed services. *USENIX Symposium on Operating Systems Design and Implementation (OSDI)*, pp. 367–380, 2006.

[159] D. Purandare and R. Guha. An alliance based peering scheme for P2P live media streaming. *IEEE Trans. Multimedia*, **9**(8):1633–1644, December 2007.

[160] H. Xie, Y. R. Yang, A. Krishnamurthy, Y. Liu, and A. Siberschatz. P4P: Provider portal for applications. *ACM SIGCOMM Computer Communication Rev.* **4**: 351–362, October 2008.

Index

Printed in the United States
by Baker & Taylor Publisher Services